THE STORY OF
HUMAN ERROR

THE STORY
OF
HUMAN ERROR

EDITED BY

JOSEPH JASTROW

Essay Index Reprint Series

BOOKS FOR LIBRARIES PRESS, INC.
Freeport, New York

First Published 1936
Reprinted 1967

LIBRARY OF CONGRESS CATALOG CARD NUMBER:

67-30219

PRINTED IN THE UNITED STATES OF AMERICA

PREFACE

THE COMPLETION of the circuit that converts an idea into a project is often a matter of circumstance. It was so in regard to the present undertaking. I was much impressed, in visiting the Century of Progress Exposition in Chicago in 1933, with the magnificently staged demonstration of the progressive rôle of science in human affairs. But it also impressed me then ocularly, as it had done reflectively years before, that the story of the triumphs and their stages omitted the even more significant story of errors on the way.

The central conviction responsible for the undertaking is that the vital story of scientific thought is one of the surmounting and correcting of error. In the long historical vista the high-light, constructive moments are few and far between: Pythagoras, Aristotle, Archimedes, Roger Bacon, Galileo, Newton, Harvey; and then the modern era when the supremacy of scientific thinking became established under the leadership of a succession of great minds. From Darwin on, the march of science has accelerated by leaps and bounds. Through biology, anthropology, and psychology we became conscious of the course of our intellectual ascent. A new complexion was given to the story of man when James Harvey Robinson made the significance of history to be "Mind in the Making."

From the present peak of scientific achievement the career of science invites a survey of the departures and misemployments on the way. Professor Eric T. Bell comments in the same conviction.[1] Speaking of Pythagoras, he conjectures that *if* instead of following the lure of that Greek sage, we "had continued along the harsher road of scientific experiment,"

[1] In his penetrating and hilarious *Numerology* (1933); and *The Search for Truth* (1934), jocosely profound.

we should not have had to wait until 1581 for "Galileo to start the scientific age"; and *if* Archimedes had not been butchered in 212 B.C. but had been accepted as the prime guide instead of Plato, then by 100 A.D. Julius Cæsar might have sent the "first and second cruising squadrons of the Roman Air Fleet in a two-hour flight from Ostia [to Athens] with incendiary bombs and cholera germs," and by 300 A.D., "for lack of brains or science to conduct its affairs intelligently, civilization would have ended and *Homo sapiens* returned to its ancestral trees."

But the dismal powers of science are no more in point than its potential benefits. In either case the might-have-been conjecture makes pivotal the errors and lost opportunities. When the rare chances of intellectual emancipations were missed, there ensued long stagnant periods of unprofitable treadmill intellection, which somehow—and that, to our emancipated minds, is the not easily intelligible part of the story—was more congenial to the intelligentsia.

The table of contents indicates the plan of the book. The two major hemispheres of investigation are World and Man; and where man enters, the perspective of significance acquires an added dimension. As history recedes—like the retinal images of objects as they are more remote from our point of vision—the greater emphasis is given to modern errors, which are nearer and seem larger. Only such applications have been included as depend closely upon fundamental ideas. Institutions and codes, churches, courts, states, schools, engineering, affairs generally, fall outside the canvas. The content and perspective of presentation are justified by the yardstick of human interests, both for understanding and for welfare.

I express my obligation to my contributors, and particularly to Professor Robinson, to whom I took my project in its unformed stages and whose cordial encouragement spurred it along. This acknowledgment is now a tribute to his memory. Other distinguished exponents of science whose names do not appear in the volume gave me helpful advice.

A dozen and more minds, representing as many disciplines,

will inevitably each approach his topic and the general thesis from his special interest. The resulting variety of presentation reflects the contemporary picture without detracting from its unity of theme. My contributors have been patient with my suggestions and coöperative in their consideration. The value of the volume rests upon the expertness of their authoritative presentations. I am likewise indebted to the publishers for their sponsoring of the venture from its beginnings as a scheme to its consummation as a book.

JOSEPH JASTROW

CONTENTS

INTRODUCTION

PART I

WORLD

Section I

THE COSMIC REALM

CHAPTER I

CHAPTER II

CONTRIBUTORS

Joseph Jastrow

Emeritus Professor of Psychology, University of Wisconsin. Born in Warsaw, Poland, 1863, A.B., University of Pennsylvania, 1882, A.M., 1885; Ph.D., Johns Hopkins University, 1886, the first doctorate conferred in psychology. Professor of Psychology, University of Wisconsin, from 1888 to 1926. Was in charge of the section of psychology at the World's Columbian Exposition, 1893; lecturer at Columbia University, 1910. Since 1927 resident in New York. Lecturer at the New School for Social Research. Conducted a syndicated column, "Keeping Mentally Fit," for four years; broadcasts on psychology through the National Broadcasting Company. Frequent contributor to scientific, psychological, educational, and popular journals. Past president of the American Psychological Association. Author of *Fact and Fable in Psychology*, *The Subconscious, Character and Temperament, The Qualities of Men, The Psychology of Conviction, Keeping Mentally Fit, Piloting Your Life, Managing Your Mind, The House that Freud Built, Sanity First, Wish and Wisdom, The Life of the Mind* (in preparation).

James Harvey Robinson

Late Professor of History, Columbia University. Born in Bloomington, Illinois, 1863. A.B., Harvard University, 1887, A.M., 1888; Ph.D., University of Freiburg, 1890; LL.D., University of Utah, 1922; L.H.D., Tufts College, 1924. Taught European history at the University of Pennsylvania and Columbia University, 1891 to 1919. Acknowledged founder of the modern method in history. A founder of the New School for Social Research, New York. Author of many books, including *The New History, Mind in the Making,* and *The Humanizing of Knowledge*. Wrote the article "Civilization" for the latest edition of the *Encyclopædia Britannica*. Died, 1936. His contribution to this book was his last writing for publication.

Harlan True Stetson

Research Associate in Geophysics, Harvard University. Born in Haverhill, Massachusetts, 1885. A.B., Brown University, 1908; A.M., Dartmouth College, 1910; Ph.D., University of Chicago, 1915. Instructor at Dartmouth, Northwestern University, and Assistant Professor of Astronomy at Harvard University, 1916 to 1929. Then appointed director of the newly established Perkins Observatory at Ohio Wesleyan University. In 1933 resumed his connection with Harvard University. Has participated in eclipse expeditions in California, 1923, Connecticut, 1925, Sumatra, 1926, Norway, 1927, Malaya, 1929, Maine, 1932. Author of *Thermoelectric Methods in Photographic Photometry, Manual of Laboratory Astronomy, Man and the Stars,* and *Earth, Radio and the Stars*. Distinguished for his contributions to astronomical research, notably to the applications of astronomy to problems of the earth, and is well known for his popularization of astronomy.

KIRTLEY FLETCHER MATHER

Professor of Geology, Harvard University. Born in Chicago, 1888. B.S., Denison University, 1909; Ph.D., University of Chicago, 1915. Was successively instructor in geology in the University of Arkansas, Queen's University (Canada), Denison University, and Harvard University. Was connected with the U. S. Geological Survey and worked as geologist in Bolivia and Argentina. Served as captain in the O.R.C. Is a fellow of the Geological Society of America and of the Royal Geographical Society. Author of numerous reports on geological surveys, and in popular presentation of *Old Mother Earth, Science in Search of God*, and *Sons of the Earth*. Editor of the Century Earth Science Series. Among wide educational interests, is the Director of the Summer School of Harvard University, and President of the Adult Educational Council and of the Boston Center for Adult Education.

JOHN BARGER LEIGHLY

Associate Professor of Geography, University of California. Born Locust Grove, Ohio, 1895. A.B., University of Michigan, 1922; Ph.D., University of California, 1927. American-Scandinavian Fellow at Stockholm, 1925-1926. Social Science Research Council Fellowship, 1929-30. Field assistant in Land Economic Survey, 1927, and special assistant, Kentucky Geological Survey. Special interests in physical climatology, meteorology, and the mechanics of streams. Author of numerous scientific papers and reviews.

WILLIAM FRANCIS GRAY SWANN

Director of the Bartol Research Foundation of the Franklin Institute. Born at Ironbridge, England, 1884. B.Sc., University of London, 1905, D.Sc., 1910; A.M. (honorary), Yale University, 1924. Served as demonstrator in physics at the Royal College of Science and in the University of Sheffield. Came to the United States in 1913 and was associated with the Carnegie Institute of Washington; then became professor of physics at the University of Minnesota, University of Chicago, and Yale University. Past president of the American Physical Society. Among recent activities are the devising and checking of instruments for flights into the stratosphere. Contributions to technical periodicals many and important. Author of *The Architecture of the Universe*, a brilliant survey addressed to the general reader.

ERIC TEMPLE BELL

Professor of Mathematics, California Institute of Technology. Born at Aberdeen, Scotland, 1883. A.B., Stanford University, 1904; A.M., University of Washington, 1908; Ph.D., Columbia University, 1912. Became Professor of Mathematics at the University of Washington, and since 1927 has been at the California Institute of Technology at Pasadena. Writes in an exuberant style, keenly humorous, in discussing weighty subjects, as appears in his books on *Numerology* and *The Search for Truth*. Author of many other books, both popular and scientific.

CHARLES ALBERT BROWNE

Chief of Chemical and Technological Research, United States Department of Agriculture. Born North Adams, Massachusetts, 1870; A.B., Williams College, 1892, honorary Sc.D., 1924; M.A. and Ph.D., University of Göttingen,

1901; honorary Sc.D., Stevens Institute of Technology, 1925. Has been successively connected as Research Chemist with the Agricultural Experiment Stations of Pennsylvania and Louisiana, and the U. S. Bureau of Chemistry and Soils. Author of *Handbook of Sugar Analysis* and of numerous contributions to scientific and popular journals. Editor of *A Half Century of Chemistry in America*. President of the History of Science Society and a distinguished contributor to the history of science. Was awarded gold and silver medals at the St. Louis Exposition, 1904.

HOWARD MADISON PARSHLEY

Professor of Zoölogy, Smith College. Born Hallowell, Maine, 1884. A.B., Harvard University, 1909, A.M., 1910, Sc.D., 1917. Author of *Bibliography of the North American Hemiptera-Heteroptera, Science and Good Behavior, The Science of Human Reproduction*, 1933. Is a frequent contributor to and reviewer of the general literature of biology in its bearings upon sociological problems.

HOMER WILLIAM SMITH

Professor of Physiology, New York University, College of Medicine. Born in Denver, Colorado, 1895. A.B., University of Denver, 1917; Sc.D., Johns Hopkins University, 1921. Held research positions with the U. S. Public Health Service, Harvard Medical School, American Museum of Natural History, Mt. Desert Island and Bermuda Biological Stations. Professor of Physiology at the University of Virginia, 1925 to 1928. Technical work particularly on the physiology of the kidney. Is known to a large public as the author of *Kamongo* (Book of the Month Club selection in 1932), dealing in the form of fiction with the African lung-fish, upon which he has made extended metabolic observations; and *The End of Illusion* (1934), a humanistic defense of the scientific doctrine of determinism, likewise presented in the form of fiction but with the locale shifted to Malaya.

CHARLES JUDSON HERRICK

Professor of Comparative Neurology, University of Chicago. Born in Minneapolis, 1868. B.S., University of Cincinnati, 1891, Sc.D., 1926; M.S., Denison University, 1895, Sc.D., 1926; Ph.D., Columbia University, 1900. Taught zoölogy at Granville Academy, Denison University, and Ottawa University (Kansas); has held his present position since 1907. Is a foremost authority on neurology; his contributions to the literature are voluminous, and his texts widely used. Has been an influential interpreter of the neural basis of life and the general concepts of behavior in all its aspects, notably in *The Thinking Machine* and *Brains of Rats and Men*.

RALPH LINTON

Professor of Anthropology, University of Wisconsin. Born in Philadelphia, 1893. A.B., Swarthmore College, 1915; A.M., University of Pennsylvania, 1925; Ph.D., Harvard University, 1925. Was associated with the Bishop Museum, Honolulu, and the Field Museum of Natural History, Chicago. Was a member of archæological expeditions to New Mexico, Guatemala, Marquesas, Tuamotos, Tahiti, and Madagascar. Has written extensively in reports of these expeditions and has published a comprehensive introduction to anthropology entitled *The Study of Man*.

GEORGE MALCOLM STRATTON

Professor of Psychology, University of California. Born in Oakland, California, 1865. B.A., University of California, 1888; M.A., Yale University, 1890; Ph.D., University of Leipzig, 1896. In the faculty of the University of California 1896-1904, Johns Hopkins University 1904-1908, and since 1908 in his present position. Past president of the American Psychological Association. Has lectured at the Universities of the Philippines, Columbia, Yale, Chicago, Tokyo, and Kyoto. During the World War he served as Major in the Air Service, U. S. Army, being president of the Aviation Examining Board in San Francisco, and Head of the Psychology Section, Medical Research Laboratory, at Mineola, Long Island. Author of *Experimental Psychology and its Bearing upon Culture, Psychology of the Religious Life, Theophrastus and the Greek Physiological Psychology before Aristotle, Developing Mental Power, Anger, Its Religious and Moral Significance, Social Psychology of International Conduct, International Delusions,* and of articles in various psychological journals.

HARRY ELMER BARNES

Born in Auburn, New York, 1889. A.B., University of Syracuse, 1913, A.M., 1914; Ph.D., Columbia University, 1918. Taught history and sociology at Syracuse University, Columbia University, Barnard College, Clark University, Amherst College, Smith College; lecturer in the New School for Social Research, New York. On the staff of the New York *World-Telegram.* Widely known as a brilliant writer and forceful speaker, his contributions have a wide range through the historical and social sciences. The most notable among his many books is the *History of Civilization.*

HOWARD WILCOX HAGGARD

Professor of Physiology, Yale University. Born in La Porte, Indiana, 1891. Ph.B., Yale University, 1914, M.D., 1917. Apart from an association with the U. S. Bureau of Mines, has been continuously connected with Yale University. During the World War he served as captain in the Chemical Warfare Service. Has made notable contributions in his researches on respiration and metabolism generally, and is known to the general public for his vivid and popular presentations in *Devils, Drugs, and Doctors, The Lame, the Halt, and the Blind, Mystery, Magic, and Medicine, The Doctor in History,* and (jointly with C. C. Fry) *The Anatomy of Personality.*

ABRAHAM MYERSON

Professor of Neurology, Tufts College Medical School. Born in Yanova, Russia, 1881. Emigrated in early youth to this country and took his medical degree at Tufts College in 1908. Is a well known authority in psychiatry. is the author of *Inheritance of Mental Diseases* and has contributed widely to the study of feeble-mindedness and the psychic side of criminality. His more popular presentations are *The Foundations of Personality, The Nervous Housewife,* and *When Life Loses Its Zest.* The breadth of his interests is indicated by his recent volume on *Social Psychology,* the first such survey by a psychiatrist.

ILLUSTRATIONS

THE STORY OF HUMAN ERROR

INTRODUCTION

THE PROCESSION OF IDEAS

By Joseph Jastrow

> Every work of science great enough to be remembered for a
> few generations affords some exemplification of the defective
> state of the art of reasoning of the time when it was written;
> and each chief step in science has been a lesson in logic.
> —CHARLES S. PEIRCE.

THE MODERN OUTLOOK

NEO-MODERN MAN, surrounded by the comforts and
benefits of science, with triumphs of invention at call,
is apt to accept his heritage without reflection or gesture
of gratitude. In a day's work and an evening's play, he hears
by telephone and radio, sees motion-pictures and telepho-
tographs, making remote peoples and events as familiar as
neighbors. He learns week by week what experts have seen
through ultramicroscope and supertelescope, by X-ray and
by light beyond the spectrum. The daily wizardry of his life
exceeds by undreamt spans of wonder the fabled marvels of
old. He reaches the metropolis through a mile-long tube
under a river, is lifted to offices a hundred floors above the
street; travels fast and far on electrified rails in luxurious
air-conditioned coaches in stream-lined trains; and flies still
faster on scheduled airways across continents.

The neo-modern mind takes still more for granted when
its accepts unreflectively and as its assured birthright present-
day knowledge and the techniques for arriving at it. Such
insensitive complacency may well be disturbed by tracing

the slow, groping steps—through false leads, futile quests, quagmires of ignorance, thickets of superstition, obstructions of dogma, ineptitudes in reasoning, interspersed here and there with the ventures of pioneering minds—by which our intellectual forebears prepared the way for our present commanding, enlightened estate.

The story of the triumphs of man the thinker should not crowd out the yet more instructive career of man the blunderer. An increasing price has been paid for the emergent control and insight. The responsible reflective should know something of the story of human error, something of the procession of ideas that set the themes for the irregular march of thought through forgotten centuries, something of the pain and labor of the cultural mind in the making. That story of stories begins with Neanderthal Man and proceeds through Nineveh and the Nile, through Athens, Rome, medieval Europe, Dark Ages and Renaissance, heralding the modern era with its swift-coming discoveries culminating in the Machine Age. It is not the story of occupation and dominion that is most significant, but that of the cargo of ideas, of gold and dross alike, that were carried in the caravans of migration and intercourse of peoples. Mind in the making follows no straightforward progression; its many wanderings in the quest for truth compose a cyclopedia of error and vain solutions far more than orderly annals of successful advance. There is in some measure a parallel chronology in the procession of empire and in the calendar of thought; yet the critical events of the two sequences are shaped by different moments. The accepted premises in the search for truth determine the quest. They give a distinctive set to the ancient, the medieval, the modern, and the neomodern mind. Each comes to the inquiry with a different equipment, speaks another and a growing language. When we hear cited the stock scholastic discussion of how many angels can dance on the point of a needle, we are struck not merely by the incongruity of angels and needles, but by the state of mind and interest that could raise and argue the question; yet minds then and now and through earlier and intervening ages had the same fundamental endowment.

Such shifts in the temper of consideration—major shifts and minor ones—no less than advancements in data and logical standards, place and date the milestones in the procession of ideas. Comparing ourselves with former intellectual generations, we find that we no longer ask the same questions; that the questions we ask, the data we collect and seek, the concepts we develop, the techniques we employ for solution of our problems, the conclusions we reach, are widely different in temper as well as in content, differ as markedly in approach as in conviction. The structures of the brains of men show no such decisive contrasts as do their employments; it is a progress in heritage far more than in capacity. Often the strands of thought-kinship of old and new can be traced only through large concessions. Civilization is an issue of learning and logic; the primary datum is what men put their minds to and how.

Neo-modern thought is the qualified and privileged heir of all the ages of scientific thinking; still more, it represents the thorough reconstruction of that heritage. From our point of vantage we contemplate the record of the ancestral explorations of the phenomena of the world, of the nature of man and his place in it, and trace the course of error on the map of science. We are impressed by how much has been outgrown, disowned, corrected, and superseded, yet retained as foundation for structures that stand. In the life sciences the effective center of our outlook is as recent as Darwin, and it has shifted since. Every province of knowledge has been reorganized by the wisdom of the moderns. It is our twentieth-century privilege to enter confidently into our possessions, recognizing our obligation to the past while boldly following the lead of the present.

THE EMPLOYMENT OF INTELLIGENCE

A standard psychological experiment tests the rat's ability in learning to run varied patterns of mazes in the search for food or escape. The results show the rodent's procedure to be one of *trial and error*. Rats learn to avoid wrong paths which lead nowhere and to take correct turns by gradual profits of direct experience. Human learning proceeds both similarly and far more complexly. It is the human privilege to plan and map the solution, to consider it fore and aft. It is also the human liability to form irrational attachments to false turns and blind alleys and to ignore experience in consequence of a maze of predilections. The progress of the many depends on the occasional success of a wiser "rat" to escape from the contagion of the mental habits of his fellow-men. In that the wisdom of the few becomes available to the many, there is progress in human affairs; without it, the static routine of tradition continues. The rats of today, even those which have enjoyed the—doubtless to them meaning-less—advantages of a training in a psychological laboratory, are no wiser, no better informed, than those which gnawed the corn in Pharaoh's granaries. Yet Professor C. J. Herrick presents vividly the common qualities of the *Brains of Rats and Men*.

Our theme converges upon the employment of intelli-gence—to what uses man puts the sapience which since the days of Linnæus has become his zoölogical appellation. The capacity to reflect, to analyze, to organize, to review the stages of mental progression, is the decisive quality; it pro-ceeds upon the democratic dissemination of the results, which is education. The exceptionally apt at the process become the *intelligentsia* of their day.

The procession of ideas wends its uncertain way, subject to the vicissitudes of circumstance, awaiting the direction of pioneering leaders. There are, however, moments of critical consequence in its course, when the stream of thought ran full and high. One such golden period stands unexcelled and establishes a classic kinship across five-and-twenty centuries. The flair for reflection of the ancient Greeks created an epoch in the history of thought. The recognition of its double stem then became articulate. The one element is the tendency to accept conclusions and devise solutions at the instigation of urges, of the personalized motives and satisfactions that men live by; such understanding of human mind-ways became—*longo intervallo*—Psychology. The other is the critical examination of the validity of the process, the analysis of methods of procedure, the program of confirmation—evidence, proof, test, prediction, consistency; that critical expertness developed into the discipline of Logic. Jointly they were applied to the problems of the world and life, from which emerge the several sciences—virtually all one science of Nature. From the ways of knowledge spring the ways of life; civilization, in Santayana's pithy phrase, combines passion and policy.

Combined with that insight was another of equal value— the advance to the natural from the supernatural explanation. A great step was taken when the flash of lightning and the peal of thunder were explained, not as the wrath of the gods, but as the operation of physical forces under atmospheric conditions. It was a far-distant step in the same procession when Franklin sent his key along a kite and established the kinship of the lightning bolt and an electrical discharge. The Greek sage and the experimenting American philosopher, despite their distance in time and training, are brain-brothers in science. The varying patterns of fusion of logical and psychological elements characterize the advance of knowledge. The present project reconstructs the byways in the natural history of thought; selected scenes from that long-range panorama may be pieced together as the procession of ideas.

THE ANTECEDENTS OF SCIENCE

The natural history of error goes back to the earliest stages of thought. So foreign to our ways of belief are the workings of primitive mentality that we scarcely think of its products as errors—only as crudities of mind in the making. Yet they belong to man's authentic intellectual past and have left their mind-prints on the sands of time.

The prescientific mentality of primitive man is conditioned by the undividedness of his universe—of his view of it. All events and their loosely and vaguely observed "befores" and "afters" are of one undifferentiated order; and that order is *psychic,* patterned after the most intimately known of antecedents and consequents, man's "inside" experience of his fears and angers, his hopes for favors and benefits to come from the powers that be. Into what we regard as forces of Nature, he projects the same friendly and antagonistic feelings as arouse his own activities. He animates and animizes Nature—woods and weather, stars and sun and moon, sea and land, animals and plants. With little insight into physical operations, he creates a world of psychic magic, peopled with gods. He invents ceremonies to propitiate their dispositions, to avoid their ill-will, to atone for his transgressions in offending the powers of his world; yet he has as well an uncertain, feeble notion of error checked by experience.

Primitive man's dominant beliefs and life-serving customs may all be interpreted from this position; this *Weltanschauung* stands as the great, prehistoric antecedent of science—richly false. Its massive impress upon habits of mind remains, from the days of its universal prevalence to its sporadic survivals in our midst. This tendency, when it persists in a sophisticated culture, is called unscientific or superstitious. Until the emancipation from primitive mentality was fairly on the way, beliefs worthy of the name of science could hardly develop. Beliefs formed in the same temper and under much the same outlook continued even among savants until the scientific renaissance of the early seventeenth century. Current popular mentality still bears the impress of the ancestral primitive mind; the folk-mind flows congenially with the ancient stream. The doctrinal mind, while yet affected by naïve premises, wrought its way to more disciplined thinking.

That a learned science can be founded on primitive assumptions appears in the invaluable illustration of the heroic error of astrology. The primitive personal intrusion is the intrepretation of cosmic events as portents and heralds of human fate. The study of the heavens with this intent developed astrology. It required the study of the celestial positions and movements for interest in their laws alone to establish astronomy. Not only *motive* but the *level of logical method* shapes inquiry; the beliefs of astrology proceed on the folklore level, elaborated by learned doctrine; the conclusions of astronomy are framed—with whatever measure of error and imperfection—on the scientific level of investigation.

The simple *portent* stage falls well within the comprehension of the primitive mind. As disturbances of the regular order of events, eclipses were regarded by untutored minds as foreboding evil. They accounted for the eclipse realistically as a monster biting a slice out of the moon or sun. Belief led to practice: alarmed, men made hideous noises to frighten the intruder away. If they sought confirmation, they found it in the observation that in due course the menace withdrew and the celestial orb resumed its proper appearance.

Following a primitive ideology, astrology devised an intricate system of seemingly profound but logically baseless premises. By the accident of assigning the names of certain gods to certain planets, and by the further, equally arbitrary, assignment of human traits among these gods according to the canons of mythology, the planet-deities or deified planets, through analogies of relations in their celestial positions, were assigned an influence upon the fates of men, all concentrated at the moment of birth. For long and unprofitable ages this pseudology absorbed a goodly share of intellectual occupation, and—yet more strangely—it still survives among the untutored, or among those who take lightly their loyalty to science or repudiate their heritage in reserved areas of their convictions. Error and futility embalmed give the appearance of vigorous longevity. The horoscope remains as the blue-ribbon exhibit of the misuse of intelligence.

Astrology proceeds in the temper of a total view of Nature and its operations which prevailed through survival and revival up to the advent of the modern era. As celestial positions are matters of observation out of reach of control, the central purpose of astronomy was to interpret—to collate and map movements and relations and, in astrology, to interpret their influence on the personal affairs of men. Alchemy and chemistry employ and manipulate experimental procedures, deal with tangible things and workable processes, with the purpose of command and control. The illegitimacy of alchemy lies only in the direction of that control toward a specific human desire. Both astrology and alchemy study with a predesignated intent, instead of investigating to determine the laws of matter and processes for that end alone —though ever with a view to application. Alchemy and chemistry overlap; in the course of his manipulations the alchemist accumulated much useful along with more spurious information. Yet the alchemical motive insidiously spread its falsifying influence over the investiture and experiments of the laboratory; it vitiated the temper of inquiry, misdirected it even more than it introduced errors of procedure and conception.

The quest of alchemy was a formula to convert the baser metals like lead into the nobler, more valued ones like silver and gold. The philosophy of alchemy assumed elements and essences and quintescences, distillations and tinctures and "spirits" directing the fanciful processes carried on in retorts and cucurbits and alembics. The appeal was to an occult and magic universe. It has graduated beyond the naïve, primitive world of myth and fairy-tale, wherein a magic lamp or a magic carpet or a magic word works wonders; it proceeds upon the more advanced notion that Nature has her secrets, that the veil may be rent as in a vision and the true essence of being stand revealed. Hence the search for the "philosopher's stone," which would serve at once as a potent alchemical transmuter and a rejuvenating elixir, conferring wealth and youth and vigor by one complete and supreme efficacy. Alchemy became the dominant expression of the magical-mythical concept of Nature, continued in waves of recurrence in the esoteric pursuits of occultists still challenging the ways of science.

As an issue of magical manipulation, exceeding ordinary human powers and calling upon the supernormal, which the Church interpreted as the diabolical, there arose the distinction between the legitimate "white" and the forbidden "black" arts. The Friar Bacon and Dr. Faustus legends turn on this theme. The magic-mystic concept of Nature and its operations dominated thought even as it moved toward the naturalistic view. It appears notably in the practice of early medicine; alchemist and physician were often one, and both in part occultists.

The necromantic logic proceeded thus: "Here is a plate of lead on which is engraved the symbol of a planet; and beside it a leaden flask containing gall. If I now take a piece of fine onyx marked with the same planet-symbol and this dried cypress-branch, and add to them the skin of a snake and the feather of an owl, you will need but to look into one of the tables given you to find that I have only collected various things in the elementary world which bear a relation of mutual activity to Saturn, and if rightly combined can attract both the powers of that planet and of the angels with which it is connected." Plainly there was a long road to travel for the occultist to become a scientist—for long ages a road of compromise and confusion.

That the same order of theory and practice continued well along into the days of scientific advance appears in the career of Jerome Cardan (1501-1576), who adhered to a strange mixture of valid knowledge and gross superstition. Mathematician and physician and naturalist, Cardan was independent enough to protest against the slavish following of authority and to emphasize the importance of observation. Yet he was so credulous that every howling dog was an omen; that he heard murmurs in his ear when others were speaking of him; that he saw on his finger an image of a bloody sword when his son was arrested for murder, which disappeared when the son was executed. When he explained that he was fond of river crabs because his mother, while carrying him, ate many of them, he was giving learned sanction to popular error. When he treated his patients on the theory of heat and cold and the four temperaments, he was following the current heritage; in purging the brain by anointing the coronal suture with ship's tar and honey, or prescribing a diet of tortoise soup, distilled snails, barley-water, and asses' milk, he was following a hodgepodge of theory, analogy, and fancy—the notion being that snails and tortoises, as cold and torpid creatures, would reduce hot, active fever. Consider that picture of an eminent savant-physician and the knowledge and practices of his successors of three centuries later—the laboratories, clinical facilities, and, most of all, the concepts of the processes of health and disease—and the story of science appears in its dominant plot, the elimination of error.

That primitive concepts survive even under fairly advanced conditions of learning appears in the survival of the principle of sympathy and sympathetic magic as advocated by Giambattista della Porta (1543-1617), like Cardan a leader as well as misleader of the science of his day. "There is a deadly hatred between the colewort and the vine. The vine with its tendrils clings to everything else, but shuns colewort as if told that her enemy were at hand." His prepossession vitiates observation: "If colewort is seething in the pot and you put ever so little vine into it, it will neither boil nor keep its color."

Sympathetic cures dominate primitive medicine, continue through medieval prescriptions—notably and curiously in Saxon remedies—and appear in Renaissance days in the celebrated "powder of sympathy" of Sir Kenelm Digby (1640). This powder or salve was to be applied to the weapon that inflicted the wound, whereupon the wound would heal. Charms and spells were compounded upon the same concept, in which a remote and figurative relation was brought into the category of cause and effect.[1]

Both primitive and developed medicine held that eating the heart of a lion would give courage, while partaking of deer would induce timidity, though it might by the same influence confer swiftness in flight. In the classic treatment, from Empedocles to Lucretius, the qualities of attraction and repulsion appear in more abstract forms, serving at once as philosophical concepts metaphorized as love-forces and hate-forces, operating even among atoms. It requires a further step in objectivation to reach a physical concept of cause and effect. Such was the muddled mixture of superstitions, of folklore, and of doctrinal errors that flourished before the great modern awakening, for which Harvey's discovery of the circulation of the blood (1629) may be accepted as a critical date. From then on, the scientific profession, mainly through the emphasis upon observation, moved away from speculative and fanciful explanation. Pre-Harveian and post-Harveian science assume different positions; it is such shifts in the intellectual temper, such advances in the uses of mind, that mark the decisive stages in the elimination of error.

The story of human error follows the objective trail of the development of knowledge and the successive views of the "nature of things" in the chief divisions of learning; it follows the subjective trail of the trends in human thinking that lead astray. The scientific habit of mind has a long and varied struggle with psychological trends in belief and with rival methods of reaching opinion. In consequence, the story of science as a pursuit of knowledge and solutions is as ancient as reflection; but that of the establishment of the scientific method under the dynasty that now reigns is

[1] A related, though not the same, notion appears in the doctrine of signatures, by which the appearance, color, taste, or other quality of a plant indicated its uses and benefits. These beliefs are considered in my *Wish and Wisdom* (New York and London, 1935), Chaps. i, vii, and xix. As an argument by analogy, see page 31.

short and recent. Yet in the antecedents, as in the course of human invention, all the fundamenal stages appear. The basic inventions of the lever, the screw, the rudder, the wheel, the oar, the spear, the arrow, the hammer, the scraper, the drill, fire by friction, and scores of others were all made in prehistoric days—with vast improvements and a store of entirely novel principles contributed since. Similarly, the fundamental logical processes were sensed and practised in early reflection. The story of error inevitably emphasizes tendencies that were in their prime when science was young; yet they have their counterparts still. Modern errors appear in a closer, more detailed perspective; in the historical perspective we look upon what is near with a magnified interest, which appears in the chapters on the various sciences that follow. The present essay attempts to supply for them a broader background.

FETTERS OF KNOWLEDGE

The vicissitudes of belief share with all mundane affairs the subjection to circumstance. The course of discussion is not completely free; knowledge does not proceed untrammeled. The laboratory is a late comer on the human scene; the scepter, the battle-field, the arena, the mob, tribunals for heresy, the stake, are far older as moulding instruments of belief, and more direct and effective. The powers that be acquire a vested interest in the acceptance of belief and establish rigorous censorship. The hemlock for Socrates, martyrdom for Giordano Bruno, forced concealment of his discoveries for Copernicus, forced recantation for Galileo, persecution, ostracism for a long line of victims of authority, whether of Church, State, or School! Still further, the battle for truth encounters prejudice, popular misunderstanding, and the persistence of tradition. Innovation in ideas has as thorny a road to travel as has nonconformity in conduct. Few weighty topics of reflection remain beyond the influence of the temporal interests. Science has had to resort to the locked cell and the secrecy of a suspected guild; academic freedom remains an issue even in our enlightened society. Science has not always been a peaceful pursuit.[2]

The several chapters of this book touch upon the dramatic incidents in this long-enduring and still active conflict. It requires no more specific reference. Interpreted in terms of motive, there emerges as a basis of the conflict the maintenance of the dignity of man. As the bare insistence of dogma and authority, there is little arresting in the tale of intolerance. Thus was the world created; such is man's place in creation; this is man's proper career and destiny. If science concludes otherwise, it must be prepared to meet the opposition of the Church or other constituted authority. It is the same order of conflict as obtained or still obtains within opposing faiths; its dismal records appear in persecutions for heresy, in the annals of witchcraft, in denunciations of evolution and the proscription by legislatures of its teaching in the schools.

The dominion of dogma and the sway of authority prevail not alone by the exercise of constituted power; they exist by virtue of their psychological function in settling things and of their social function of uniting vast multitudes in a common faith. Logically, settling things by authority is just a false method of arriving at conclusions; its practical utility and psychological satisfyingness override this objection. Creeds, gospels, texts, are safe and stable — assured props of faith and secure conviction. Doubt is uneasy and troublesome; security is equally a biological and a psychological asset. It is ever easier to accept than to investigate; intellectual indolence and inertia favor the rule of authority. The scientific ideal is inevitably a late arrival in the cultural complex.

By the same psychology arises hero-worship; appreciation

---

[2] Influential as these obstacles of circumstances are in the historical panorama, their contribution to the analysis of error is ever a variation of one theme: the conflict between rival authorities in establishing and enforcing belief. The most comprehensive survey of dogmatic error in the sciences—a classic still—is written from this point of view: Andrew D. White, *A History of the Warfare of Science with Theology* (New York, 1896).

turns to veneration and assumes that genius can never go wrong. Once accepted as authoritative by Church or by science itself, the word of an Aristotle or a Galen becomes law, and the criterion of truth that of agreement with the dicta of the master. The deference to authority, as well as the absorption in dialectic, becomes a habit of mind, defeating inquiry. The Dark Ages were such by mental habit even more than by neglect of learning; they were unprogressive and obstructive through the pursuit of false patterns of reason. The lack, not of good minds, but of good methods accounts for the stagnation. The method of science has potent rivals in the fixation of belief.

The intrusion of Church-sponsored moralistic and dogmatic doctrine appears saliently in the zoölogical realm. It was held that originally, in the state of Paradise, lion and lamb lay down together; spiders were harmless as flies; "none attempted to devour or in any wise hurt one another" (Saint Augustine). The "fierce and poisonous animals were created for terrifying man in order that he might be made aware of the final punishment of Hell" (Venerable Bede). "Before Adam's sin there was no death and therefore neither ferocity nor venom" (John Wesley). The serpent was then degraded to go upon its belly; having previously no "serpentine form" (Richard Watson, a contemporary of Wesley). Saint Augustine classified animals as useful, hurtful, or superfluous—the last not necessary to human service, yet completing the design of Nature. Luther removed flies from the superfluous to the noxious class, as they were sent by the Devil to vex him in his meditations. A naturalistic view had to contend with prepossessions thus derived.

Other mental attitudes, mainly general credulity, lent credibility to such interpretations. The standard zoölogy was itself a fusion of sound observation and imaginative, dramatically conceived creations without benefit of fact. The result was a fabulous zoölogy, likewise moralized. It appears in the bestiary called *Physiologus* which enjoyed great popularity in the Middle Ages and beyond. The factual and the fabled consorted amicably. Combinations of lion-fish, gorgons, sala-manders, centaurs, unicorns, mermaids, were all acceptable.

A doctrinal intrusion was the anatomical deduction that since. Eve was formed of one of Adam's ribs, men should have one less rib than women. The report of Vesalius from actual dissection that men and women had the same num-ber of ribs was strongly resented. For the most part, dogma and verification had no contact. When the theologian gave himself full sway, even as late as the eighteenth century, it was laid down, as by Dr. John Lightfoot, Vice-Chancellor of Cambridge, that "of clean sorts of beasts there were seven of every kind created, three couples for breeding and the odd one for Adam's sacrifice on his fall, which God foresaw." Dogma, authority, congenial myth, were all closer to the mental habit than rigid observation and grubby facts.

It was not this or another group of specific doctrines that obstructed science; it was the habit of mind that prevailed in learning generally, which spread its influence over the ways of science. Such was the composite scholastic trend—in general, a forsaking of reality and a substitute for it of concepts, dogma, verbalisms, and dialectics. Over it all was the authority of the Church, claiming dominion over secu-lar knowledge and developing within the theological realm the niceties of disputation theoretically, and practically the ceremonials of ritual and symbolical interpretation. This universal pattern of intellectual occupation was not limited to the Christian tradition; it had its earlier counterpart and continuous parallel in rabbinical learning. Nor was it limited to theological and ethical controversy, though there it found its most intensive expression. It was not alone that science had to contend against the pronouncements of ecclesiastical authority; the logic of science had to make its way against the differently tempered methods and ideals of learning fos-tered by the long-prevalent absorption in theology. All this left its stamp upon the procession of ideas. The Baconian reformation was a formulation of a wholly different ideal of learning and of the direction of the intellectual pursuit.

ERROROLOGY

The present project brings into relief the study of error as an integral aspect of the history of science in one bearing, and in another as an embodiment of logical and psychological trends in the search for truth; it may be regarded as a project in "errorology"—if a hybrid term may be pardoned. Attempts at such study appeared as long ago as Roger Bacon (c. 1214-1294), who called these misleading tendencies *offendicula,* or impediments, of thought.[3] He recognized four such obstacles to inquiry: the overweight of authority, the slavery to custom, the dominance of popular opinion, and the concealment of ignorance by pretense of knowledge. Thirteenth-century psychology and mental habit seem remarkably like those of the twentieth century.

Far better known are the four *idols* of Francis Bacon (1561-1626), which are derived from shrewd observation of the thinking habits of both scholars and laymen. Bacon opposed strongly the heavy hand of tradition and commented on the "vanity" of some "moderns" of his time to "attempt to found a system of natural philosophy on the first chapter of *Genesis*." Though Macaulay's charge may hold, that Bacon wrote of science like a Lord Chancellor and preached better than he practised, he must be given credit for an orderly, if at times cavalier, recognition of the profitable procedures of science.

Bacon's initial idol of the *Tribe* covers popular foibles in the thinking process—primarily the insistence that Nature is as the human mind would have it—but also embraces such common fallacies as selecting favorable instances and rashly generalizing. "The human understanding is like a false mirror, which receiving rays irregularly, distorts and discolors the nature of things by mingling its own nature with it."

That is sound psychology for all time. The idol of the *Cave* or *Den* is that of the individual cast and limitation

---

[3] I have used this phrase and illustrated its scope with reference to our own standards of thinking in *Managing Your Mind* (New York, 1932).

of outlook, making men non-understanding and intolerant of others' ways. It is the personal or class idol of prejudice and bigotry. The idol of the *Forum* or *Market-Place* recognizes that opinions operate socially, as "it is by discourse that men associate." The undue deference to public opinion —in our pragmatic days, the idol of the band-wagon and the box-office—belongs here, as also the confusing effect of the "counters of thought" (words), as confusing today as in Bacon's time. The idol of the *Theater*—which is both stage and platform—is particularly directed to the contentions of the scholastic contingent. It applies to doctrines fostered by authority, tradition, vested interests, and to false notions introduced by "worlds of their own creation after an unreal and scenic fashion." The bulk of scholastic and ecclesiastical doctrine was so much conceptual and verbal scenery and screenery—a windmill version of reality against which doughty academic knights tilted; though otherwise accoutred, they also survive.

In modernizing the idols, shaping them more closely to their psychological origins, I have suggested a group of three *subjective* and three *objective* idols.[4] First there are the idol of the *Self*, the projection of the subjective upon Nature; of the *Thrill*, the favoring of the dramatic and romantic; of the *Web*, the spinning of imaginative data.[5] These spring directly from Psyche's realm. In the setting—in that belief is a social, institutional affair—there are the idols of the *Mass*, the undue deference to popular opinion; of the *Mold*, the restriction to one's own class-cast of mind and outlook; of the *Cult*, bondage to dogma and 'isms. These are resultants of the milieu of opinion. Impediments or inexpertness

[4] In *Managing Your Mind* (New York, 1932).
[5] In *The Psychology of Conviction* (Boston, 1918) I have illustrated its vicissitudes. To Henshaw Ward the all-inclusive idol is the idol of the THOB, thinking out an opinion and believing it. His book on *Thobbing* presents a composite of these several feelings in the several sciences in their current temper. With all our advance in standards of believing, much thinking is so infused with thobbing as to vitiate its validity. It is abundantly clear that correct thinking is a difficult art and an uncertain venture. No education is complete without instruction in mental management.

in the thinking process and errors in scientific exploration derive from the same habits of mind; they hark back to the same traits in human psychology that detract from man's qualifications as a logician. The scientist is also affected strongly by his human liabilities. If to err is human, the redempton from error is no less a privilege of man's estate.

LOGICAL PROCEDURE

Science moves on in parallel columns of advance. Setting the line of march is the more rightly oriented, more critically analyzed concept. These are ever wider surveys of fact, better organized knowledge, directed by more refined, improved methods of inquiry. Hence the inclusion in the present essay of an account of logical procedure; for errors in logic are as significant as those in fact.

The logic of science, no less than science itself, is a product of master minds. Like the content of science, command of method grows by experience and then yields to formulation. It has been concisely stated by the master mind of Charles S. Peirce.[6] There are three major orders of *inference,* inference including all varieties of steps from premises to conclusion. Each has its distinctive formula and service. Prompted by curiosity, knowledge begins with the *results* of observation. Gathering is always selection; what one observes reflects the total mental habit, determines both content and conclusion. The personal equation permeates thinking —for better or for worse.

When thinking is on explanation bent, it frames a guess or theory, in logic called *hypothesis.* When thinking aims to discover the rules under which things happens, it *generalizes,* and that is *induction;* when thinking applies what rules have been found, the step is *deduction.* It is all in the service of getting order and understanding of the world and adjusting to it.

[6] Charles S. Peirce, *Chance, Love and Logic,* edited by Morris R. Cohen (New York, 1923), pp. 131 *seq.*

EXPLANATION

Logic sets forth the several steps and types of procedure. If I begin with *results* (of observation or experiment) and know (or suspect) the *rule*, I infer a *case;* an *hypothesis* is my conclusion. An hypothesis may be anything from a guess —even a wild one—to a highly organized, technically stated theory. Myths are imaginative explanations; Kipling's *Just So Stories,* explaining how the elephant got his trunk, or the Dakota legend accounting for the short white-underneath tail of the deer, are poetic versions of what in form is an explanation.

An apt instance of an hypothesis is a diagnosis. The physician notes symptoms (*what* he notes depends on his schooling); he knows that clinical experience has established such a disorder as typhoid fever; on these premises he diagnoses his patient as a *case* of typhoid. That is the framework of his procedure. Other symptoms would lead to a diagnosis (hypothesis) of pneumonia or tumor. Working under other rules, determined by general concepts of the ways of Nature —and that is ever the vital consideration—his remote predecessors, making no sharp distinctions, concluded that disease was a punishment for the patient's neglect of sacrifice to the gods. Or they, like the children of today, by a naïve hypothesis *explained* a blister on the tongue as a *case* of telling a lie. To account for it as a case of slight infection through germs is a scientific explanation in a highly advanced stage of knowledge. These two hypotheses are logical leagues apart, as contrasted as the ox-cart and the automobile in the procession of invention. Yet in bare logical procedure they are alike. They are both ventures in determining cause and effect; that in turn implies a general philosophy of the nature of things.

From primitive peoples to the tribe of Einsteins, Millikans, and Jeans, minds are on explanation bent. The *status* of explanation, the *logical level* of the thought of the explainer, mark its place in the procession of ideas. Of that the bare form of the logical pattern tells nothing; but to the student of scientific progress and its deviations in error, the grade of the explanation forms the nub of the tale. How the several sciences develop and employ hypotheses is too long a story for summary here; it will appear in the contributions that follow. The initial step starts with a *fact* or a group of observations; these suggest a *theory* of explanation, and that in turn is verified by a further fact or consequences of the hypothesis; this second step is *deductive*. It may be taken by way of verification or in application of the theory. Hypothesis-making is running ahead in the course of solution and then going back to check it; it is such in advanced stages of science.[7]

A complex set of observations led Pasteur to the conjecture that a germ was responsible for suppuration; the deduction followed that if a wound were kept scrupulously clean, it would heal without the formation of pus. This was only one section of the evidence that led to the comprehensive germ theory of disease, in turn applied and confirmed. The formation of hypotheses and their testing often fall far short of this ideal procedure. In many investigations an approximation to it must be accepted.

Logical procedure is shaped by the measure of *control* of the data. Most controllable, and thus subject to experiment, are such of the processes of physical and chemical action as may be brought into the laboratory. The range of experimentation with living organisms—from microbe to man—has advanced prodigiously in modern techniques; the establishment of Pasteur's hypothesis depended upon it. A science the data of which are (with slight exceptions) not under control, which is dependent upon such observations as Nature affords, which employs hypotheses as its major

---

[7] All the sciences make use of hypotheses, but each in distinctive fashion. In that hypotheses or theories are so readily formed and may be baseless, speculative, or fantastic, they come under suspicion, even under a ban. The legitimate protest is against unsupported or falsely devised hypotheses. Newton's *hypotheses non fingo* meant not that he did not employ hypotheses, but that he avoided fictitious ones by deriving them from the data at hand and verifying them by further observations.

procedure, is geology. The story of the formation of the earth can be only hypothetically reconstructed. The hypotheses of the serial geological epochs and of the successive glacial periods are founded upon and verified by wide ranges of observation of their annals in stone; but very little may be observed in operation. The nebular hypothesis is of the same order. The time-clock cannot be set back; but so confident is geology of the chronology thus hypothetically constructed that it is applied to certify the age of human remains. The earlier hypotheses formed to explain fossils ranged from thunder-stones to tryouts of the Creator, to shells dropped by returning Crusaders, to vestiges of creation. In a scientific hypothesis fossils are stages through æons of time of biological development, in a series with living forms and embryological stages of growth. They support the theory of evolution, which in its cumulative strength is so firmly established that it is the master-key of biology.

Without the use of hypotheses science would be lame, halt, and blind, and knowledge only empirical. We should be collecting information, observing effects or benefits, but without insight into their *modus operandi*. Under the guidance of false hypotheses we may follow wrong paths and miss opportunities. A critical sense of value in theories, an ingenuity in proving them right or wrong, is the quality of the high-grade scientific mind.

GENERALIZATION

When from his stock of observations the observer notes wherein specimens agree, he attempts a *generalization* and discovers a rule (*induction*). By repeated observations of instances he may reach the valid rule that Scandinavians are (generally) fair-haired. If he happened to know a few Scandinavians who proved untrustworthy and on this basis should conclude that *all* (or most) of that race, or that tow-headed persons generally, possessed that undesirable quality, he would be generalizing, but generalizing rashly. Many prejudices are rash generalizations. But generalize we do and must, constantly arriving at rules which hold more or less; once having the interest of belief in this rule or that, or committed to it, we note conformities and ignore exceptions. Most rules, which are also known as *laws* or *principles,* are not absolute or universal, not uniformities but preponderant regularities; reduced to measurement or enumeration, they yield *statistical generalizations,* serving scientific conclusions.

As with all factors of logical thought, it is the level and sweep, the scope and importance of generalizations that fix their place. To bring apparently unrelated observations under one inclusive generalization is a scientific achievement. When Newton brought the fall of the apple and the forces that hold the celestial bodies in their orbits under one great law, he achieved a mighty generalization, completely out of reach of an ordinary mind. An hypothesis is a proposition; to establish and strengthen it requires *verification* by way of *deduction.* In investigation, in attempting the solutions of problems, in penetrating a stage beyond the frontiers of science, explaining, generalizing, applying, go hand in hand.

The supreme treasury of science is the knowledge of Nature's laws, her tablets of myriad commandments. They are formulated as generalizations, products of the skills of mind in induction. Without rule all would be chaos and the mind stand in bewilderment. Law and order find their reflection in human affairs.

APPLICATION

As experience accumulates, the individual and the race gather a considerable collection of cases and rules; that composes the mental equipment which is applied both to interpret and to produce results. Similarly occupied, though far less reflectively, is a child playing with blocks when it discovers the rule of gravity and the rule of balance, and the cases under them, and builds accordingly. When once we have learned a great number of rules, the problem assumes its typical form: namely, which rule applies to which case. *What* sets of rules and cases we carry in mind, and *how* and *when* we apply them (which implies that we are alert enough to think of their application when needed), that composite mental quality stamps our logical capacity. To possess knowledge and a fair logical equipment is one qualification; to use it rightly is another.

Applications may be momentous, as that of the electromagnetic principle (Faraday) to the construction of motors or, in other and independent variations, to the telegraph and the telephone. The logic of application stands forth as clearly in simpler instances as in momentous ones which by virtue of their consequence are the more dramatic. James Watt was far from the first to observe that steam could lift the lid from a kettle; he is notable for applying the idea and harnessing the power. The concept of electric power generated a giant among inventions. The story of applied science—engineering in all its branches—falls outside the scope of this enterprise, though it is plain that the story of scientific thought is reflected in the procession of inventions. Invention illustrates the logical procedure of application. Applicational projects follow rapidly upon the discovery of a new principle and may develop, as in the use of the X-ray in surgery, in unexpected directions.

Such is one great employment that keeps many minds busy. Another arises is a converse situation which we call a *problem*. The solution of a problem by fitting the right principle—or the most effective, convenient, or economical one—to the situation at hand is an equally common operation illustrating the logic of application. As a slight instance, assume that my problem is to attach a hook to an enameled-iron surface. I know that this is not a proper *case* for driving a nail (impossible) or drilling for bolt or screw (inconvenient) ; I think, and recall the principle of suction. This is a case for that principle. I apply it by fixing the hook in place by means of a rubber cup in which it is set. Or again, there is the Egyptian problem-story of the placing of the huge, heavy obelisks on their stone pedestals. When hoisted by means of a derrick as high as it would go, the obelisk still swung below the level of the base. Then someone shoulted, "Wet the ropes!" The moisture shrank the ropes, and the obelisk was swung into place.

Familiar though it be, it may be well to illustrate the logical relation of principles to inventions. The principle underlying the motion-picture is typical—the *stroboscopic* principle. It reads that *when the eye is given a properly sequenced series of momentary glimpses of an object (or its picture) with proper interruptions to prevent fusion of after-images, the eye will compose these rapid and related glimpses into the appearance of a moving object.* For that is actually how the mind behind the eye sees (infers) movement. The first stroboscope, called a *zoetrope,* later *kinetoscope,* was a circular hat-box of cardboard with slits in the vertical sides and a strip of appropriate drawings on the inside of the box, which were seen through the slits, as the box was rotated on its axis. From that simple toy of our grandfathers has sprung this indispensable ministration to knowledge and pleasure, and the investment of billions of dollars. Every improvement depended upon further principles, such as those involved in recording sound and synchronizing eye and ear.

A far less momentous but equally instructive example, deriving likewise from the use of the eyes, is the *stereoscopic* principle, which a century ago led Sir Charles Wheatstone to devise a stereoscope, in its present form improved by Sir David Brewster and Dr. Oliver Wendell Holmes. This principle reads that *an impression of depth,* or the third dimension, *results from giving each eye its own view of an object and a convenient means of combining the two retinal images.* It may be as truly said of the invention of the instrument that it was a proof of this principle as that the principle led to its application in the instrument. The historic fact seems to be that there was no intention to invent a depth-seeing apparatus (as there was to record movement), but to determine on what basis we see objects as solids. The principle has again become important in X-ray stereoscopic photographs, which enable the surgeon to locate a fracture or foreign object in the body in its depth from the surface.

The logical procedure is simply formulated, but the scope of its application is indefinitely varied; such is the logic of application in science. Pure science pursued for the insight into principles and operations has led to even more momentous advances than the direct search for applications.

DEDUCTION

Practical appliances are not the only examples of deductive steps. In the history of thought another deductive process plays a prominent part, that of deducing further consequences from laws (rules) or facts (results) already reached. This type of deduction is the mainstay of disciplines that frame their own premises, such as mathematical propositions, which end typically with a Q.E.D. Deduction is then bringing out explicitly conclusions contained implicitly in the premises. The general formula reads: If A, then B, and if B, then C, and so on in equation upon equation. Reduced to elementary form, deduction concludes that if John is taller than James, and James is taller than Julius, John is taller than Julius—which is a valid though not very succulent inference. But so complex are the implications of "propositions" in geometry, trigonometry, algebra, or calculus that the development of these instruments of scientific method becomes an elaborate technique. Relations are reduced to abstractions in terms of which equations are solved and the solutions retransformed into practice; thus are bridges built with the stresses in each member accurately calculated.

The principles involved may be formulated abstractly in equations or derived empirically. One can draw an ellipse by using a loop of string longer than the distance between two fixed pins over which it is placed, and describing the curve with a moving pencil held taut against the string, which forms a triangle with the two pins. This device may come first and the equation of the ellipse be deduced therefrom; or the equation may be first determined and a device invented to embody it. Empirical knowledge—rule of thumbs —antedates abstract knowledge—rule of head. There are today calculating machines for all sorts of purposes—cash-registers, slide-rules, multiple-correlation calculators, saving not physical but mental (logical) labor.

Mathematics, the "queen of the sciences," is not in the naturalistic, descriptive sense a science at all, but the sovereign *method* of science, so far as it is applicable. The relations of factors involved in the phenomena of the exact sciences—and in the inexact or variable realms as well—are so complex that the quantitative technique of measurement and the resulting equations form the indispensable calculus for their management. The high development of mathematics, without which modern science could not have arrived, is a tribute to the importance of method; mathematicians are specialists in the development of methods of handling quantitative relations involved in scientific pursuits.[8]

In such sciences as astronomy, the data of which are necessarily expressed in mathematical terms, the entire logical procedure is thereby determined. Inductions and deductions proceed by equations, and their hypotheses are mathematically as well as observationally confirmed. Thus, at one stage, Johann Kepler, still under the bondage of the "perfection" notion of the circle, tried in turn no less than twenty-two hypotheses to fit the observed positions of Mars into an orbit before he finally hit upon the ellipse; a modern mind, free from the older assumptions, would have tried

[8] "Relationology" would be an inclusive term for the method of reducing the relations of things to symbols combined in equations. The *algebra of logic*, as recently developed, is substantially that; it is a specialized medium of scientific disciplines. All elaborate thought requires a highly developed language. A further contribution to relationology is the method of correlation, for which see page 29.

this at once. Centuries later the brilliant verification of an hypothesis based upon elaborate calculations resulted in the discovery of Neptune—predicted independently by two mathematicians. Without the technique of mathematics the human mind could not handle the relations involved in cosmic movements, or in physical, mechanical action, or in construction. Mathematics adds many a cubit to the stature of science. It creates a quantitative world, which not only is represented by Bureaus of Standards, but lies at the basis of recipes and formulae in science generally. It is not, as was anciently assumed, that the cosmos is mathematical or that God geometrizes, but that the human mind has acquired a mathematical form of apprehension.

FORMAL LOGIC

In the educational realm, the central place of deductive discipline was for centuries occupied by formal logic, ruled by the syllogism. Inaugurated by Aristotle, it developed in scholastic dialectic as the daily diet of student minds; it lost its hold only with the reinstatement of inductive logic by John Stuart Mill (1806-1873), a pioneer in the study of the logic of science. Students' minds were drilled in logical squirrel-cages in converting syllogistic "figures" and "moods"—an exercise even less useful than cross-word puzzles.

The standard (and boresome) deductions of the logical text read: "All men are mortal. Socrates is a man; hence Socrates is mortal." Or, as applied to yourself: You will go the way of all flesh, and may wisely prepare for it by making a will—and hardly require a syllogism as a reminder. But so useful an application did not enter the scholastic horizon; the student was supposed to be contented with the thrill of recording that Socrates was mortal, and of recognizing by virtue of what order of syllogism he became so. The supreme virtues of logic are not assailed by its ready misuses. If minds are so trained or tradition is so set that minds prefer the sawdust to the lumber of which useful structures may be built, that is no proof of the worthlessness of architecture. The logical architecture of science, properly conceived, remains its glory.

It was a profound error, at least a barren misdirection of intellectual acumen, that set generations of students to sharpen their wits on deductive subtleties out of relation to the vital problems of Nature's rich universe. By a similar confusion they substituted grammar for literature; the student was taught to parse instead of to interpret. It is unfortunate that scholastic deductive logic came to be regarded as the model of logical procedure. An appreciation of the indispensability of logic in scientific pursuits, and as a criterion of valid results, has arrived only within our own, or at best the preceding, generation; it is still most inadequately disseminated. The greater proportion of current errors—including those in economics and politics—result from a weak sense of the logic of evidence. The error is peculiarly disastrous in the humanized sciences—psychology, anthropology, sociology—and in certain phases of medicine, in which the data are not susceptible of exact formulation, but are no less subject to the same logical dominion. It is the hope of modern ideals to bring to the knowledge of human relations the benefits of scientific thinking.

CAUSE AND EFFECT

From another approach, the generic logical problem in a world so bewilderingly full of numbers of things is to trace in mazes within mazes the connections of cause and effect. In that pursuit, as already indicated, the general notion of how causes and effects are related determines the enterprise; it may proceed in the manner of philosophy or theology, myth, or science. So long as such remote and fantastic cause-and-effect relations as those connecting planetary conjunctions and human traits and fates were credited, any right relations of the conditions affecting human qualities were excluded; no progress, only increased confusion, could result from continued following of that false clue.

The entire procession of ideas takes its impress from the currently entertained notions of cause and effect. Once committed to a naturalistic causology, and ever more extensively and intensively, the progress of science was assured. But issues remain: the general controversy between materialism and idealism, the specific controversy between vitalism and mechanism in biology, are conflicts of view in regard to the spheres of causation that may legitimately be called upon in accounting for observed phenomena. Can mind be reduced to matter, in the sense that mental action can be expressed in terms of brain physiology?

Anyone who could prepare a complete treatise or give adequate rules for the correct tracing of cause and effect in any one of the hundreds of problems engaging the best minds of today would be a genius indeed. We are surrounded by effects and are working by ever more and more refined methods to determine their causes; and other researchers are trying all manner of agencies (causes), not at random or by unregulated speculation or venture, but along the clues of ascertained principles, to observe effects, ever with the goal that some of them may prove useful, and with confidence that increase of knowledge is a worthy pursuit— again not indiscriminately.

What is the cause of the tides; why the huge tides of the Bay of Fundy and the modest ones in New York Harbor? How far does heredity account for human differences, and how does heredity operate? What is the cause of dementia præcox? What is the function of the frontal lobes of the brain? A thousand and one major and minor problems are under investigation, and in each the method of pursuit is as characteristic as the solution reached.

As an instance of the devising of supplementary method to detect more precisely—and with applicability to less manageable material—what "things" are related, and to what extent and regularity, the method of *correlation* may be cited. In this formula, an invariability of relation is indicated by 1.00; no relation at all, or pure chance, by 0; and an intermediate relation by some number between. If the rule of relation for human beings were the taller the brighter, height would be positively and highly correlated with intelligence; and if the rule read the taller the duller, negatively so; if tallness and brightness are totally unrelated, the correlation would be o. For many such relations formulae have been elaborated and tested, and the degree of their reliability likewise measured. It has been shown, for instance, that children resemble parents as closely as they do their brothers and sisters, and that identical twins "correlate" far more highly than siblings. While the study of relations gives no final clue to cause and effect, it serves as a datum in that determination. The modern scientist must be an adept in many methods.

LOGIC AND PSYCHOLOGY

Psychology investigates and describes how the psyche in all its phases works, including how the human mind tends to think. Logic prescribes how the mind should think to discover and interpret things as they are and thus to control their operation. There are *natural* thought-patterns (psychological) and *acquired* thought-patterns (logical). Profiting by the lessons of the past, we shall not repeat the scholastic error of substituting the forms into which inferences can be fitted for the operations as they occur. That we can reduce inferences to syllogistic terms does not prove that in our thinking we syllogize. The great workers observed the rules of grammar, but grammar does not make literature.

Holding to a naturalistic interpretation, we turn to primitive and early thought to observe what thought-patterns they preferably employ and how logic would classify them. The prevalent early thought-pattern proceeds by what logic calls *analogy.*[9] It is the inference—often loose and vague—of a further degree of resemblance from a noted resemblance. All argument by analogy is weak at best and treacherous

[9] I have presented "The Natural History of Analogy" in an essay included in *Fact and Fable in Psychology* (Boston, 1900).

besides; yet it is universally employed. The doctrine of "signatures" exemplifies it, and that of correspondence in occult thought likewise. When walnuts were prescribed for brain trouble because the hard shell is remotely like the skull and the soft and convoluted kernel again remotely resembles the brain, the pattern of such thinking is by analogy. The mandrake was assigned high therapeutic qualities because its bifurcated roots remotely resembled the human form, and the myth arose that it issued a moan when torn from the ground. The weakest of all arguments logically is the most prevalent pyschologically. In general type the argument by analogy is inductive; it is used in scientific thinking, if at all, as a suggestive hint for inquiry. Analogy furnishes a neat instance of the relations between psychological and logical trends.

ERRORS IN THEORY AND PRACTICE

There is an element of paradox in the advance of science through logical steps in defiance of the reminder that one cannot by taking thought add a cubit to the mental stature, the significant fact being that generations have done so progressively. The method thereof is the point of interest. The situation brings to mind the argument to prove the impossibility of building a tunnel: for the masonry is needed to support the vault of the excavation, and the excavation must be made before the masonry can be placed. In practice a length is excavated and a section of masonry placed, and then another length—if necessary, with temporary shoring. Thus theory—which is view and insight and a push ahead—and practice—which is experience and construction—compromise and coöperate, and the tunnel proceeds to completion.

This dual dependence is reflected in the diversities of minds; thus arise theorizers and practitioners, the one apt in plan, direction, and design, the other in assembling and placing bricks of facts. But the parallel fails in that data have not the uniformity of bricks, but arise out of an interest and in a setting that gives a fringe of meaning to a nucleus of observation. For the very recognition and use of a fact implies selection. Theory issues the summons. The "theory"-suggestiveness of a fact and the "fact"-suggestiveness of a theory enter into and compose the values of the logical campaign by which new territory is won or the conquered area more thoroughly explored. Thus science builds its tunnels.

Errorology includes the correction of errors that prevail in regard to logical procedure and the relations of its components. The precept first to secure facts and then to frame theories is wise within its limitations, but it is too simple to cover the scope of the adventure in inquiry. Visitors to a strange country who go as military, economic, or ethnological observers in part select their facts, as they pack their kits. A map of boundaries is in itself a mingling of geographical and political alignments; a geological map has a different appearance from one of commercial products. Unprejudiced as the scientific observer should be, he is not blandly neutral or as indifferently registering as a camera. He comes to his survey with chart and compass and technique; and that includes a critical and disciplined view of the nature of fact and theory and their relations, as applied to the terrain of exploration.

Gross errors in the concept of "facts," which ignore the setting of theory in which facts emerge or ignore that a belief in a theory may *create* the alleged facts, are found only in extreme instances. Trained minds, within the scope of their training, are immune to such misapprehensions, though they appear constantly in popular thought. Yet a recent distinguished physiologist, Charles Richet, argued in regard to so radical an innovation as "ectoplasm" that the proper procedure is to admit the "fact" of plasmic emanations from human mediums and explain it as best we can. He ignored that he himself contributed the "plasmic" theory to a "phenomenon" which might and was until then posed as a "spirit-force"—itself a wild, unscientific hypothesis—

or accounted for as a fraudulent performance. Which is the fact? that is the issue. The provenance of "facts" is a vital factor in their status. "Facts" come thickly encrusted with theory, like the evidences (facts) of witchcraft, and decline when the state of mind responsible for the theory loses its hold. Or if this seems a far-fetched instance, its parallel may be found within the more exact domain of physics. The chemist Karl von Reichenbach announced the discovery (theory) of a totally new physical force. "Od" on the basis of "facts" which were actually the subjective impressions of neurotic subjects, though recorded with all the (spurious) objectivity of a laboratory protocol. A right sense for fact is indispensable to scientific pursuits.

When fact and theory proceed in proper dependence and within the accredited rules of procedure, there still remains a suspicion of theory by the practical-minded and a charge of the futility of mere fact-accumulation by advocates of theory. Theories without facts or based on uncritically selected facts are vain, and facts without theoretical interpretation, blind. The appearance favors the facts. When master and dog go out for a stroll, the dog runs ahead of the master, suggesting that the master is following the dog; in reality, the dog is keen to any turn of the master, yet once it is taken, he goes along confidently, as though the route were of his selection. Actually the dog is following the master. Theory in the broad sense—a vision of the larger planned procedure —is the decisive step.

There is a measure of truth in the dictum that science is organized common sense; but it is far truer that the organization far exceeds the reach of common sense. Common sense would pronounce the earth flat and accept the vault of the horizon as a reality. Correcting common sense is the business of science; common sense is a stabilizing influence in checking the extravagances of theory, but is itself responsible for errors past and present.

Truly, there is vouchsafed to the scientific mind no gift of supernal transcendent revelation of another order than applies to the humblest employment of reason. Science, like many a common and false solution, is guesswork; but the distinction of science is no less illustrious, in that it turns upon the critical acumen by which a guess becomes a valid working hypothesis.

However scientific the intent, however versed in knowledge, resourceful in hypothesis, the explorer may be, if he lacks the technique of logic and the sobriety of outlook that such adherence confers, his course is precarious. This judgment may be applied, with all respect, to the theories of Freud, which have so consequentially affected current thought. Framed in the temper of science, their genuine insight cannot offset their extravagance of theory and lamentably weak sense of logic.

In as much as the theories of our predecessors proved variously defective and fallacious, so, it may be argued, are our guesses no better than those of old, and in turn they will go the way of all knowledge. Theories rise and fall as better, truer theories replace them; yet it is unwarranted to conclude that science is truth for a day. The essential truths endure; the stepping-stones within the same circle of validity have a permanent service, for they bear the stamp of a more rigid discipline.

Likewise is it idle to proclaim that there is nothing new in rational experience under the sun; or to ignore that the imaginings of a Lucretius or the dreams of a speculative Roger Bacon and the orders of inquiry that have actually produced the flying-machine have little in common. It is still less warranted to find any parallel in goal or method between the alchemists' hollow dream of transmutation and the recent accomplishments of the physicists. Such identification ignores the logic of procedure.

TECHNICAL ERRORS

We may dismiss with a bare statement the comprehensive observation that the errors of modern science, subsequent to the recognition of the logical framework to which all science conforms, are technical errors. The cruder errors, like the basic discoveries and inventions, belong to the earlier stages of science; they are no longer in and of our outlook. Ineptitudes in the management of the logical machinery of investigation, of interpretation, of proof, constitute the technical errors of science that are considered in their several provinces in the contributions that follow. When these are examined comparatively and analytically, they yield general conclusions illustrative of the vicissitudes of ideas. The divisions of the sciences represent convenient categories of specialized interests; Nature is one. All the sciences have passed through similar stages of logical evolution; all have encountered, though in varying measures and aspects, the prepossessions that the human mind brings to the quest for truth.

The use of intelligence is the highest privilege and the deadliest menace of humanity. By the increase of understanding and organization, man has largely transferred the control of his destiny from Nature to his own decisions. It is a sobering or a despondent reflection that the inability to enlist intelligence in the service of an enlightened, tolerant dispensation may annihilate—as if by a ghastly super-gas-bomb of malign stupidity—the cumulative heritage of thought and enterprise. On the one hand the brilliant triumphs of logical discipline; and, in dismal contrast, the undisciplined warrings of nations over fleshpots and messes of pottage; the resources of Nature under superb orderly dominion, and the contentions of men under unruly strivings of warped judgments. The atmosphere of laboratory and council-chamber seem so alien as to deny their common origin in sapience. In that perspective the story of human error in the sciences may be lowered in significance, compared with the Herculean task of bringing intelligence to correct the follies of ambitions of empire and tyrannical control. When, in the twentieth century, master minds of science may become hounded political exiles, and scientists of distinction sacrifice their loyalty to truth to remain under the protection of perverted authority, the shock compels a revaluation of pen and sword. To follow the story of human error in the intolerances of government is almost too discouraging an undertaking; yet rationalism and sanity have a common cause. A life of reason remains a worthy and a possible goal, despite the impression which this volume countenances that the story of man might well be entitled *Our Misbehaving Minds.*

It is its purpose to incorporate into the story of thinking a consideration of the obstructions, the false leads, the ineptitudes along the slow, irregular course that forms an integral portion of our intellectual heritage, thus restoring the perspective of the epic of Man the Thinker.

PART I
WORLD

Section I
THE COSMIC REALM

CHAPTER I

ERROR IN ASTRONOMY

By Harlan T. Stetson

THE STORY *of error in science may well begin with astronomy, for astronomy is the oldest of the sciences—in a certain sense a parent of them all. Primitive man, in observing the star-strewn skies, naturally formed a homocentric interpretation of the universe around him. Trusting implicitly in appearances, his early attempts to picture the universe inevitably involved gross errors. Not until some sense of the law and orderliness of Nature replaced man's credence in magic and caprice, and not until the mind of man was able to divorce itself from the error of the homocentric concept, was it possible for astronomy to make distinct progress. Slowly, step by step, by trial and error in ideas and theories and in interpretation of observations, the human mind has been led out of the primitive wilderness of misconceptions into a world which views the universe in the light of modern astronomy— with new depths of discovery unfolding to every increase of power in the telescope.*

From the vantage-ground of the astronomy of today we can perceive the mistakes of thinking and of judgment in the views of yesterday; and it is indeed remarkable that on such a series of badly laid foundations the present formulations of the universe should have been constructed. The possible errors in modern astronomy need be slightly considered, since with constantly changing conceptions and added observations many of the present-day theories, such as that of the expanding universe, must remain uncertain. The advances of another generation will pass fairer judgment on the rôle of error in the concepts of the astronomy of today. Our story, however, should make clear how, through logical methods and improved accuracy in thinking, the probable error of judgment is slowly being reduced; that at least we may feel confident that the facts of observation form a body

39

of knowledge which is unchanging through the years, whatever may be the changes in interpretation that time may bring.

NAÏVE ASTRONOMY

WHEN ONE READS the astounding facts of modern astronomy and learns of galaxies upon galaxies already revealed, with more yet to come when the great 200-inch telescope for the California observatory is completed, curiosity asks, "How has modern science been able to achieve such a stupendous disclosure of the universe around us?"

It was from man's wonder about the stars that the first idea of natural law and order, which is the basis for all science, came into existence. Paradoxical as it may seem, his chief source of error was the tendency to take Nature at her face value, to accept appearances as the warrant of reality. Yet one without astronomical knowledge can readily reënact much of the drama of human error in gaining a knowledge of the universe, by going into the open on any clear night and looking into the vault above him. By a strange illusion, such an observer seems to be rather centrally located on a flat disk of earth, overspread by a hemispherical dome bedecked with scintillating points of light which we call stars. If the observer lingers long enough, he will discover that the spangled dome of sky appears to be slowly revolving about him. Constellations rise in the east, pass slowly across the sky, and sink in the west, as though a sphere were being majestically turned by some mysterious spirit hidden beyond the celestial blue.

In this simple picture appears the first set of errors in astronomy, which had to be outlived before any progress could be made towards understanding of the constitution of the stars and their location. What primitive man saw—as we see—and accepted as the reality was a flat earth, himself at the center, with a celestial sphere rotating about him and carrying the sun, moon, and stars, as though these objects were attached to its inner surface.

Perhaps the first great disillusionment came when some ambitious traveler ventured far north or south from his home. As he receded from either pole of the sphere, he would see that certain stars near this pole gradually described lower and lower arcs until they were lost beneath his horizon. If our early observer lived in the northern hemisphere, as most probably he did, he would find likewise that to the south new stars appeared above his horizon as he journeyed southward. Thus he would make the first important discovery, that the appearance of the sky changed with his location on the earth. Reasoning from such facts as these, the early Greek astronomers came to the conclusion that they must be traveling on a sphere—not on a flat earth. Thus came the first correction of the erroneous conception of the position and movement of the earth in cosmic space.

We next encounter in the same early stage of reflection on the celestial world a totally different order of error, not of observation but of interpretation, derived not from the firmament but from human psychology: namely, the belief that spirits, deities, agencies, associated with the skies, have influence upon men's lives. In some respects perhaps this was a most fortunate error, for it gave stimulus to astronomical study at a time when interests and instincts were primitive and personal. Spurred by the desire to interpret the aspects of the skies as they shaped the fortunes of men, the early observers noted carefully the appearances of the heavens to anticipate the secrets of coming events.

Much of early astronomical knowledge was acquired in the service of astrology—a differently focused interest. Not until centuries elapsed, until man came to perceive his apparently insignificant place in the cosmic realm, did the absurdities of early astrological thinking give way to the desire to discover truth for its own sake.

THE PLANETS

Conspicuous among the brighter objects of the night-time skies were certain brilliant stars whose peculiar movements

indicated that they did not belong to the same category as
the so-called fixed stars, which night after night, year after
year, faithfully retained their positions among their neighbors in the revolving celestial sphere. The ancients recognized seven roving objects—the sun, moon, Mercury, Venus,

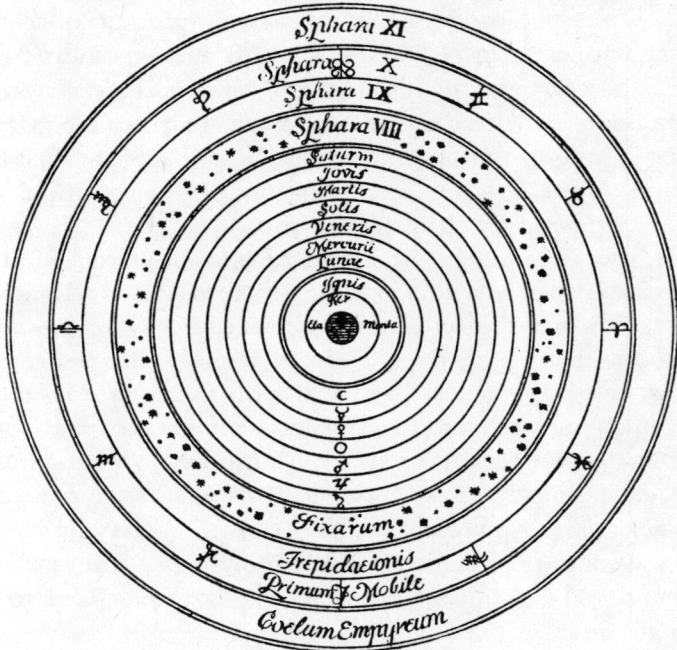

THE PTOLEMAIC SYSTEM

From Harlan T. Stetson, *Man and the Stars* (New York, 1930). Courtesy of
the McGraw-Hill Book Company.

Mars, Jupiter, and Saturn; by reason of their peripatetic behavior, these objects were named *planets,* which means wanderers. Continuing to mistake appearances for reality, they
assumed that these bodies must be revolving about the earth
against the dome of heaven. So complicated were the curves
that the planets described in the sky that the early philos-

ophers resorted to all sorts of mechanical analogies to explain their movements.

A fairly satisfactory approximation to the movements of the planets (not including the sun and moon) could be based upon the assumption that each planet revolved on the

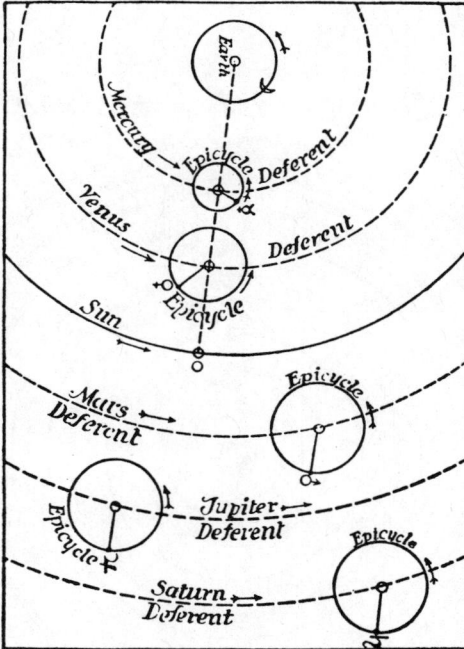

PLANETARY EPICYCLES

From Harlan T. Stetson, *Man and the Stars* (New York, 1930). Courtesy of the McGraw-Hill Book Company.

periphery of a circle called an epicycle, the center of which traversed the circumference of a larger circle known as the deferent. At the center of all these larger circles was the earth. Thus the system is known as the *geocentric system.* Receiving the official stamp of approval of Claudius Ptolemy in his epoch-making book *The Almagest* (about 150 A.D.), this early Alexandrian cosmology long since became known as the Ptolemaic theory. But the insensitive planets knew nothing of Ptolemaic contrivances to govern their motions and impishly persisted in wandering off their prescribed courses—much to the consternation of the Greek astronomers. Ptolemaic philosophers repaired their tottering theory by the addition of epicycle upon epicycle, as Milton said,

> With centric and eccentric scribbled o'er,
> Cycle and epicycle orb in orb,

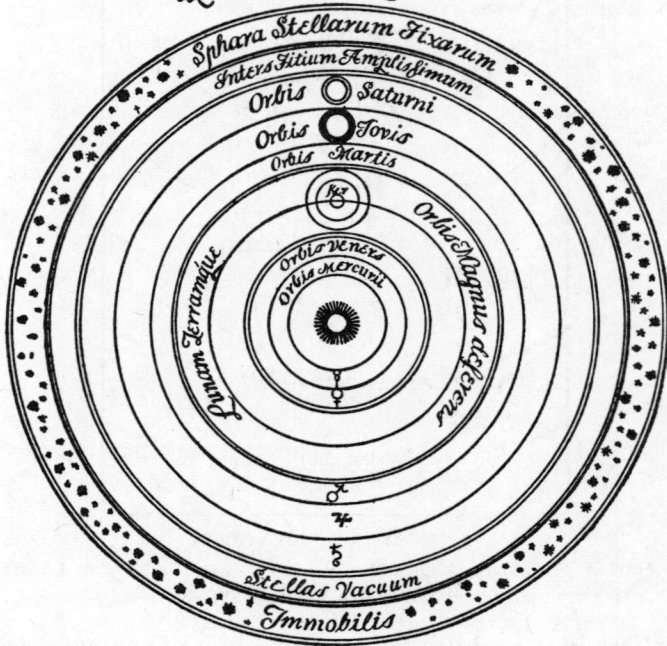

THE COPERNICAN SYSTEM

From Harlan T. Stetson, *Man and the Stars* (New York, 1930). Courtesy of the McGraw-Hill Book Company.

a complication which for a time served to improve their calculations for the predicted planetary positions. But with a system based upon error they could not forever postpone the downfall of the geocentric theory, although it dominated astronomical science for fourteen centuries.

THE COPERNICAN SYSTEM

It was Nicolaus Copernicus (1473-1543), a Pole, who first dared to refute the errors of his predecessors in his celebrated work *De Revolutionibus Orbium Cœlestium*. The book was published just before the death of Copernicus in 1543, and

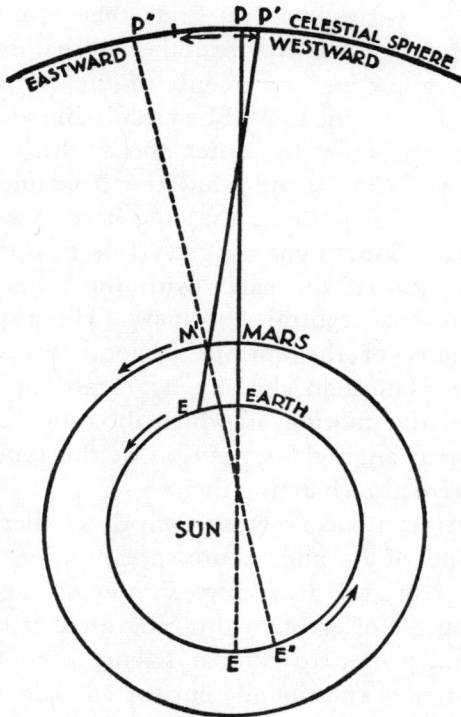

THE RETROGRADE MOTION OF A PLANET

It will be observed that the outer planets must always retrograde when both planets are on the same side of the sun. From Harlan T. Stetson, *Man and the Stars* (New York, 1930). Courtesy of the McGraw-Hill Book Company.

he was thus released from witnessing the havoc that followed. One of the tragedies of human thinking is the inertia that resists change in ideas once they are accepted. The more closely emotional response is entwined with error, the more difficult does it become to change one's thinking. The accepted cosmology of Ptolemy had become fundamental in the concept and plan of creation in medieval Church dogma; hence the painful readjustment to the Copernican doctrine.

Copernicus, through researches in the documents of the Alexandrian school, had learned that certain Greeks—among them Aristotle, Pythagoras, and Eratosthenes—had conceived not only that the earth was round, but that so far as the effect on the apparent movements of the sun and planets was concerned, it might indeed be circulating about the sun, rather than itself being the center about which the sun and planets revolved. So, abandoning the fundamental notion of the geocentric hypothesis that the earth was the center of all creation, Copernicus assigned this distinction to the sun and argued that the earth with the other planets revolved about this central luminary. He explained the simple geometry of the curious periodic backward movements of the planets in the sky by regarding them as resultants of relative motion, in which the observer on a moving earth was changing his position at the same time that the planets were also changing theirs.

The Copernican theory was promptly challenged, chiefly on the ground of its unorthodox presentation of the Creator's design. Here was a striking example of one of the most dangerous sources of error in thinking, in that the ecclesiastic zealots endeavored to combat science with theology, to reason with tenets emotionally supported. Did they not insist that the "earth hath He fixed that it shall not be moved"? Some few, however, with a glimmer of scientific reasoning, objected to the new doctrine on the ground that if the earth moved in an orbit about the sun, as Copernicus would have them believe, the stars would appear to be seen in different directions in the sky during the course of the year as the earth completed its revolution. This was really a very sound objection, and one which could not be met in Copernicus' day because of the lack of any precise instruments to measure so small an apparent change in the direction in which a star is seen as is caused by the earth's motion in its orbit. Not until nearly three hundred years later was the first so-called *parallax* of a star measured. Parallax, in astronomical nomenclature, is the term used to represent the difference in direction in which a star is seen from the circumference of the earth's orbit as compared with its position relative to the sun at the center of the earth's orbit. To this objection against his theory Copernicus is said to have replied, perhaps more wisely than he knew, that the stars were so far away that the variations in their apparent positions due to the earth's turning about the sun were too small to be detected.

PROGRESS IN PLANETARY OBSERVATIONS

With the stage thus set for the development of new ideas in astronomy and for the inevitable conflict with the old, a Danish astronomer, Tycho Brahe (1546-1601), was inspired to gather further observational material which was to help to decide the outcome of the battle of error against truth in the evolution of the planetary theory. With a grant from Frederick II of Denmark, Tycho set up on the Isle of Hveen an elaborate observatory which he called Uraniborg, the castle of the heavens. Here he developed instruments more precise than any that had been constructed hitherto. Night by night he carefully measured the positions of stars and planets, hoping no doubt that if the Copernican doctrine should prove true, he would be able to detect small changes in the motions of the stars corresponding to the revolution of the earth about the sun. In any event he was charting the positions of the planets in a way which should serve to determine with greater precision the course of the orbits they traversed.

Whether Tycho was politically shrewd in deference to ecclesiastical authority and therefore held to the error that the earth must be the center of the universe, or whether it was his own mental inertia that made it difficult for him to embrace the novel doctrine of the heliocentric theory,

remains a matter of opinion. He was responsible for in-
venting a hybrid theory which subsequent events were to
show was the result of a marriage of error and truth—the
error that the earth was actually the center of the universe,
and the truth that all the planets revolved about the sun.
Obviously, Tycho did not accept the idea that the earth itself behaved like any other planet. He therefore postulated
that the moon and the sun revolved about the earth as the
center of creation, but that the planets Mercury, Venus,
Mars, Jupiter, and Saturn revolved about the sun; he thus
held to the idea of the epicycle advanced by Ptolemy, but
made the several planetary epicycles concentric with the sun.
The Tychonic system holds slight place in the history of
astronomy, but the thousands of observations Tycho made
with his new instruments became a monumental contribution to the development of true ideas of planetary motion
and made possible the work of his successors.

KEPLER'S SOLUTION

Tycho Brahe had in his observatory a young German
assistant by the name of Johann Kepler (1571-1630), and it
was to Kepler that the great Danish astronomer bequeathed
his observations and the unsolved problems to which they
gave rise. Kepler, attacking the problem of planetary motion with a mystical mind, at first followed the same fallacy
held by his predecessors—that since the Creator was a *perfect* being, and the circle was the only *perfect* geometric
figure, *therefore* the planets must move in circles. In vain
did he try to fit his master's observations into orbits of a
circular nature—equally vainly whether the earth or the sun
was postulated to be at the center. With one bold stroke
Kepler ventured to forsake the ways of traditional thinking
and try other forms of curves for the orbits of the recal-
citrant planets. At length he tried the *ellipse,* and found that
he could represent planetary motion far more accurately than
had yet been done in the history of astronomy. He announced the first fundamental law of planetary theory—that

each of the planets revolves about the sun in an elliptical orbit with the sun at one of the foci.[1]

Further examination of Tycho Brahe's masses of observations led to two more empirical laws. The first of these states that the speed of each planet at any point in its orbit is such that a line joining the planet and the sun will sweep over the same area in equal units of time. This means that when the earth is, for example, farthest from the sun in July, it moves more leisurely in its slightly elliptical track than it does in January, when its smaller distance requires that it speed up its motion in order that its radius vector shall describe the same area in the course of a day that it does at the earth's greater distance during the summer months. Fortunately for those of us who live in the northern hemisphere, the nearer approach of the earth to the sun in January means that our winters are less cold and of somewhat shorter duration than they would otherwise be—than, in fact, they are in the southern hemisphere.

Kepler's third important law relates the speed of the different planets to their respective distances from the sun. With a high degree of approximation, the square of the time that a planet requires to make one complete revolution about the sun, divided by the cube of its mean distance from the sun, is a constant. To Kepler's mind this conveyed an idea of rhythm and harmony in Nature which led him to call this principle his Harmonic Law.

Now that we command the mathematical theory of planetary motion, it seems the more surprising that Kepler should have been able to formulate such accurate conclusions in these three laws; to him they were entirely empirical, for Kepler had no notion of the fundamentals of mechanics, which had to await the discoveries of Newton. In the third,

---

[1] "He [Kepler] accomplished this by his incomparable energy and courage, blundering along in the most inconceivable way (to us), from one irrational hypothesis to another, until, after trying twenty-two of these, he fell, by the mere exhaustion of his invention, upon the orbit which a mind well furnished with the weapons of modern logic would have tried almost at the outset."—Charles S. Peirce, *Chance, Love and Logic* (New York, 1923).

though here with no serious effect upon the truth, an alien
esthetic notion was again intruded into what otherwise was
a scientific observation.

GALILEO'S REFORMS

Contemporaneously with Kepler lived the Italian Galileo
Galilei (1564-1642), justly called the father of experimental
science. More than any other representative of Renaissance
Science, Galileo broke the shackles of traditional thinking
that bound men to error in their search for truth. Until the
day of Galileo little was known concerning the fundamental
principles of motion. If a body moved, it was because something or someone moved it; hence it had been postulated
that spiritual forces were necessary to keep the planets going
in their courses about the sun. Galileo supported the thesis
that the inertia of motion and the inertia of rest were fundamentally the same—that a body set in motion would continue moving unless it were stopped by some object or opposing force, just as a body at rest would not start moving
unless an external impelling force acted upon it. Such unorthodox thinking brought Galileo into constant conflict
with the philosophers of his day.

Probably the most spectacular performance staged by
Galileo in disillusioning men's minds of tradition was the
classic demonstration at the Leaning Tower of Pisa. The
familiar tale [2] has a permanent place in the story of error.
Aristotle had stated that bodies fell at speeds proportional
to their weights, and the ideas of Aristotle were held in such
reverence that, like the law of the Medes and Persians, they
could not be changed. Boldly challenging this position, Galileo appealed to Nature to uphold his conclusion. Carrying
with him a massive hundred-pound shot and a shot weighing one pound, Galileo trudged his way up the stone stair-
case of Pisa's inclined tower. Poising the weights upon the
balustrade until he had the attention of the crowd beneath,
he pushed off the two shots together. Together they fell all

[2] See also Chapter IV, "Error in Physics."

the way, and together they struck upon the turf. Yet jeering savants returned to the philosophy of Aristotle, read again his statement that the speed of falling bodies was in proportion to their weight, and repeated, despite the evidence, "Galileo is wrong!"

More trouble was in store for Galileo when in 1610 he directed his newly contrived telescope toward the sky. Although the exact origin of the telescope is in doubt, it is established that Galileo was the first to apply it to the stars and to appreciate its significance for astronomy. With his magic "optick tube" he found the surface of the moon to be rough and rugged, pitted with hundreds of craters of what appeared to be extinct volcanoes. The immaculate sun he found besmeared with spots which drifted across its surface, betraying a globe in rotation and thus at least furnishing an analogy for his belief in the rotating earth. Galileo discovered that Jupiter was accompanied by four satellites revolving about the planet, thus adding additional members to the solar system. This discovery brought additional storms of protest from the dogmatists of his day. The accredited doctrine argued that as there were seven days in the week and seven churches in Asia, so the number of planets was *necessarily* seven. If the number of planets were increased, then the whole system was doomed. "Moreover," argued Francesco Sizzi, "the satellites are invisible to the naked eye and therefore can have no influence on the earth, and therefore would be useless, and therefore do not exist."

Still more acute were Galileo's later troubles. Venus appeared in his small telescope to show phases like the moon, proving that the planet was shining by borrowed light from the sun and fulfilling an old prophecy of Copernicus that "God is good and some day he will show the phases of Venus to somebody." The price Galileo paid for his acceptance and advocacy of the Copernican doctrine was humiliation and the condemnation of the Church. His mind, his temperament, and the temper of the day were such as to engender the bitterest hostility of an ecclesiastical régime whose theology, wedded to medieval cosmology, long retarded the progress of science. How strange, indeed, that the zeal of the sponsors of spiritual thinking, whose purpose it is to lead man outward and upward in his search for truth to the highest achievements of which the human mind is capable, proved a stubborn obstacle in the path of progress.

NEWTON'S GREAT INDUCTION

With the initiation of the scientific attitude of mind by Galileo's contributions, astronomy moved rapidly to the status of a mathematical science. The termination of the age-long conflict over the planetary system came with the enunciation of the law of universal gravitation by Sir Isaac Newton (1642-1727), whose contributions to science belong equally to physics and astronomy. The traditional story that the fall of an apple led Newton to evolve his gravitational theory may or may not be based on historical foundation. His great induction was the embodiment in a single statement of a law which applies accurately and equally to the fall of an apple and the movements of celestial bodies, such as the moon, which literally falls to the earth while sweeping forward in its orbit.

It was a long step from the initial conception which the apple may have suggested to the complete formulation of Newton's law of gravitation: "Every particle of matter in the Universe attracts every other particle with a force which varies directly as a product of the masses of the particles concerned, and inversely as the square of the distance between them." But could the earth and the moon, relatively large spherical bodies, be considered to behave as particles? Before he answered this question in making general the application of his law, Newton sought to convince himself through mathematical reasoning that any spherical body may be considered as having all its mass concentrated at its center. This was essential to his major deductions.

Observational error played a further part in delaying somewhat the completion of Newton's gravitational theory.

As a datum for applying his law of inverse squares to prove that the same force holds the moon in its orbit as acts upon falling bodies on the earth's surface, Newton needed to know the distance to the moon. It was common knowledge that the moon's distance was about sixty times the radius of the earth; but until the careful survey of Jean Picard in France (1671) the accepted value of the earth's diameter was about 15 per cent too large. Using this erroneous value, Newton's assumed distance to the moon was in corresponding error. The calculation of the speed of falling bodies from the rate at which the moon fell toward the earth agreed so poorly with measured speeds of bodies falling at the earth's surface that Newton, we are told, was so disheartened that he abandoned for many months the completion of his publication. The story goes that when word was received of the true value of the earth's radius, obtained from Picard's new measurements on the meridian near Paris, and Newton revised his calculation with the new value for the moon's distance, he was so emotionally moved to find an exact accord in the place of the former discrepancy that he called an assistant to write for him the final figures of his computations.

STELLAR PARALLAX

Not infrequently has the science of astronomy been advanced through the curious events in human experience that we sometimes call accidents. James Bradley (1693-1762), for a time Professor of Astronomy at Oxford and later the Astronomer Royal at the Observatory at Greenwich, hoped that the progressive improvements in the construction of the telescope and other astronomical instruments would make possible the measurement of parallax, which, as already defined, is the small annual displacement of the position of a star due to the motion of the earth in its orbit about the sun. With the accumulation of hundreds of observations of star positions, Bradley made a thorough search to determine whether there was not some small periodic displacement of twelve months' duration that would reflect the annual revo-

lution of the earth, as had been predicted from the days of
the Copernican theory. He was not long in reaping a reward.
He found that not only the nearer but the farther stars all
shifted their positions by small but significant amounts. In
the part of the sky directly over the earth's orbit they described small *circles* of 20″ of arc in radius. Stars near the
plane of the earth's orbit oscillated back and forth in *straight
lines,* while stars in the region of the sky intermediate between the pole of the earth's orbit and its plane described
ellipses, the half major axes of which were again 20″ of arc.
It was always the same amount for every star, whether dim
or faint, and therefore presumably whether far or near. This
could not have been true if what Bradley thus had discovered
was parallax, for the *nearer* stars would then shift their positions by a *greater* amount, just as the nearer objects in the
landscape viewed from a moving train shift more conspicuously with the passage of the observer than do the hills on
the distant horizon. No, Bradley had not measured parallax;
but he had discovered something very different from that for
which he had searched—the so-called *aberration of light,* a
displacement due to the combination of the velocity of the
earth in its orbit with the velocity of light from the distant
star.

At that time Ole Römer had just announced the finite
velocity of light as calculated from observations of Jupiter's
satellites. He had previously deduced a fair value for its
velocity from his observation that light consumed some eight
minutes in coming from the sun to the earth, a distance of
93,000,000 miles. Utilizing Römer's value, Bradley was able
to show that his discovery had led indeed to an observational
proof that the earth revolved about the sun in Copernican
fashion, and that it was moving at a speed of 19 miles per
second—a value quite consistent with the velocity arrived
at by other lines of evidence. Thus, while disappointed in
not succeeding in determining the parallax of any star, Bradley contributed, as so often occurs in science, a by-product
of his efforts equally significant in value. It was not until

1838 that the first successful measurements of stellar parallax were made.

NEW PLANETS DISCOVERED

Up to this point we have traced the story of error in the development of astronomy through the principal solutions of the problem of planetary motion, as finally unraveled by Newton, and through the actual demonstration of the orbital motion of the earth about the sun. Aside from the discovery of the satellites of Jupiter and Saturn, which came with the invention of the telescope, not a single new planetary member of the solar system was revealed until 1781. On the night of March 13 of that year, William Herschel (1738-1822) discovered an outlying planet beyond the orbit of Saturn, a planet first named after Herschel but later called Uranus. The frontiers of science were being pushed back. Here was another giant planet swinging around the sun at a distance twice as great as that of Saturn and obeying the same laws of gravitation as controlled the earth and the moon.

With the discovery of Uranus, Herschel, already famed as a musician, began his career of fame in astronomy. He was, however, by no means the first observer to have seen Uranus; for here again appears a comedy of errors. The planet had actually been observed *nineteen* times before that memorable night in March, 1781. Flamsteed had entered it in his catalogue of stars in five distinct places. It was observed as a star—not as a planet—also by Bradley, Johann T. Mayer, and Pierre Charles Lemonnier. Had Lemonnier been more prompt in reducing the records of his observations, wherein no less than twelve positions of this body were recorded, he would have had the honor of the discovery. As it is, all honor fittingly goes to Herschel for the identification of the planet Uranus.

But the discovery of Uranus brought new difficulties. Careful observations for months and years disclosed discrepancies in its predicted positions. These gave rise to the query whether possibly the law of gravitation, which had held for all the other planets, was not completely applicable to the orbit of Uranus. Two mathematicians, John C. Adams in England and U. J. J. Leverrier in France, with unshakable faith in the law so thoroughly demonstrated, endeavored to vindicate the very law Uranus appeared to violate as pointing to the existence of another planet lying far outside the orbit of Uranus, which, through its own gravitational attraction, was a disturbing influence in the irregular behavior of the planet discovered by Herschel. With surprising precision the position of the undiscovered hypothetical planet was *predicted* independently by the English and the French scientist. The planet Neptune was actually discovered just where predicted, by John G. Galle at his observatory in Berlin on the night of September 23, 1846, one of the most brilliant confirmations in all science.

Likewise based upon the mathematical predictions of two American astronomers, Percival Lowell and William H. Pickering, was a second remarkable discovery of a new planet, Pluto. The perturbations of Neptune and Uranus led Lowell and Pickering to assume an unknown planet X beyond Neptune and to calculate its hypothetical orbit. Planet X was photographed by C. W. Tombaugh at the Lowell Observatory at Flagstaff, Arizona, in 1930, fourteen years after Lowell's death, and designated Pluto. The orbit of the new planet agreed remarkably well with that calculated by Lowell. Later, however, the investigations by E. W. Brown of Yale indicated that the small mass of Pluto could not have affected Uranus to the extent required. It was then determined that slight errors in the early observations of Uranus led to the remarkable coincidence of the actual and predicted positions of Pluto. It was an extraordinary contingency that a chance error happened to be in the right direction and amount to lead to the discovery of an unknown planet.[3]

[3] The story is admirably told in the *Biography of Percival Lowell* (New York, 1935) by his brother, A. Lawrence Lowell, Ex-president of Harvard University, with a critical discussion of the astronomical technicalities by Henry Norris Russell.

Another remarkable instance of an important discovery resulting from an unsuccessful effort to determine parallax is connected with Herschel. Herschel resolved to search for a small displacement of fixed stars which should reflect the annual motion of the earth about the sun, by comparing the position of a bright star week by week and month by month in relation to the position of a closely adjacent faint star, presumably much farther away from the earth. Herschel had no difficulty in discovering thousands of such pairs of "double stars." He expected to find that the brighter star of a pair—assumed from its extra brilliance to be much nearer than the other—would be observed to move with respect to the fainter star in a tiny circle or ellipse during the course of the year, because, as the earth moved in its orbit about the sun, he would be directing his telescope towards the star from a slightly different position in different months. Herschel was not in error in his theory or in his search for this motion, but he was in error in thinking it could be detected by that method. For today we know that many of Herschel's double stars are *binary systems,* in which the fainter star, usually the less massive, circulates about the brighter star in periods of many years, often of centuries. We, of a later day, can appreciate Herschel's initial bewilderment when he found that it was the faint star that apparently moved with respect to the bright star, and that it completed its circuit in a much longer time than a single year. He was very quick, however, to catch the significance of his new discovery and to realize that there were stellar systems as well as planetary systems; that although he had not achieved the expected demonstration of the earth's motion about the sun, he had come upon a far more striking demonstration— the demonstration that Newton's law of gravitation was valid for stellar systems far, far beyond the limits of our own planetary range.

THE ORIGIN OF SIDEREAL ASTRONOMY

Herschel added to these achievements a systematic study of the distribution of stars based on actual counts of their numbers in selected fields of his telescope. From these so-called star gauges he deduced a theory of the construction of our universe. Considering the limitations of astronomical knowledge in his time, it is surprisingly in accord with our present-day picture of the galactic system. Herschel likened the universe in shape to a grindstone, with the sun and the solar system situated nearly in its central plane. He accounted for the fact that we see more stars in the direction of the Milky Way by the hypothesis that we are then looking through the long dimension of the universe, which naturally includes more stars. The paucity of stars at the poles of the Milky Way, or—to preserve the figure—in the direction of the axis of the grindstone, indicates that the thin dimension of the universe lies in that direction. Assuming that the stars are equally distributed, this view would include a much smaller number than when we look towards the rim of the "grindstone" from our somewhat central position. Little did Herschel dream of the enormous extent of the galaxy.

In scanning the skies Herschel was often astounded by certain areas almost completely lacking in stars. Some of these, circular in shape, were relatively so small that he referred to them as holes in the heavens, through which man looked into the great beyond where there were no stars. In the light of modern astronomy this interpretation appears to be very much in error; for we have strong reason to believe that these dark areas are caused by occulting matter situated on this side of the star-strewn background beyond. The presence of many such dark areas has made the modern astronomer conscious of the danger of interpreting the numbers of stars seen in given areas as representing the stellar population in those regions. In fact, it appears entirely possible that gigantic clouds of dark absorbing matter—cosmic dust and fogs of tenuous material—dim the light of distant stars so as to confuse our interpretation of the dimensions of the universe in many directions.

Among the peculiar objects that puzzled Herschel were the so-called *star clusters* and *nebulae*—colonies of stars compactly gathered together, some in the form of a luminous nebulous mass like that of a distant street-lamp seen through a thick London fog. Because of their cloud-like nature, thousands of these objects have been catalogued as nebulae, the word *nebula* literally meaning a cloud. With the increasing power of Herschel's larger instruments he soon found that many of these objects resolved themselves into thickly compacted clusters of stars, and he began to wonder whether, if only we had telescopes powerful enough, we should not be able to resolve all of these diffuse nebulosities into star clusters.

STELLAR SPECTRA

The answer to Herschel's question could not be given until, shortly after the middle of the nineteenth century, a new instrument was introduced into astronomy—the spectroscope. This instrument, a product of the chemical and physical laboratories, was not at first enthusiastically received by conventional astronomers, who thought of the problems of astronomy as limited to such as could be solved from observations of position and motion. Astronomers who first applied the spectroscope to their telescopes in an endeavor to analyze the light from distant celestial objects—as the chemist would examine hot incandescent gases in his laboratory—were looked upon as departing from the tradition of an astronomical observatory. At length, however, those astronomers who became sufficiently versed in physics to understand the manipulation of the spectroscope and the interpretation of spectra were the creators of a new science recognized today as *astrophysics*—known in the latter part of the nineteenth century as "the new astronomy."

To the spectroscope we owe important revelations in our knowledge of the constitution of the sun and stars. By its aid we have been enabled to determine the physical and chemical constitution of every astral body that emits light enough to make possible the analysis of its spectrum. Spectroscopic analysis has demonstrated that matter everywhere throughout the universe is the same, that practically every element known on earth is found among the stars, and that we ourselves are of such materials as stars are made of. How tragic would have been the error had the spectroscope been left out of astronomy, and yet how slowly was this unconventional instrument introduced into the observatories of yesterday! Today physics, chemistry, and astronomy are jointly concerned in the supreme quest for the ultimate nature of matter.

It is interesting to note that, having contributed to astronomy a means of studying the physical and chemical nature of the stars, the spectroscope should also make possible the study of their motions without waiting the long intervals of time required by the older methods of positional astronomy. For just as the blast of the whistle of a locomotive is at a higher pitch while the train is approaching and falls immediately in pitch when receding from the observer, so, if you please, the vibration of light as recorded in the lines of the spectroscope shifts a little higher in the octave when a star is approaching the earth and apparently slows down in its vibrations—moving the spectral line towards the lower or red end of the spectrum—when the star is receding from the earth.

A universal application of this principle, known from its first discoverer as *Doppler's principle,* has made it possible from a single observation of a star's spectrum to determine the relative velocity between the earth and the star and to remove all doubt as to whether the star is receding from or approaching the earth. From extensive observations of thousands of stars with the use of the spectroscope, we have come to discover that our own sun is moving at the rate of 400,000,000 miles a year among a group of nearer stars within our galactic system, and that our whole galactic system is itself in rotation about a common center. This same instrument has answered Herschel's dilemma about the nature of the nebulae, for it has taught us that some nebulae are indeed aggregations of stars, while others are enormous whirls of glowing gas incandescent through their own luminous fires, and still others are clouds of gas electrically excited by the presence of near-by stars.

GALAXIES

The universe of Herschel, bounded by the Milky Way, has grown in size in modern astronomy. Galaxies upon galaxies now are recognized, so that our universe of yesterday takes a secondary place among the spiral nebulae which strew the sky by millions. Less than a generation ago astronomers were debating the size of our own galactic system and whether or not these spiral nebulae were part of it. The major diameter of our whole Milky Way system was defined by Simon Newcomb, a leading astronomer of a preceding generation—now voted into the Hall of Fame—as probably about 2,000 light-years. From a study of the fluctuations of light from certain types of variable stars, it has become possible indirectly to determine the distances of some of the star clusters of our own galaxy. Harlow Shapley less than two decades ago astounded astronomers with a computation that our galactic system could not be less than 300,000 light-years in diameter, large enough presumably to include the spiral nebulae themselves.[4] More recently the application of the same law of the relation of the period of a cluster-type variable star to its luminosity brought the startling result from the 100-inch telescope at the Mount Wilson Observatory that the Andromeda Nebula is approximately three times as far away as even this unthinkable diameter of our galactic system. Other spirals yielded even

[4] This estimate first appeared in a paper in the *Astrophysical Journal* in 1918, reprinted in *Contributions of the Mount Wilson Solar Observatory*, Carnegie Institution of Washington Publication No. 157, p. 1. A later estimate of not less than 200,000 light-years appears in Professor Shapley's *Flights from Chaos* (New York, 1930).

greater distances, so that an old idea of Herschel, that there are island universes, was again revived. Application of the spectroscope to the determination of the velocities of these many spiral nebulae has revealed that for the most part the lines of the spectra of the spirals are shifted considerably towards the red, indicating that they are receding from us at a rate of hundreds and even thousands of miles per second. Does this mean that the whole universe, which comprises a galaxy of galaxies, is expanding in all directions? If so, there must have been a limited time since it began its expansion; and if so, must we look into a future when the universe will have dissipated into even greater tenuosity than its present state? The mathematical theories of such geniuses as Einstein, De Sitter, and Le Maître would indeed seem to offer some theoretical basis for believing that such is the case. The observational evidence of lines shifted towards the red in the spectra of the spiral nebulae would appear to be confirmatory of the theoretical deductions of mathematicians.

THE DESTINY OF THE UNIVERSE

Our survey has reached modern times. Our very proximity to the picture of the universe now being painted, and to the artists who are sketching its details, makes perspective difficult. To trace the story of error in our modern thinking may require the coming of another day in astronomy. We may well be on our guard, however, knowing that so many times Nature has tricked scientists into acceptance of appearance for reality. Theories of the expanding universe, however rigid may be the mathematical deductions, must be founded upon certain premises and postulates. But, one asks, do not the observational data of the spectroscope show beyond doubt that the distant galaxies are receding? Yes, if our only possible interpretation of the shifting of the lines to the red is that based on Doppler's principle. If, however, through some mechanism in the distant sources of light, or through the medium through which that light is transmitted to earth, the vibrations are slowed down, then we might well have a falling pitch of the light-wave, or a shifting of the spectral lines towards the red, quite independently of Doppler's principle, which depends upon the recession of the distant stars.

The history of error in our ideas of the evolution of the solar system and of the age of the planet earth on which we live, while well within the province of astronomy, may be left to the geologist to relate. Through error and its correction we have progressed in our knowledge of the universe towards an approximation to truth in the realm of astronomy. The story makes scientists tolerant towards conflicting theories that touch the frontiers of knowledge, confident that out of confusion truth will ultimately emerge.

CHAPTER II

ERROR IN GEOLOGY

By Kirtley F. Mather

GEOLOGY *is popularly conceived as the study of rocks and minerals, metallic and non-metallic ores, fuels and water-supplies, fossil animals and plants. It is all that and more, for it is nothing less than the history of the earth and its inhabitants. The geologist is constantly asking how things came to be as they are. When he is concerned with the discovery and utilization of the natural resources upon which our physical prosperity depends, he must find out how the iron ore or petroleum or building-stone came into existence or was concentrated in bodies suitable for economical exploitation, in order to guide the search for additional deposits or the processes of recovery from the known deposits. When he is concerned with the surface features of the earth, he must learn how soils are formed and transported, how rivers behave, how glaciers sculpture the land, how waves and currents erode the shores of sea and lake or build new lands along the strand. His descriptions of the things he sees are almost invariably only the first step towards an explanation of those things in historical terms.*

Men first became interested in geology because of spectacular events like floods, earthquakes, and volcanic eruptions, or because of their discovery of beautiful crystals, unusual stones, or curious fossils. Gradually that interest expanded to include ordinary as well as extraordinary objects and events. Gradually also the demand for suitable building materials, for gold and jewels, for the great variety of metals, salts, and fuels useful in industry and commerce, stimulated the search for the natural resources hidden in the earth's outer shell, and that in turn awakened a new interest in this particular branch of science.

Obviously, errors in interpretation and causal explanation are much more likely to occur than errors in description, although frequently it has been faulty observation that has led to erroneous interpretation. It is therefore not surprising that the history of geological science is liberally sprinkled with naïve and ludicrous explanations of natural phenomena. In the pages that follow, the attempt has been made to select the most interesting and significant of these errors from each of the more important subdivisions of the science. Taken as a whole, this chapter should serve to acquaint the reader with the major steps in the still incomplete process of unraveling the tangled threads of earth history.

I T SEEMS to be impossible to discover any logical point at which to begin the story of geological error. The scanty records of the dim beginning of man's endeavor to understand the earth upon which he dwells and to comprehend the ways in which its features have been shaped reveal the gradual emergence of knowledge from superstition; but there is no date that can be selected as the birthday of the science. Observation and experiment were occasionally the basis for ideas even in the days of the cave-man and the Greek scholars. Uncontrolled speculation and authoritarian dogma still persist in our century, the alleged age of reason.

Error in geology has resulted largely from inadequate observation and faulty experiment, but it has also been a consequence of subservience, either conscious or unconscious, to the dicta of theologians and philosophers. This mixture of intellectual ills has produced a long series of quaint, naïve, and often ludicrous ideas, but usually there is a gleam of truth in the midst of the error. Rarely can we say that a particular geological theory is all wrong or all right; various shades of gray are far commoner than either black or white in the geological color-scheme.

THE EARTH'S INTERIOR

Take for example the curious ideas that men have entertained concerning the structure of the earth beneath us. It was to be expected that those whose intellectual heritage was drawn from the Mediterranean region should have early concluded that below the solid surface of the earth there must be a place of eternal fire. The doctrines concerning hell, so cruelly orthodox in medieval Christianity, were obviously supported by everyday observations. The floods of lava, the clouds of ashes, and the intensely hot gases ejected from Etna and Vesuvius certainly came from the earth's interior. How the fire at the center of the earth was kindled and how it was maintained presented a difficult problem to those whose interest in "how," the most scientific of all questions, was engendered by their interest in "what" and "where," "why" and "when."

Frascatus had a neat answer.[1] The earth stands at the center of the universe; consequently its center is the center of the whole universe. Both are spherical in form. The sun, moon, planets, and fixed stars all shoot out rays of heat which tend to concentrate at the center of the universe, as the spokes of a wheel meet in its central hub, so that the center of the earth must be the hottest of all places in the universe. Some think that these rays find their final goal at the earth's center, others that when they reach the center they are reflected back along the same course or at an acute angle; but whichever view be true, these rays, colliding with one another and rubbing against each other in the narrowed space about the earth's center, by their friction generate fire, which is nothing else than excess of heat. The heat thus derived from celestial sources will be maintained so long as the universe endures.

Such ideas were widely current during the sixteenth and seventeenth centuries, quaint though they appear in the light of modern science. There were, however, a few medieval scientists who even then considered them fantastic. Agricola, for example, contended [2] that "the heat was derived from the combustion of beds of coal, bodies of bitumen, or,

---

[1] Gabrielis Frascati Brixiani, *De Aquis Returbii Ticinensibus* (1575); quoted by F. D. Adams in "Origin and Nature of Ore Deposits," Bulletin of the Geological Society of America, Vol. 45 (1934), pp. 375-424.

[2] Georgius Agricola, *De Ortu et Causis Subterraneorum* (1546); quoted by Adams, *loc. cit.*

in some cases, masses of sulphur which existed beneath volcanic centers, and that these were ignited by intensely
heated vapors which in their turn derived their heat from
friction which was set up within the gaseous mass itself or
by its contact with the walls of narrow spaces through which
it was forced when in rapid movement within the earth."
He implied that the subterranean fires were localized beneath volcanic vents, rather than widespread throughout the
whole interior of the earth. This explanation of internal heat
in terms of combustion of coal and other inflammable materials was quite generally accepted in later years and was
promulgated by Werner and his followers even as late as the
opening of the nineteenth century.

It soon began to appear, however, that high internal temperature was not a local phenomenon confined to volcanic
regions. So far as available records indicate, the first definite
observations concerning the temperature in the deeper por-
tions of mines were made in 1740 by Gensanne in the lead
mines of the Vosges. By the end of that century it was known
that wherever observations were made, there was a notable
increase in temperature with depth below the surface. Evidently, human beings were living on the graciously cooled
periphery of an intensely hot planet.

This was, of course, happy news in the nineteenth century. Under the influence of the Laplacian nebular theory
of earth origin, concerning which more will be said later,
the scientific authorities in the first half of that century accepted as fact the idea that the earth was essentially a ball of
molten rock material surrounded by a comparatively thin
crust. Cooling had progressed from the gaseous to the liquid
stage, and at last a sufficient thickness of solid rock had provided the necessary environment for life.

But how thick was the crust? For decades the popular
answer was a thickness somewhere between 14 and 75 miles,
a trivial fraction of the earth's radius. Shortly after the
middle of the nineteenth century, however, it began to be
recognized that the earth is an extremely rigid body. If it

were essentially a liquid sphere encased in a comparatively thin surficial shell of solidified rock, gigantic tidal waves would necessarily take place within its body as a result of the attraction of the moon. George Darwin effectively disposed of the popular and long-standing error concerning the interior of the earth by demonstrating that tidal ebb and flow could be resisted only by a crust of enormous thickness: at least 2,000 miles of solid rock must intervene between the surface and any liquid nucleus. This computation has recently been confirmed by modern seismology, although the precise nature of the central core of the rigid earth is even yet unknown.

THE FALLACY OF THE NEPTUNISTS

Quite naturally the earlier geologists devoted much more attention to the rocks that could be seen at the surface of the earth than to those which are concealed from view in its deeper interior. For decades a bitter controversy was waged concerning the origin of the various kinds of rock in European mountains. One group, ably led by Abraham Gottlob Werner (1749-1817), argued stoutly that all rocks were "precipitations and depositions formed in succession ... from water which covered the globe." The minerals (included in those days among the general designation "fossils") "which constitute the beds and strata of mountains were dissolved in this universal water and were precipitated from it." [3]

The followers of Werner, appropriately labeled "Neptunists," were vigorously opposed by the "Plutonists," among whom James Hutton (1726-1797), "a private gentleman" of Edinburgh, was the recognized leader. The plutonic doctrine explained such rocks as granite as the result of the deep-seated crystallization of molten material, certainly far removed from water in its nature, and suggested that basalt was formed from lava flows and could not be a precipitation

[3] A. G. Werner, *New Theory of the Formation of Veins*, translated by Charles Anderson (Edinburgh, 1809).

from the sea. Curiously enough, it was around the origin of basalt that the battle was waged most furiously. Many were the learned dissertations which proved with equal finality that the extensive basaltic flows of Germany and France were or were not the result of the former presence of the element under the jurisdiction of the god Neptune where today are lofty peaks and widespreading plateaus.

One of the most telling blows which did much to demonstrate the error of the Wernerian school was delivered by Rudolph Raspe (1737-1794), a German mineralogist and archæologist who found it necessary to emigrate to England and who was doubtless the most notorious rogue ever connected with the science of geology. His fame as the greatest liar of all time is well established, for he is credited with being the author of *The Adventures of Baron Munchausen*! In science, however, he deserves great credit for his careful observations and accurate descriptions of the basaltic lavas and associated volcanic rocks in the Habichwold. His conclusions concerning their volcanic nature are now universally accepted; the Neptunists have disappeared completely from geological society.

THE MEANING OF FOSSILS

But the abandonment of the Neptunist doctrine that all rocks originated as precipitations in the sea should not lead to the inference that no rocks came into existence in that way. Even in the midst of the error there was a modicum of truth. Yet that truth emerged only after a long struggle. Nobody knows who first noticed the fossil shells of marine animals among the sedimentary rocks of mountainous uplands, but such evidences of the former presence of the sea where now is lofty land were obviously most puzzling. The simple explanation that crustal upheaval had lifted the sea floor to heights of many thousands of feet above sea-level involved mental readjustments which were not easy to make. Handicapped by the notion that the earth had been created by Divine fiat in its present completed form, and with wholly inadequate concepts of the duration of time involved in earth history, the medieval philosopher-scientists were hard put to it to explain such anomalies.

Avicenna (980-1037) suggested that fossils had been generated in "the bowels of the earth" by the mysterious creative force that continually strives to produce the organic out of the inorganic, and that they were "unsuccessful attempts of nature, the form having been produced but no animal life bestowed." [4] For three hundred years, beginning in the fifteenth century, scientific circles were frequently enlivened by disputations concerning the nature of fossils, and numerous clever suggestions were proposed in the midst of arguments of which the net result was to pour more heat than light upon the subject. Fossils were the abortive results of the germination of living seeds or eggs carried in mists blown from the ocean and lodged in crevices in the rocks. Or they were regarded merely as illusory sports of nature, possibly even implanted in the ground by the Creator with the meretricious purpose of setting a trap for those foolhardy individuals who essayed to eat of the fruit of the Tree of Knowledge and made bold to probe the unfathomable secret of how the world was made.

Despite the fact that, near the close of the fifteenth century, Leonardo da Vinci (1452-1519), the world-famous artist and architect, announced that the fossils he had seen while constructing canals in North Italy were the remains of marine organisms which had actually lived where they were found at a time when the sea covered all that region, error still reigned in the minds of the orthodox intellectuals. In 1557, Fallopia of Padua, one of the greatest anatomists of his time, described the fossil teeth of elephants as earthy concretions and the mineralized shells from Volteranno as the result of fermentation and exhalations from the earth. Michele Mercati prepared good illustrations of fossil bivalves, ammonites, and nummulites in the museum of Pope

---

[4] Karl A. von Zittel, *History of Geology and Paleontology* (London, 1901), p. 13.

Sixtus V, and these were published between 1717 and 1719 in the *Metallotheca Vaticana*. He named the fossils according to Pliny, and after long discussion of their probable source came to "the conclusion that they took origin under the influence of the stars." [5]

Lister and Lhuyd in England and Lang in Switzerland clung to the naïve absurdities even while making truly valuable contributions to the advance of knowledge. Martin Lister (1638-1711) laid down the important principle that the different sedimentary rock formations might be distinguished by their particular fossil content, and illustrated living and fossil bivalves and snails side by side to demonstrate their resemblance. But at the same time he insisted that the fossils were merely poor imitations of the real forms and must have been produced in the rocks by some unknown cause.

The last of this long line of believers in magic among those who have retrieved the records of prehistoric life development was Johannes Beringer, a professor in the University of Würzburg, whose scholarly career collapsed most tragically as a result of the pranks of his heartless students. In 1726 he published a remarkable paleontological monograph entitled *Lithographia Würceburgensis* in which he illustrated a miscellaneous collection of true fossils from the Triassic rocks of North Bavaria, mixed with stone replicas of sun, moon, stars, and Hebraic letters, all of which were soberly described as having grown in the rocks. "As a matter of fact, his students, who no longer believed in the Greek myth of self-generation in the rocks, had placed artificially concocted forms in the earth, and during excursions had inveigled the credulous professor to those particular spots and discovered them! But when at last Beringer's own name was found apparently in fossil form in the rocks, the mystery was revealed to the unfortunate professor." [6] The distracted man tried frantically to purchase and destroy all the copies of his book, but without avail. In 1767 a new edition was

[5] Zittel, *op. cit.*, p. 16.
[6] Zittel, *op. cit.*, p. 18.

published, and the book is now preserved as a scientific curiosity.

One of Beringer's most interesting contemporaries was the Swiss naturalist Johann Scheuchzer. By the time he was ready to prepare his natural history of Switzerland, the organic origin and historical significance of fossils had become rather generally recognized. But very promptly the religious forces of the day claimed that all fossils were vestiges of the Biblical creation and represented the creatures that had been interred during the Noachian Deluge. The "Diluvialists" were numerous among the geologists of the seventeenth and eighteenth centuries and were, of course, warmly supported by the Church. Scheuchzer accepted the doctrine wholeheartedly and sought eagerly for every possible proof of its correctness. His *Querulae Piscium,* or *Complaint of the Fishes,* published in Latin early in the eighteenth century, is a fantastic volume. It is illustrated by many expensively engraved figures of various fossils, most of which are described as the remains of fishes which through no sin of their own had been destroyed in the world catastrophe and imprisoned in the rocks ever since. It also contains what is probably the first picture of ichthyosaur bones. But these vertebrae, found by Scheuchzer and his friend Langhans in the "inclosure of the gallows" at Altorf, are described as human bones, relics of "the accursed race destroyed by the flood."

Late in his life Scheuchzer found at Oeningen some fossil bones which he described as "the bony skeleton of one of these infamous men whose sins brought upon the world the dire misfortune of the deluge." To the fossil he accordingly gave the name *"Homo diluvii testis."* The bones are, however, those of a large Tertiary amphibian ancestrally related to the salamander! They are now on display in the Teyler Museum at Haarlem and bear the label *Andrias scheuchzeri,* in honor of their Swiss discoverer. Scheuchzer's engraving of this "Witness of the Flood," as also that of the ichthyosaur vertebrae of Altorf, was reproduced in the famous "Copper

Bible" as positive proof of the literal accuracy of the biblical record.[7]

Somewhat similar motives to those which stimulated the Diluvialists have persisted even into the twentieth century.

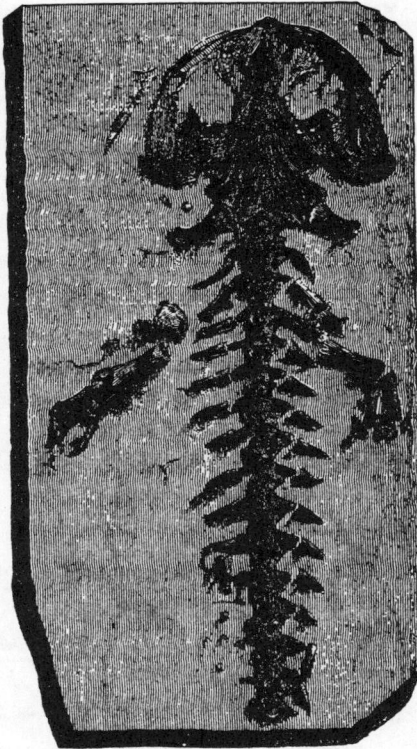

HOMO DILUVII TESTIS
From H. Alleyne Nicholson, *The Ancient Life-History of the Earth* (New York, 1899).

There is little doubt that the imposing array of prehistoric animals and plants, whose fossil remains have been discovered by thousands of geologists and arranged in chronologic order, provide for the average layman the most compelling of all the arguments that have been advanced in favor of the

[7] S. W. Williston, *Water Reptiles of the Past and Present* (Chicago, 1914), pp. 107-9.

GEOLOGICAL CHRONOLOGY

		Periods and Epochs	Geological Events (North America)	Life
CENOZOIC ERA	Tertiary	**RECENT** About 15,000 years ago		Man
		PLEISTOCENE Culmination of mammals followed by extinction of many groups About 1,500,000 years ago	Continental glaciation	
		PLIOCENE Animals and plants much like those of modern times About 15,000,000 years ago	Cascade and Coast Ranges	
		MIOCENE Widespread appearance of modern families of mammals About 30,000,000 years ago	Marginal seas	Flowering Plants / Mammals
		OLIGOCENE First Anthropoids. Extinction of archaic mammals About 40,000,000 years ago	Widespread terrestrial deposits	
		EOCENE Mammals become dominant, all modern groups represented About 50,000,000 years ago	Mountain glaciers	
		PALEOCENE First great differentiation of Mammals About 60,000,000 years ago	Rocky Mountains	
MESOZOIC ERA	Late Mesozoic	**UPPER CRETACEOUS** Extinction of Dinosaurs, Mosasaurs, Plesiosaurs, Ichthyosaurs, Pterosaurs and Ammonites. First Placental and Marsupial Mammals About 105,000,000 years ago	Great continental seas	
		LOWER CRETACEOUS First abundance of plants with conspicuous flowers and fruits About 120,000,000 years ago	Terrestrial deposits Marginal seas	
	Early Mesozoic	**JURASSIC** First Birds and Modern Fishes. Ammonites, Ichthyosaurs, Pterosaurs, Plesiosaurs and Dinosaurs abundant About 150,000,000 years ago	Sierra Nevada Great continental seas in Western America	Reptiles
		TRIASSIC Culmination of Ammonites and Amphibians. First Dinosaurs, Mammals and Modern Corals About 180,000,000 years ago	Terrestrial deposits in the East	

principle of organic evolution. If, therefore, the popular acceptance of that principle is to be undermined by anyone who believes it to be antagonistic to religion, he must destroy the cogency of the argument based on the succession of fossil forms. Within recent years the charge has been made, frequently in sermons and addresses, occasionally in books and periodicals, that geologists have reasoned in a circle. Having assumed evolution "from amœba to man," they arrange the rocks in order from oldest to youngest on the basis of their fossil content—the more primitive and generalized the animal remains, the older the rocks, and vice versa. Then they announce that because the fossils of the more lowly

GEOLOGICAL CHRONOLOGY—*Continued*

			Periods and Epochs	Geological Events (North America)	Life	
PALEOZOIC ERA	Upper Paleozoic	Carboniferous	**PERMIAN** Last of the Old Corals, Trilobites and Eurypterids. Spread of Insects, Amphibians and Reptiles About 225,000,000 years ago	Appalachian Mountains Glacial Epoch Draining of the seas from the continents	Cone-Bearing Plants	Amphibians
			PENNSYLVANIAN First Reptiles, Insects, Centipedes and true spiders. Great Coal Forests About 270,000,000 years ago	Period of crustal unrest, alternation of marine and terrestrial conditions over wide areas		
			MISSISSIPPIAN First Amphibians and Ammonites. Sharks, old Ganoids, Crinoids, Blastoids and Brachiopods abundant About 300,000,000 years ago	Widespread shallow seas on continent Acadian Mountains	Fern-Like Plants	Fishes
	Middle Paleozoic		**DEVONIAN** First Arthrodires, Lung fish, Ganoids, Sharks, fresh-water Mollusca and Land Plants. Goniatites and Ostracoderms abundant About 345,000,000 years ago	Broad shallow seas on the continent during Silurian and Devonian		
			SILURIAN First Scorpions, Goniatites and Fish-like animals. Eurypterids and corals abundant About 375,000,000 years ago	Taconic Mountains	Marine Plants	Marine Invertebrates
	Lower Paleozoic		**ORDOVICIAN** First Corals, Crinoids, Starfish, Bryozoans, Pelecypods, Eurypterids and Ostracoderms. Graptolites, Cephalopods and Trilobites abundant About 435,000,000 years ago	Broad shallow seas on the continent		
			CAMBRIAN First well-preserved fossils. Trilobites, Brachiopods and Gastropods abundant About 540,000,000 years ago	Narrow seas within the borders of North America		
			PROTEROZOIC ERA First fossils, all obscure, indicate existence of algae and worms About 1,000,000,000 years ago	Killarney Mountains Glacial Epoch Algoman Mountains		
			ARCHEOZOIC ERA History obscure, life existed but no fossils yet found About 1,800,000,000 years ago	Laurentian Revolution		

creatures are found in the older rocks, whereas those of more highly specialized animals occur only in the more recent formations, the evolution of life is demonstrated. Thus goes the argument of those who would preserve the revealed truth of a divinely inspired Bible against the diabolical attack of the evolutionists.

Obviously, the indictment would stand were its fundamental premise not an error. The fact is that the relative sequence of the rocks is determined in the last analysis not by their fossil content but by their physical relationships. The relative ages of all the great systems of sedimentary formations were correctly determined long before the epoch-making contributions of Darwin. Fossils are the key to correlation, not to chronologic sequence. The idea that geologists reason in a circle with regard to fossils and time relations is itself an error which needs to be scotched whenever it lifts its head.

THE IDENTIFICATION OF FOSSILS

Once the true nature and real significance of fossils have been comprehended, there remains the difficult task of reconstructing an accurate picture of the animals of the past from the fragmentary remains or scanty relics that are available. It is not surprising that erroneous inferences have occasionally resulted from the attempt to interpret the record in the rocks. The story of the fossil footprints in the Triassic red beds of the Connecticut Valley is an almost perfect example of the gradual emergence of truth from the mists of error.

These remarkable fossils were first called to the attention of the scientific world in 1836 by Professor Edward Hitchcock (1793-1864) of Amherst College. His opinion concerning them had been solicited by Dr. James Deane of Greenfield, to whom Dexter Marsh of Deerfield had gone for an explanation of the strange imprints he had observed on some flagstones being laid in front of his house. It was obvious that they were footprints preserved from time immemorial in the solid rock, but whether of man or beast he and his neighbors could not decide. Apparently it soon became the popular opinion that they were the tracks of birds, some of which must have been of gigantic size, imprinted in the soft, yielding stone when the world was in the making. Hundreds of specimens were unearthed in a score of quarries throughout the length of the Connecticut Valley southward from the northern boundary of Massachusetts. Many of the best of these are now on display in the museum of Amherst College and in other museums the world around.

Hitchcock described and figured tracks which he regarded as having been made by at least seven different species of biped animals. The only known bipeds that could have made them were birds: most of them were certainly the footprints of three-toed animals, and many of them showed that the toes terminated with claws. To these supposed bird tracks he gave .the name *Ornithichnites*, signifying stony bird tracks. Five years of further study enabled him to increase the list of species to twenty-seven, which were described and figured, natural size, in the *Final Report on the Geology of Massachusetts*, published in 1841.

Thus far Hitchcock had found no proof that any of the tracks were made by quadrupeds; but many of the footprints were so similar to the tracks of reptiles that he designated

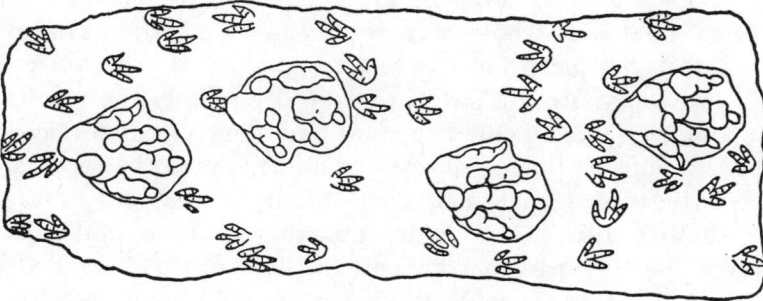

FOSSIL FOOTPRINTS OF THE CONNECTICUT VALLEY
Two species of Professor Edward Hitchcock's "ornithichnites," from his *Ichnology of New England* (1858). About one thirty-second natural size.

them *Sauroidichnites,* or tracks resembling those of saurians. To the others he now applied the name *Ornithoidichnites,* or tracks resembling those of birds. The next year, however, he announced the discovery of the first "certain evidence that any of the numerous tracks upon the sandstone of the Connecticut Valley were made by a quadruped."

By this time interest in the footprints had spread abroad; Lyell, Owen, Murchison, and other leading geologists of Europe announced their opinions. Murchison, for example, after considering the enormous size of the tracks that must have been made by the New Zealand moas, "confessed that the gigantic bones from New Zealand, evincing as they did most unequivocally the existence even in our own times of birds as large as any required by the American footmarks, had removed his skepticism, and that he had no hesitation in declaring his belief that the *ornithichnites* had been produced by the imprints of the feet of birds which had walked over the rocks when in a soft and impressible state." [8] This seems to have been the generally accepted view among geologists for at least twenty years, despite the fact that, one after another, certain of the footprints supposed to have been made by bipeds were revealed on newly discovered slabs of stone as the tracks of quadrupeds, and hence must have been produced by reptiles rather than birds.

At last, in 1860, Roswell Field, a farmer of Greenfield, Massachusetts, on whose property some of the earliest known specimens had been found, for the first time in print voiced his own opinions. He gave briefly and modestly the results of his experience in collecting and observing and announced it as his opinion that the tracks had all been made by reptiles. That the animals that made them usually walked on two feet he admitted, but he contended that they could as well have walked on four had they chosen. In proof of this he added: "We find tracks as perfect as if made in plaster or wax, which, to all appearances, as to the number of toes and the phalangial or lateral expansions in the toes, agree perfectly well with those of living birds, and still we know, by the impressions made by their forward feet, that these fossil tracks were made by quadrupeds." [9]

Field's conclusion was right. Beginning in 1893, the petrified bones of slender dinosaurs have been found in the Triassic rocks of the Connecticut Valley in a sufficiently good state of preservation to permit the restoration of their skeletons and give an adequate concept of their body forms and locomotor habits. Some were quadrupedal, but many walked in a semi-erect posture upon their hind legs alone. There is

[8] G. P. Merrill, "Contributions to the History of American Geology," *Report of United States National Museum for 1904* (1906), pp. 625-33.
[9] *American Journal of Science,* 2d series, Vol. 29 (1860), pp. 361-63.

now no question that the fossil footprints of that region were made by dinosaurs and their reptilian kin. Once more a scientific argument has been settled, not by polling the opinions of experts, but by obtaining additional information.

THE ORIGIN OF GLACIAL DRIFT

And now we must return for a moment to the Diluvialists whom we temporarily abandoned a few pages back, for they played an important rôle in the comedy of errors from which emerged the truth concerning the Great Ice Age. Strewn across the landscape of the northern United States, Canada, and northwestern Europe are innumerable boulders, large and small, of a great variety of rocks, many of which are wholly different from the bed-rock of the locality where the boulders occur. These "erratics," together with the accompanying deposits of clay, sand, and gravel, have long been appropriately known as "drift." The material had evidently been spread broadcast over hills and plains during some comparatively recent episode in geologic history. Quite naturally the Diluvialists claimed it as just another bit of evidence of the world-wide catastrophe recorded in Biblical annals as the Deluge. Even today such débris is still designated in an occasional scientific treatise as "diluvium."

What seems to be the earliest reference to the drift in North America is found in a letter from Benjamin DeWitt published in 1793 in the *Transactions of the Philadelphia Academy of Science*. DeWitt had noted the great variety of rocks on the shore of Lake Ontario. "Now, it is almost impossible," he wrote, "to believe that so great a variety of stones should be naturally formed in one place and of the same species of earth. They must therefore have been conveyed there by some extraordinary means. I am inclined to believe that this may have been effected by some mighty convulsion of nature, such as an earthquake or eruption; and perhaps this vast lake may be considered as one of those great fountains of the deep which were broken up when our earth was deluged with water, thereby producing that confusion and disorder in the composition of its surface which evidently seems to exist."

Such statements are quite characteristic of the mental attitude of the Diluvialists of a century ago, both in America and abroad. Attributing the diluvium to the Noachian catastrophe in vague and general terms is all very well, but when one considers the details of the mechanics whereby boulders as big as a house are transported overland for scores or hundreds of miles, even the most orthodox has his doubts. Perhaps "earthquakes or eruptions" must be summoned to the aid of the flood-waters. The suggestion was even made that during the episode of Divine punishment, survived only by Noah and his fellow-passengers in the Ark, the rotation of the earth was suddenly stopped; in consequence the ocean waters were catapulted forward and swept irresistibly over the highest lands with velocity adequate for the task of transporting the huge boulders to their resting-places.

But closer scrutiny revealed another obstacle to the theory that flood-waters were responsible for the "drift." Peter Dobson, a Connecticut cotton manufacturer, writing in 1825, described boulders which were unearthed at Vernon, Connecticut, during excavating preliminary to the erection of a cotton factory, as "worn smooth on their under side as if done by their having been dragged over rocks and gravelly earth in one steady position." They also showed scratches and furrows on the abraded parts. He could account for these marks only by assuming that the boulders had been worn when, suspended and carried in ice, they were propelled forward in the water. The drifting icebergs off the Labrador coast, he thought, might well illustrate the conditions and methods of the process.[10]

This explanation received the hearty approval of Sir Roderick Murchison in his anniversary address before the Geological Society of London in 1842. Once more the ortho-

[10] *American Journal of Science*, 1st series, Vol. 10 (1826), pp. 217-18, and Vol. 46 (1844), pp. 169-72.

dox Diluvialists were at ease. Icebergs from the northern seas were visualized as playing tag all over the interior of Europe and North America when the Deluge was at its climax. Not only were the stones embedded in the icebergs worn and scratched, but also the solid rocks of hills and valley slopes were grooved and polished where the bergs ran aground or dragged bottom while en route.

There were, however, a few doubting Thomases who were skeptical concerning the ability of icebergs to do the Herculean tasks that the Diluvialists demanded of them. Many an ardent mountaineer had noted the morainic ridges beyond the margins of the glaciers in the ice-bound valleys of the Alps, and a few, such as Kuhn in 1787 and De Saussure in 1803, had correctly interpreted them as evidence of a formerly greater extent of the ice in that region. There could be no question in the mind of an observant Alpinist that glacial ice was competent to transport huge erratics as well as vast quantities of finer débris, and that it could also scratch, furrow, or plane the rock floor over which it moved. The only wonder is that the suggestion that glacial ice was responsible for similar phenomena at places remote from the Alps was so long delayed.

In 1832, Bernhardi, an obscure professor of forestry, first proposed the glacial theory to account for the drift in northern Germany. He made the suggestion "that the polar ice once reached clear to the southernmost edge of the district which is now covered by those rock remnants; that this, in the course of thousands of years, gradually melted back to its present extent; that, therefore, those northern deposits must be compared to the walls of rock fragments which surround almost every glacier at varying distances, or, in other words, that they are nothing other than moraines which that enormous ice-sea left behind on its gradual withdrawal." [11] But Bernhardi's brilliant deductions passed unnoticed, and it was not until Louis Agassiz (1807-1873) cham-

[11] A. Bernhardi, *Jahrbuch für Mineralische Geognosie, Geologie und Petrefaktenkunde* (Leonhard), Vol. 3 (1832), pp. 257-267.

pioned the idea that it was given even the courtesy of
adverse criticism.

In 1840, Agassiz demonstrated to the satisfaction of his
scientific colleagues "that the entire massif of our Alps has
been covered by an immense sea of ice, from which great
projections descended to the edge of the surrounding low
country." [12] His further contention that "at the end of the
geological epoch which preceded the upheaval of the Alps,
the earth was covered with an immense sheet of ice . . . which
extended to the south as far as the phenomenon of erratic
blocks, surmounting all the inequalities of the surface of
Europe as it existed then, filling the Baltic Sea, all the lakes
of northern Germany and Switzerland, stretching beyond
the borders of the Mediterranean and the Atlantic Ocean
and covering even all of North America and Asiatic Russia," was not so favorably received. And, indeed, he had
included altogether too much territory, as we now know.

But it was not the area of Agassiz's ice-sheet that raised
the storm of protest; it was the very idea of widespread
refrigeration. The "glacial nightmare" became at once the
issue in many a heated discussion. Was this not proposing
a catastrophe far more unlikely and hypothetical than the
Noachian Deluge to which all minds had long been ad-
justed? Agassiz met the challenge in the only scientific way.
For twenty years he studied the earth's surface at every
opportunity, first in Europe and later in North America.

Gradually he amassed such a body of data concerning the
distribution of morainic débris and erratics and the location and nature of striated and ice-marked surfaces of bed-
rock that even the Diluvialists, if they lived long enough,
were convinced. Probably the one most influential bit of
field data was that found near Edinburgh in Scotland,
where a tablet may be seen today commemorating the event.
To this place Agassiz brought several of the most dogmatic
defenders of the iceberg theory. An overhanging cliff displayed at its base deep longitudinal scratches underneath

[12] Louis Agassiz, *Etudes sur les Glaciers* (Neuchâtel, 1840).

a jutting ledge. Obviously no iceberg could have left its autograph in such a recess. Only glacial ice spreading over the land could have pushed into the hollow below the overhang to sculpture the bed-rock and engrave its surface. The glacial theory was validated.

Still a major error remained. Agassiz and his contemporaries for a long time thought and spoke in terms of a polar ice-sheet, spreading southward from the ice-bound "roof of the world." It is now known that the ice-sheets of the Great Ice Age were not polar in origin but had their centers in Scandinavia, Quebec, northwestern Canada, and elsewhere, in latitudes twenty to thirty degrees southward from the North Pole. From these centers they spread in all directions, northward as well as southward, influenced by local climatic conditions and topographic features.

THE ORIGIN OF THE EARTH

The desire to know how the world was made is certainly as widespread as human geography and probably as old as human history. Creation stories are an essential part of almost every body of folklore. Throughout most such stories there runs the same fundamental theme. In the beginning there was a primordial substance, generally described as more or less chaotic, and some sort of Maker or Creative Being. At first, both were at rest; then, out of the stillness came motion which in some way produced light, and out of the light came all created things. After creation evil entered; after evil came the deluge. But in the midst of the deluge the mountain-top appeared, and upon the ruins of the Old Earth the New was constructed.

Even in the heyday of Greek scholarship there was no improvement on these philosophical speculations. Not until the early part of the seventeenth century was there any serious attempt to replace such subjective imaginings by inductive reasoning based upon quantitative observations of the vestiges of creation. René Descartes (1596-1650) blazed a new trail when he announced his famous principle of the constancy of the amount of motion or "momentum" in the universe. If that principle is accepted, it becomes possible to reason backward from effects to causes. Only thus can any scientific answer be discovered to the question *how* the world was made.

More than a century elapsed, however, before Immanuel Kant (1724-1804) published in 1755 the conception that the ordered universe might have been produced merely by mechanical forces acting upon a vaporous chaotic mass. It is quite likely, though, that Pierre Simon Laplace (1749-1827) developed his Nebular Hypothesis of earth origin entirely unaware of Kant's contribution to this subject. Be that as it may, it was in 1796 that the first wholly scientific explanation of the origin of the earth was published. Laplace was the first to cut loose entirely from religious dogma and vague references to "attractive and repulsive forces" or to "whirls of movement." His nebular hypothesis seemed so adequate that for almost a century after his death it was accepted as orthodox scientific lore. Only within the last few years has it been generally discarded as erroneous.

Laplace called attention to the fact that all six planets known to astronomers of his day move around the sun from west to east in almost the same plane, that all moons or satellites then known move in similar direction around their controlling planets, and that the sun rotates in the same direction upon its own axis. As a matter of fact, not until some such information as that was available could any real attempt be made to construct a scientific explanation of the origin of the earth. No amount of geological knowledge could be adequate for such a task. The earth is a member of the solar system, and astronomy must come to the aid of geology whenever we probe back to the beginning of earth history.

But by the latter part of the eighteenth century astronomers had gathered sufficient data to make it apparent that the solar system could not have been the result of mere chance, but must have been organized under the control of some general principle which determined the remark-

able orderliness of its several parts. Laplace therefore suggested that at some time in the past all the matter in the
solar system—sun, planets, and moons—had constituted a
gigantic nebula of extremely rarefied and fiery hot gas,
shaped like a discus and whirling in the midst of space.
As the gaseous nebula cooled, it necessarily contracted,
and in consequence its velocity of rotation increased, in
accordance with the principle of constant momentum. At
last the centrifugal force developed around the edge of the
whirling disk must have equaled the force of internal gravity which kept the nebula intact. Further cooling and shrink-
ing would result in the separation of a ring of gas which
would continue to revolve in the same direction and at the
same rate as that of the entire nebula at the time of its
separation. This process would be repeated until rings had
been formed in sufficient number and at appropriate distances from the center of the nebula to provide for each
planet in the solar system. Each ring would become aggregated into a ball, which would perpetuate its revolutional
velocity and would rotate forward because the outside of
each ring must have been spinning faster than its inner side.
Each planetary sphere of gas would proceed to cool and
shrink in the same fashion as the parent nebula, and some
of them would drop rings of gas from their equators to
form the several moons which now revolve around certain
of the planets.

Thus Laplace explained all the facts known in his day concerning the positions, relations, and motions of the various
members of the solar system. Forward rotation, uniformity
in direction of revolution, regular increase in revolutional
velocity from outermost to innermost planet, would all
be a consequence of the simple principles under which his
gaseous nebula was transformed into the solar system as it
cooled and contracted.

The earth thus began its history as a ball of incandescent
gas having a diameter at least twice as great as the distance
from earth to moon. As the gas cooled and condensed, after

the moon ring had been dropped, it would begin to liquefy, and in time the surface of the liquid sphere would solidify to form the rocky crust through which the liquid interior would occasionally exude during volcanic eruptions. Such ideas were eagerly welcomed by nineteenth-century geologists as entirely in harmony with prevailing concepts concerning the origin of igneous rocks and the condition of the earth's interior.

Progressive cooling of a rapidly aging earth seemed particularly appropriate to those who ascribed the coal of the far-off Carboniferous Period to tropical climates prevalent throughout North America and Europe. The demonstration that glacial climates had been widespread in middle latitudes during the Great Ice Age, in whose "shadow" we are living today, fitted nicely into the mental picture. But as the nineteenth century waned, many geologists and astronomers were looking with growing scepticism upon the Laplacian nebular hypothesis. Glacial deposits were found in pre-Cambrian rocks, a half billion years old. The plant fossils of the Carboniferous coal-measures were identified as indicative of comparatively cool rather than tropical climates. Serious discrepancies between the motions of several members of the solar system and the requirements of the theory were discovered.

To make a long story short, it is now known that the earth and its companions in the sun's family could not possibly have been originated in the manner suggested by Laplace. The outer moons of Saturn and Uranus, quite unknown until long after the death of Laplace, revolve in the opposite direction to that required by his theory. Phobos, the inner satellite of Mars, revolves three times while Mars rotates once. Whereas the revolutional velocity of Mercury is 29 miles per second, the sun rotates so slowly that a point on its equator moves less than two miles per second. The equatorial plane of the sun is inclined five degrees from the plane of the orbits of the planets. To separate the earth ring from the Laplacian nebula that

nebula must have had 1,800 times the momentum now possessed by the earth, Venus, Mercury, and the sun.

The real difficulty is found in the fact that because of
the slow rotation of the sun in contrast to the swift speed
of the planets as they circle it, the momentum possessed by
the system as a whole is concentrated in the planets. About
one-seventh of one per cent of the mass of the entire solar system carries 97 per cent of its total moment of momentum.[13]

During the first two decades of the twentieth century,
T. C. Chamberlin and F. R. Moulton developed a wholly
different concept of the origin of the earth and gave to
the world the Planetesimal Theory. To explain the curious
distribution of momentum within the solar system they suggested the idea of a dynamic encounter between the sun
and a passing star. If another star approached closely to the
sun, gravitational atraction might cause the ejection of solar
material which would acquire angular momentum at the
expense of the energy of motion of the star as it passed by.
The ejected material might subsequently be organized into
planets, satellites, and other bodies pertaining to the solar
system.[14]

According to this view, the earth is an offspring of the
sun, but has inherited some of its characteristics from an
unidentifiable star which contributed to its birth, a couple of
billion years or so ago. This concept of a biparental origin
for the planets, resulting from the close approach of the
sun and a star, is generally accepted among astronomers and
geologists today. There are, however, many differences of
opinion concerning the precise details of the encounter and
the ensuing events.

Under the Planetesimal Theory, the earth is pictured as
having been built up from a countless number of tiny bodies
—the planetesimals—by a process of accretion. Jeans and
Jeffreys, on the other hand, maintain that the earth is the
result of the condensation of a fragment of a filament of

[13] T. C. Chamberlin, *The Origin of the Earth* (Chicago, 1916).
[14] T. C. Chamberlin, *The Two Solar Families* (Chicago, 1928).

extremely hot gas drawn from the sun during the encounter with the other star.[15] The differences in the juvenile history and internal constitution of the earth are of great significance to geologists, but not enough facts are at present available to permit the rendering of a verdict in favor of either variant of the theory.

Indeed, there are serious obstacles to the unqualified acceptance of any of the ideas concerning the origin of the earth that are now being entertained. Russell, for example, closes his recent masterly survey [16] of the subject with the apt though disappointing suggestion that he may have succeeded only a little better than Browning's philosopher, who had "written three books on the Soul

> Proving absurd all written hitherto
> And putting us to ignorance again."

Certainly, when our thoughts are turned towards so complicated and obscure a problem as that of earth origin, it is important for us still to keep an open mind, remembering that here if nowhere else in science the latest episode in the story of error is not necessarily the last.

15 Harold Jeffreys, *The Earth,* 2d edition (Cambridge, 1929).
16 Henry Norris Russell, *The Solar System and Its Origin* (New York, 1935).

CHAPTER III

ERROR IN GEOGRAPHY

By John Leighly

OUR INTIMATE DEPENDENCE *upon the earth—the source of our sustenance, the scene of our labors, and the final recipient of our mortal dust—has been an obstacle to the easy acceptance of it as a simple object of disinterested observation. Geography has had at all times to struggle against subjectivity in judging the qualities of the different parts of the earth. When it has yielded to this subjectivity, the earth has claimed attention as the cradle, the nursery, or the dwelling-place of mankind. Viewing the earth in this light, every people has tended to overestimate the inconvenience of living in unfamiliar parts, to set up for places a scale of values in which the home region occupies a high, if not the very highest, rank. The classic example of this subjective evaluation of regions is presented below in our section dealing with the habitable zone. This kind of error, it may be remarked. has not wholly disappeared with increased knowledge: the American still receives the commiserations of European acquaintances on his having to live in an uncomfortably low latitude, and on this side of the Atlantic a chorus of indignant protest greets an attempt to colonize interior Alaska.*

Errors that have grown out of objective curiosity concerning the earth, which occupy the larger part of the present chapter, display a greater variety than those which stem from deep-rooted preference for the familiar. There are, first, the honest errors of ignorance, illustrated below by the efforts made to assign definite dimensions to the earth and the lands on it before the technique of the necessary measurements was adequately developed. Others derive their peculiar qualities from the fact that, because of the great area of the earth's surface relative to what one is able to observe personally in an ordinary lifetime, much of one's knowledge of it is gathered at second hand. Every commentator or map-maker who takes an inclusive view of the earth exposes himself to the danger of making erroneous assertions, verbal or cartographic, concerning the parts he does not know personally. Only the most cautious can resist the temptation to treat the unknown as if it were familiar or to extend the lines on his map beyond where they have been accurately surveyed.

There is no evident regularity in the distribution of land and water, mountain and lowland, over the earth. Yet the human mind has been extraordinarily unwilling to accept their distribution as irregular. Confronted with a multitude of facts—though not by all the known or knowable facts—concerning the earth, man's mind has repeatedly tried to reduce them to order by means of one of its favorite tricks, the ascribing of complicated phenomena to simple causes. Was there uncertainty, in the sixteenth and seventeenth centuries, whether the southern hemisphere consisted mainly of water or of land? Then conclusions were drawn from considerations of symmetry of the two hemispheres, and the learned world soberly concurred in the opinion of Edward Brerewood, that "the earth should in answearable measure and proportion lift it selfe and appeare above the face of the sea, on the South side of the line as it doth on the North." No generalizations such as this, however, that have aimed to predict the distribution of land and sea in unknown parts of the earth, or the relief of unexplored lands, have proved tenable.

The errors discussed below are such as have been rectified by simple observation, not by recondite scientific researches; and such as have been corrected once for all, except for details. With the passage of time and the bringing of almost the entire area of the earth's surface within the horizon of the known, the opportunities for the commission of such errors have become progressively more limited. At the same time, inquiry into the qualities of the earth's surface and the attempt to explain them on physical grounds have opened the gates to a flood of errors in the physical interpretation of terrestrial phenomena. Errors of the last-named sort, in contrast to the simple errors that can be corrected by the observations of the merest layman, form a part of the general flux of scientific thought, in which the daring hypothesis of today becomes the accepted dogma of tomorrow and the outgrown error of the day after. They are left out of consideration in the present chapter, however, in the interest of unity of kind among the errors enumerated.

THE PRIMARY ERRORS of geographic thought may be either errors in the general concepts "earth," "land," "sea," "atmosphere"—the receptacles of the isolated facts of observation or report—or errors in the facts themselves. The errors that have had influence in science have resulted mainly from ascribing to unknown parts of the earth qualities deduced from observation of the known parts, and from preconceptions as to what, on such an earth as is imagined, these unknown parts should be like.

Errors of factual detail about unfamiliar places belong to the commonest faults in our individual stocks of knowledge. It may be that in the twentieth century we sin less grievously in this respect than did our medieval ancestors, who accepted mountains of silver and gold, griffins and basilisks, tailed and dog-headed human beings, as elements of their mental pictures of remote lands. We may smile superciliously at our forefathers for being so gullible, but with a little effort we could certainly match their errors of fact from the intellectual lumber of those we meet daily.

Such errors do not really form a part of the annals of scientific thought. If we wish to form an opinion of the errors committed in thinking about the earth in the dark ages of European intellectual history, it is better to examine the deductions from the general concept "earth" then made to explain common terrestrial phenomena. Consider, for example, the manner in which the sainted Isidore, Bishop of Seville, writing about the year 600, discussed the cause of the tides. He quoted Solinus (a late Roman collector of crumbs from the rich table of classical science) as saying that there are passages in the depths of the ocean, like nostrils of the world, through which its breath is alternately inhaled and exhaled, so that the waters of the sea are made to fall and rise. He reported the opinions of others, that the sea rises and falls with the waxing and waning of the moon; and of still others, that the sun and stars draw water from the ocean to temper their heat. "But whether the waters are raised by the blowing of the winds, or rise with the waxing of the moon and withdraw as the moon retreats from the sun, only God knows, who made the world, and to whom alone its workings are known." One who resigned curiosity about Nature as completely as did the good Bishop of Seville could commit no scientific errors that can interest us today.

The geographic errors of Europeans become interesting again only after they recovered the fertile ideas of the earth held by the ancients, which came back gradually into currency between the years 1200 and 1500. The most influential item of ancient geographic writing then recovered was the *Geography* of Claudius Ptolemy, written in the second century A.D., which became known in western Europe in the fifteenth century, much later than the same author's astronomic writings.

Ptolemy's *Geography* was a poor vehicle for the transmission of the geographic lore of antiquity, but it was enough to stimulate a lively curiosity concerning the earth. It gave to the medieval European, whose learning had already been rather thoroughly dusted and aired by three centuries of the study of Aristotle, a new and attractive image of the earth. Though he might draw monsters in the blank spots on Ptolemy's map, there was no room on Ptolemy's earth for impassable oceans or other ultimate barriers to free movement on its spherical surface. It contained, it is true, unknown lands, but no utterly unknowable lands. It had definite dimensions, so that one could translate one's travels into arcs of latitude and longitude, and arcs of latitude and longitude into distance to be traveled in passing from any one place to another. It gave the European a new general concept "earth," from which he could deduce much concerning the parts of the surface of the real earth that were not known.

Ptolemy's model of the earth was the weapon by which

the real earth was conquered intellectually in the Age of Discovery. In using it heroic deeds were done: we can still render the most profound admiration to Magellan, who ventured forth on a voyage of circumnavigation with complete confidence in the accuracy of Ptolemy's earth model. But it had grave defects which only slowly and painfully could be discovered and rectified; in using it the greatest geographic blunders of scientific history were committed.

Before the two classic geographic errors derived from the recovery in the late Middle Ages of ancient geographic lore are enumerated, space may be given to a particularly long-lived blunder committed by the Greeks themselves.

THE HABITABLE ZONE

At the very beginning of their science of the earth, the ancient Greeks introduced into the system of locating places on its surface an element of confusion which has scarcely yet been eradicated from our thinking, namely, the preconception that one may judge the habitability of a place from the latitude in which it lies. It began with the generalization that is still the basis of the precise numerical location of places on the earth—that the spherical surface of the globe is concentric with the celestial sphere, and if the celestial sphere be imagined as stationary, some point on the earth corresponds to every point in the heavens. Since, however, the celestial sphere seems to rotate about the earth, it is not points—except the poles, which are stationary points in the heavens—that are transferred to the earth's surface, but circular lines, which become parallels of latitude. This elementary relation between the heavens and the earth's surface has been since antiquity a commonplace in the rudiments of earth science; at the same time, except in the strictest discussions of mathematical geography, it has been mixed with and obscured by certain anthropocentric ideas concerning the suitability of various latitudes for human habitation.

In this confusion, judgment concerning the habitability

of latitudes is the older, dating from the remote times when
the Greeks had not yet drawn the magnificent conclusion
that the earth is a sphere. They derived this judgment of
habitability from the location of their country with respect
to other lands with which they early became acquainted.
Almost at the very shores of the Euxine—our Black Sea—
begin the severe continental winters of Scythia, news of
which came early into Greek geographic knowledge. South
of the Mediterranean the deserts of northern Africa were
so near as to become known even earlier. And in the same
latitude as these deserts, but farther east, is the great dry
interior of Arabia. Our modern thinking on the subject
of climate separates fairly well the ideas of heat and dryness; but the old Greek geography put these together, because the warm season in the eastern Mediterranean is also
dry. On the other hand, the shores of the Mediterranean
itself, stretching far to the west, possessed a climate recognized as uniform with that of the Greek settlements. When,
as a result of Alexander's Asiatic campaigns, a part of Asia
in the same latitude as Greece became known, the observations made in the east did not greatly disturb the image of
a narrow habitable strip of territory situated between less
hospitable stretches to the north and the south.

It is a short and easy step from the recognition of unpleasant qualities in areas close at hand to the extrapolation of
these qualities toward more remote parts, where they may
be supposed to become intolerable. Scythia was readily interpreted as lying on the hither side of a part of the earth
so cold as to render life impossible. The summer heat and
dryness of northern Africa were certainly only a foretaste
of an utterly parched condition to be found in the lands
farther south. By such reasoning the habitable earth became
an intermediate strip to which approximate boundaries
could be set on the north and the south, but which toward
the east and the west was limited only by the all-surrounding
ocean. This accepted form of the habitable earth impressed
itself upon fundamental geographic concepts to such an

extent that we still use the word "longitude"—that is, length
—for the measure of intervals of distance in an east-west
direction on the earth, and "latitude"—breadth—for intervals
in a north-south direction. By the mere use of these words
we obey the injunction of Strabo, writing at about the
beginning of the Christian era, that when speaking of the
length and breadth of any part of the earth, regardless of
its shape, we should call its "length" its dimension drawn
parallel to the length, its "breadth" the dimension parallel
to the breadth, of the whole habitable earth. Though today
we place these dimensions in relation to a spherical surface
which has neither length nor breadth, we use the terms invented for the purpose of referring any part of the earth
to a whole having the shape of a belt, which even on a
sphere is defined by length and breadth.

The recognition of the spherical form of the earth, which
may go back to the sixth century B.C., brought confusion
by transferring certain special circles from the heavens to
the earth. Before any knowledge was gained concerning
high or low latitudes, accurate conclusions were drawn as
to the lengths of day and night and the height of the sun
in the sky in all latitudes from the equator to the poles.
From observation of the heavens came the special parallels,
the tropics and polar circles, which define accurately the
belts of the earth—the "zones" of traditional mathematical
geography—within which there is a certain uniformity in
the sequence during the year of relative length of day and
night and of noonday position of the sun.

The zones formed what at first must have seemed a valuable mathematical confirmation of the half-irrational distinction between habitable and uninhabitable belts on the
earth. Older Greek geographic writings discuss the relation
between the zones and the belts of habitability. Strabo records that Parmenides (who lived in the fifth century B.C.)
was the first to divide the surface of the earth into five
zones, but that he almost doubled the area of the hot zone,
making it extend into the temperate zone. Parmenides thus

used the quality of uninhabitability rather than the bounding tropical circles in defining the hot zone, and made his uninhabitable hot zone extend northward to include the African and Arabian deserts between latitudes 30° and 35°.

Aristotle, a century later, pushed the northern boundary of the hot zone somewhat toward the south. He identified the uninhabitable hot zone with the zone between the tropics, reserving the quality of habitability to the zone between the tropic and the polar circle; yet he extended the parched equatorial belt slightly beyond the tropic. The later tendency was to push the poleward boundary of the hot zone even farther toward the equator and to free the question of habitability from dependence upon the astronomic zones.

Strabo says that Posidonius, in the second century B.C., introduced a refinement into the identification of climate by supposing that the hottest and driest belts are not along the equator, but in the vicinity of the tropics, since here the sun remains overhead longer than at the equator. We should express this latter observation by saying that the declination of the sun changes more slowly near the solstices, when the sun is overhead near the tropics, than at the equinoxes, when it is overhead at the equator. Posidonius thought that south of the neighborhood of the northern tropic the lands are better supplied with water and more fruitful than at the tropic.

As for Strabo himself, he defined very carefully the extent of the habitable belt, which for him was rather completely divorced from the scheme of astronomic zones. He located the habitable belt between the parallels of about 12° 30' and 52° 30' north latitude, leaving a narrow uninhabitable strip at the equator. Strabo was especially emphatic on the question of the northern limit of habitability. He scornfully rejected the current account of an inhabited island, Thule, supposed to lie on the Arctic circle. For him, Ierne (Ireland), which he placed north of Britain in latitude 52° 30', was the last northward limit of the habitable earth—Ierne, "where the people live miserably and like savages on account of the severity of the cold."

The early Middle Ages drew upon compilers who wrote in Latin and who transmitted a few good and many bad ideas from classical antiquity. Isidore of Seville discussed habitability in strictly schematic terms, so far as one can judge from his uncomprehending confusion of the terms "circle" and "zone." He called the hot zone and the cold zone uninhabitable, limiting habitability to the intermediate zones, both north and south. "But those who dwell near the summer tropic are Ethiopians, burned by the excessive heat."

The later Middle Ages knew Aristotle, and acquaintance with him brought order into the geographic writing done in the thirteenth century, in the time of Roger Bacon and Albert of Bollstädt, called Albertus Magnus. Albert admitted that it was not known by experience how much of the earth was habitable and how much not. Like Posidonius, he appealed to the slower movement of the sun in declination when it is above the tropics than at the time of the equinoxes in order to explain a greater intensity of heat near the tropics than near the equator; and he granted to the vicinity of the equator a "temperate" character, though not so temperate as that of the middle latitudes. In discussing the habitability of low latitudes he was chiefly concerned with the periods when the sun is at zenith. The regions between the tropics become intolerably hot, in his opinion, only at these times of the year. Albert readily admitted the habitability of a "temperate" zone in the southern hemisphere, and even the possibility of crossing the equator from one temperate zone to the other.

Among medieval writers on geographic matters, Pierre d'Ailly (1350-1420) deserves particular attention because of the influence his writings exerted on Columbus. In his *Imago Mundi* there is no evidence of any attempt to weigh the copiously quoted but conflicting statements of older authors. In general, he clings to the schematic identification of the habitable and uninhabitable belts with the astronomic zones. It was next to a stereotyped quotation on the uninhabitability of the hot zone that Columbus wrote in the margin of his copy of the *Imago Mundi* his impatient protest: *"Zona torrida non est inhabitabilis"*—"the hot zone is not uninhabitable! The Portuguese sail through it today, finding it well populated, and have a fort on the equinoctial line itself!"

Columbus' protest records the beginning of the fall of the whole doctrine that depended upon the judgment of the habitability of a country by the latitude in which it lay. Actual knowledge of the earth gained in the period of discovery robbed the doctrine of the zones, with its implications of habitability, of all significance, so far as the persons who visited lands in low and high latitudes were concerned. Among the learned, Bernard Varenius, in the middle of the seventeenth century, handled the question in a thoroughly rational manner: he used differences in latitude for what they are worth—that is, for the convenient and accurate expression of location on the terrestrial sphere; and he depended upon observation of the physical climate of a place as an indication of its habitability. Yet even after Varenius generations of school-children have had to learn the system of the zones, as expiation of the smug conviction of the ancient Greeks that the narrow strip of latitude in which they lived was the only fit dwelling-place for rational human beings.

THE NARROW ATLANTIC

The error or chain of errors that, purely incidentally, had the most far-reaching consequences was that which is eternally associated with the name of Christopher Columbus. It was, briefly, an underestimation of the width of the Atlantic Ocean—not of our Atlantic Ocean, which lies between the western shores of the Old World and the eastern shores of the New, but of the Atlantic Ocean of the ancients, which lay between the western and eastern extremities of the Old World.

Any error, of whatever magnitude, made in estimating the width of the Atlantic in Columbus' day was a scientific error, for the estimate itself was the product of scientific reasoning. Such an estimate involved the fundamental problem in the geometry of the earth's spherical surface—the attainment of a measure in units of distance of the length of an arc of a great circle (equator or meridian circle) on the surface of the earth. Once this measure was given, the length of any other line on the earth could be computed by the familiar rules of spherical geometry. This problem was solved as early as the third century B.C. by Eratosthenes of Alexandria. Eratosthenes learned that at Syene (now called Assuan) the sun is at zenith at noon on the summer solstice; hence it was clear that Syene must lie on the Tropic of Cancer. He measured the angle between zenith and the noon position of the sun on the summer solstice at Alexandria, finding it to be one-fiftieth of a circle, or, as we should say, 7⅕ degrees. He supposed that Syene and Alexandria were on the same meridian, 5,000 stadia apart. The length of a meridian circle, and hence the length of any great circle on the earth, was then given by multiplying 5,000 stadia by 50, which equals 250,000 stadia. Eratosthenes added 2,000 stadia to obtain a number divisible by 360, and so obtained 252,000 stadia as a final value for the circumference of the earth. Using the most probable length of the stadion in which Eratosthenes reckoned distances, his estimate of the circumference of the earth was 24,740 miles. According to modern measurements, the length of the equator is 24,900 miles. So near an approach to the truth cannot be ascribed to any particular accuracy on Eratosthenes' part: it was merely the result of a canceling of errors in the fundamental data.

The second part of Eratosthenes' task, as of all similar attempts made in antiquity, was to assign a dimension to the lands known to the ancients and to place them on the sphere. It was generally supposed by Greek scholars that the land body known to them was longest (in an east-west direction) at about latitude 36° north, on the parallel of Rhodes. Eratosthenes estimated that the lands had a length along this parallel sufficient to stretch through an arc of 130° of longitude. In our figure below the two upper circles represent the parallel of Rhodes as Eratosthenes computed its size and as it is according to modern measurements. In his esti-

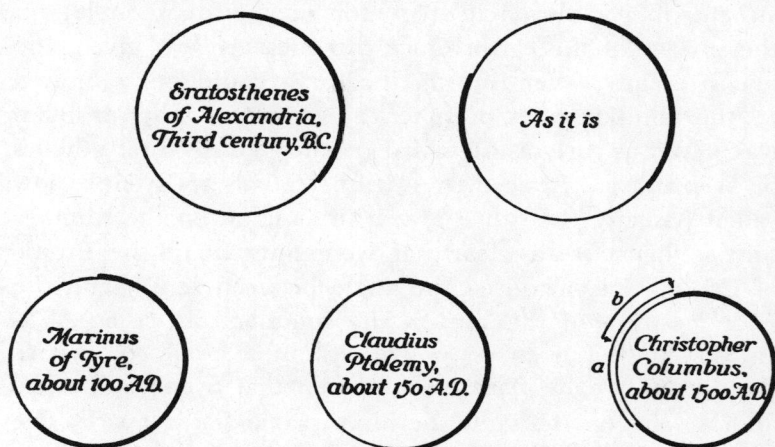

THE EARTH AND THE WESTERN OCEAN

The size of the earth and the extent of the Old World from west to east, as conceived by the geographers of antiquity and by Columbus, and as it is according to modern measurements. The heavy parts of the circumferences of the circles represent the extent of land about the parallel of 36° north. The length of arc *a* in the Columbus diagram shows the distance Columbus must have expected to sail from the Canary Islands in order to reach the eastern extremity of Asia on the parallel of 28° north latitude; arc *b*, the actual distance he traversed on his first voyage, between the Canaries and his landfall in the Bahamas. A comparison of *a* and *b* shows how confused were his notions of his position after he had crossed the Atlantic. For it was when at the western extremity of *b* that he reported his position as 135° of longitude west of the Atlantic coast of Spain.

mate of the length of the Old World, Eratosthenes was again lucky—not accurate, for he had not sufficient knowledge of the Orient to give any significance to his estimate of the eastward extension of Asia.

It was a perversion of Eratosthenes' brilliant accomplishment that ultimately led to Columbus' fateful error. It happened that his result was transmitted through Posidonius of Rhodes, who lived a hundred-odd years after Eratosthenes. Posidonius seems to have given in his writings, perhaps only as an example to show how the necessary computation should be carried out, a difference in latitude of 7½ degrees between Alexandria and Rhodes, corresponding to an estimated distance of 3,750 stadia. These data were the basis of the computation of the size of the earth accepted by Marinus of Tyre and in turn taken over along with Marinus' other geographic data by Ptolemy. Posidonius' arc of the meridian between Alexandria and Rhodes was one forty-eighth of the circumference of the earth; the total circumference was therefore 48 times 3,750 stadia, or 180,000 stadia, a third less than the dimension Eratosthenes had arrived at.

The extent of the lands of the Old World in an east-west direction was not reduced by this new dimension of the globe. By the time Marinus compiled the information that Ptolemy later used for his *Geography,* news had been received of lands lying to the east of India, the remote eastern edge of Eratosthenes' world. On the strength of these reports, Marinus estimated that the land mass extended through 225° of longitude on the parallel of Rhodes. Ptolemy subjected Marinus' data to critical sifting, and among other revisions he reduced Marinus' estimate of the east-west length of the lands. He brought this length down to half the length of the parallel of Rhodes. Marinus' and Ptolemy's views of the size of the earth and the longitudinal extent of the lands are also shown in the figure opposite. Ptolemy, the conscientious editor of Marinus' uncritical compilation, quoted Marinus' views in order to refute them, and so transmitted Marinus' dimensions to fifteenth-century Europe along with his own.

If one accepts Columbus' own evaluation of what he accomplished on his first voyage, he vindicated Marinus' dimensions of the earth against Ptolemy's corrections. That is exactly the manner in which he described his accomplishment in the letter to the king and queen of Spain written from Haiti in 1503, in which he reported the results of his fourth voyage. Indeed, he went so far as to assert that a determination of longitude he had made by observation of a lunar eclipse in 1494 had shown a difference in longitude of 135° between his newly discovered lands and the coast of Spain—exactly the difference, it will be noted, between 360°, the full circuit of the earth, and Marinus' 225° of land.

Through marginal notations in books that belonged to Columbus, we know rather accurately the sources of the ideas he held concerning the dimensions of the earth and the eastward extension of the Old World land mass. His principal source was that collection of tracts by Pierre d'Ailly, Cardinal of Cambrai, commonly called from the title of the first tract *Imago Mundi*. Columbus' notes show that he was particularly impressed by three ideas that recur in d'Ailly's writings: first, that a degree of the equator is 56⅔ miles in length; second, that there is no great distance covered by the sea between the western extremity of Europe and the eastern extremity of Asia; and third, that the parts of the earth near the equator are habitable and inhabited.

Only the first two of these ideas are of interest here. The equatorial degree of 56⅔ miles had an Arabian origin: Pierre d'Ailly borrowed it from Roger Bacon, who in turn had it from an Arab astronomer, in Europe commonly called Alfragan, whose writings were translated into Latin in the twelfth century. Alfragan was undoubtedly thinking of the Arab mile, which is longer than the customary nautical mile of Italian origin in which Columbus reckoned distances. By accepting this dimension of the equatorial degree, Columbus made the earth, already shrunken through the arithmetic manipulations of Marinus and Ptolemy, even smaller than these ancient worthies had reported it to be. (Compare the drawings in the figure on page 100.)

Pierre d'Ailly insisted repeatedly, in express opposition to numerous authors quoted by him, on the narrowness of the western sea. In an appendix to the *Imago Mundi* he summarized Ptolemy's *Geography*, mentioning the correction that Ptolemy had applied to Marinus' estimate of the extent of the lands, but without recording Marinus' dimension or stating explicitly that he accepted Marinus' opinion in preference to Ptolemy's. He continued, instead, to appeal in support of his conviction to older and less authoritative writers. Columbus possessed also a copy of Ptolemy's *Geography*, in an edition printed in 1478. This book does not bear the evidence, in the form of marginal notations, of the careful reading he gave to d'Ailly's *Imago Mundi*, from which he obviously derived the principal support for his view that the westward passage to the Indies was a short one. Since d'Ailly does not cite any specific width of the sea, however, the numerical value of the width of the Atlantic that Columbus accepted must have come from Marinus through Ptolemy's book.

In accepting Marinus' length of the lands of the Old World, Columbus paid no attention to Ptolemy's well-deserved correction. He had, instead, his own corrections to apply to Marinus, which made the gap between the western end of Europe and the eastern end of Asia even narrower than the Tyrian geographer had estimated. In the East, Marco Polo's reports indicated that the lands extended farther eastward than Marinus had supposed; and in the West the Portuguese, by their discovery of the Azores and the Cape Verde Islands, had narrowed from this side the unknown part of the circuit of the earth. From these considerations came Columbus' own numerical interpretation of the narrowness of the western sea—that the gap was no wider than 120° of longitude, a third of the length of the parallel on which he expected to sail westward.

The final circle in the figure on page 100, which shows Columbus' conception of the size of the earth and the extent of the lands, is drawn, as are the other diagrams, for the parallel of Rhodes. But since Columbus did not aim to sail westward on this parallel, but on about 28° north (the latitude of the Canary Islands), the actual distance he would have to traverse was somewhat longer than it would have been if he followed the parallel that for the ancients marked the greatest extent of the lands and consequently the narrowest part of the Atlantic.

With the adoption of a length of the equatorial degree and a value for the extent in longitude of the lands of the Old World, the scientific basis of Columbus' project for a westward passage to India was laid. It was throughout a tissue of errors, which Columbus did not have the scientific knowledge or critical acumen to detect and eliminate. It was a conceptual structure such as is characteristic of the non-scientific man, who draws from the storehouse of science such assertions, whether scientifically tenable or not, as support a conviction already arrived at by unscientific contemplation. We do not know what arguments Columbus made to the learned men appointed to examine and pass upon his proposal, nor what data they cited in opposition to his plan. If he advanced the views that he later embodied in his writings, there would have been ample reason for rejecting them. The men of learning in the Spain of his day could not possibly accept the values he would have to cite, representing the length of a parallel on the earth and the width of the Atlantic, in support of his opinion that the Orient might easily be reached by sailing into the Occident. The opinions they held were certainly those of Ptolemy; according to these, the distance through which Columbus planned to sail was far greater than the supplies he could carry on his ships would permit.

The Columbian tradition, transmitted through Columbus' son Fernando and his historian Las Casas, has given to later generations a picture of the Discoverer struggling for a new idea against intrenched unscientific conservatism. This representation of Columbus' scientific activity is certainly false; the more enlightened his critics were in the science of their day, the more firmly would they have to reject his ridiculous scientific ideas. The wonder is that he ever obtained support for his expedition. That the expedition was equipped and that he received a royal commission for his explorations was a triumph of non-scientific, presumably political and economic, considerations over the best science of the fifteenth century.

Only the wholly unpredictable existence of a New World between the West and the East of the Old saved Columbus' first expedition from being a disastrous and inglorious failure. Except for this pure accident, Columbus, if remembered at all, would have an obscure place in scientific history along with the other denizens of the lunatic fringe of science. To his death, and in the face of observations and measurements that contradicted his ideas at many points, he clung most unscientifically to the convictions at which he had arrived by his reading of the works of second-rate authors, if not by more dubious intellectual ways. His error, incomparably the most fateful geographic error of all time, had in it nothing of the heroic.

THE UNKNOWN SOUTHERN LAND

While the geographic lore of the ancient Greeks included positive knowledge of only one land mass on the earth, the Old World of our modern ideas, it left open the possibility of there being on the terrestrial sphere other lands, not connected with Europe, Asia, or Africa. Speculation on this possibility gave rise, for example, to the concept of the Antipodes, the subject of much heated theologic discussion in the Middle Ages. Perhaps it was some obscure influence of this tradition, perhaps an interpretation of some information available to Marinus of Tyre concerning the east coast of Africa and the southeasternmost parts of Asia, that led Claudius Ptolemy to put into his geographic treatise the name of an unknown land enclosing the Indian Ocean on the south. Under its Latin designation, *"Terra Australis Incognita"*—the Unknown Southern Land, or, more hopefully, *"Terra Australis nondum Cognita"*—the Southern Land as yet Unknown, it haunted maps and geographic literature for approximately three centuries, from the time of the recovery of Ptolemy's *Geography* until well into the eighteenth century. Passages to the East Indies around the Cape of Good Hope, made early in the Age of Discovery, proved a connection by sea between the Atlantic and Indian Oceans. But so potent was Ptolemy's authority that no amount of sailing on the open sea south and east of Africa, or into the Pacific about the southern extremity of South America, could do more than push the shores of the Southern Continent farther back into the unexplored parts of the southern hemisphere.

Curiously, most world maps drawn immediately after the discoveries made by Columbus do not display the Southern Land; but with the passage in 1520 of Magellan into the Pacific through the long and tortuous strait named in his honor, the Ptolemaic land leaped immediately into prominence on up-to-date maps. And naturally enough, for the lands on the western side of the Atlantic were not yet recognized as a New World, but were identified as a peninsula extending southeastward from the eastern extremity of Asia. Now Ptolemy had recorded the existence of just such a peninsula. It was merely found to extend southward farther than Ptolemy had supposed, and the ocean to the west of it found to be wider than Ptolemy had made it. These errors might easily be forgiven the Alexandrine, who after all had little opportunity to learn about the Far East. He had made this remote peninsula continuous with the Southern Land lying south of the Indian Ocean, but Magellan had found a strait between them. Tierra del Fuego, the land on the south side of the strait discovered by the Circumnavigator, was clearly the first part of the Southern Land to be discovered. When later explorers found passages about the southern end of South America that lay farther south than Magellan's Strait, they saw other lands in the vicinity that could be interpreted as parts of the elusive Austral continent.

Navigation from the Atlantic into the Pacific by way of Magellan's Strait became a commonplace in the sixteenth century, as a technical part of the consolidation of Spain's colonial empire. Spanish navigation of the Pacific was directed, however, toward maintaining connections between the homeland and the new settlements on the western coasts of America, and between these settlements and the colony in the Philippines, rather than toward discovering new lands beyond the beaten paths of colonial administration and commerce. For communication between the western coast of America and the Philippines, a convenient route lying north of the equator was early established—a route which utilized the northeast trade-winds north of the equator for the westward passage, and the westerly winds of the north Pacific for the eastward.

Working about the southern extremity of South America against the strong and steady westerly winds that blow in those latitudes, whether through the Strait of Magellan or later about Cape Horn, was generally a difficult task, even much later than the date of Spanish exploration of the tropical Pacific. Once a ship reached the Pacific, its pilot turned thankfully northward along the western coast of South America, into latitudes in which the ship could be steered out into the open ocean and westward with the aid of the trade-winds, which blow from the east in a broad belt between 20° north and 20° south latitude. Thus the winds upon which the navigators of the sixteenth and seventeenth centuries depended made the approach from the east toward the unexplored part of the South Pacific Ocean difficult.

There is, however, a second route into the unexplored area in which the sixteenth-century map-makers placed the part of the Southern Continent that extended into lower latitudes—the route from the west in the latitude of the Cape of Good Hope. Here the westerly winds of the southern hemisphere help rather than hinder. This route was in use, however, exclusively for passage to the East Indies, always a more enticing goal than any unknown land. The map on page 110 shows the state of knowledge of the lands of the southern hemisphere near the end of the sixteenth century, as this knowledge was expressed in the maps issued by the famous map-publishing houses of the Netherlands. The outline of

the Southern Land is in part hypothetical and in part incorporates observations made by navigators passing westward into the Pacific by way of Magellan's Strait and into
the Indian Ocean by way of the Cape of Good Hope. In
the quadrant between 90° west and 180° the South Pacific
is drawn as open sea in the parts that had been traversed
by Spanish ships. In the lower left of the figure is the southeastern part of the Indian Ocean, limited on the east by a
part of the Southern Land. If one compares this part of
the map with Captain Cook's map on page 111, one recognizes
immediately that this coast is really the western coast of
Australia. How it was found by some unknown mariner
on his way to the East Indies appears clearly from the practices of Dutch skippers in the early seventeenth century.
Dutch ships sailing thither found a better passage by sailing
eastward from the southern extremity of Africa to the longitude of Java and then turning northward than by setting
their course northeastward from the Cape or Madagascar.
The author of the map from which that on page 110 is drawn
placed Java (the island at the extreme left of the figure,
on which "120° E." is lettered) in the same longitude as the
coast of the Southern Continent farther to the south. This
arrangement represents the sailing practice mentioned, and
indicates that this coast was drawn from observation. Elsewhere, except at Magellan's Strait, its outline is drawn by
guess.

It was inevitable that the active Spanish navigation of
the Pacific in the sixteenth century should lead to the discovery of some of the innumerable islands that dot that
ocean between the equator and 20° south latitude. Rumold
Mercator's map of 1587 is not so well informed as to Spanish discoveries as it is concerning the Dutch. Twenty years
earlier than the date of the map, in 1568, Alvaro Mendaña
de Neyra, sailing westward from Peru, discovered the
Solomon Islands, in the far western part of the ocean, and
identified them as parts of the Southern Land. On a second
voyage, in 1595, Mendaña found the Marquesas Islands,

and farther west the Santa Cruz group, which lies somewhat east of the Solomons.

On his second voyage Mendaña had as a subordinate the
man who may without exaggeration be called the Columbus
of the Southern Continent—Pedro Fernandez de Queiros,
Portuguese by birth but long in the Spanish service. After
the return of Mendaña's second expedition, on which its
leader died, Queiros exerted himself successfully with the
Pope and the Spanish Crown to obtain support for an
expedition to the lands he had seen on Mendaña's second
voyage. His expedition sailed from Peru in 1605, and on it
he discovered the Tuamotu island group in eastern Oceania,
and much farther westward the New Hebrides, the largest
island of which he took to be actually part of the Southern
Continent. With much pomp and ceremony he took possession of the "continent," naming it *"Austrialia del Espiritu
Santo."* The island thus christened, pitifully insignificant
in comparison with what he thought it to be, is still called
by the latter part of the name he gave it. Something
approaching a mutiny among his crew caused Queiros to
return from here to America without following up his discovery. After years of weary effort to obtain support for a
second expedition to his newly discovered continent,
he died, leaving the question of the southern land unan-
swered.

Queiros did not hesitate in pleading for support of his
explorations to link his name with that of Columbus; and
there is much in his missionary zeal, his perseverance in the
face of bureaucratic obstacles, and the obstinacy with which
he insisted on the identity of the lands he had discovered that
remind one of the Genoese. His conviction of the identity
of his Austral lands was absolute: "The greatness of the land
newly discovered, judging from what I saw ... is well established. Its length is as much as all Europe and Asia Minor
as far as the Caspian and Persia, with all the islands of the
Mediterranean and ... the two islands of England and Ireland."

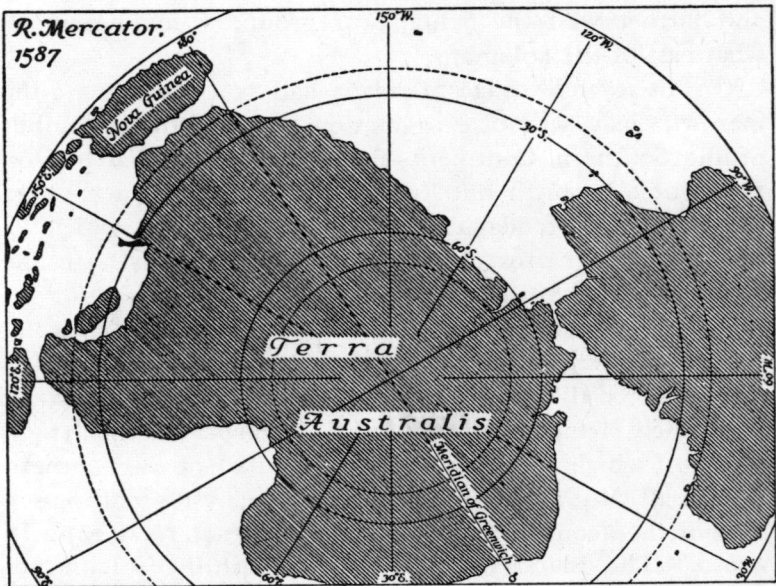

THE SOUTHERN CONTINENT AS REPRESENTED BY MERCATOR

Part of the southern hemisphere drawn from a world map by Rumold Mercator, dated 1587, showing the location and extent of the Southern Continent. An inscription on the original states that "some call this southern continent Magellan's Land, after its discoverer." The land mass at the right is South America—a name not in use at the date of Mercator's map. The island at the left side of the map in longitude 120° east is Java. The strait between New Guinea and the Southern Continent is only guessed at; other maps of the same and later dates show New Guinea continuous with Terra Australis.

Judging from what he saw in the South Sea? Only in the sense that the seeker after something intensely desired generally finds what may be interpreted as the object of his quest. He was actually judging from what he saw on some map resembling the original of the map above and from an interpretation of the island of the Holy Spirit as a part of the continent depicted on such a map. Queiros was writing in the same tradition that led the discoverer of the New World to exaggerate mightily the area and wealth of the West Indian islands.

By the time of Queiros the enthusiasm that had supported

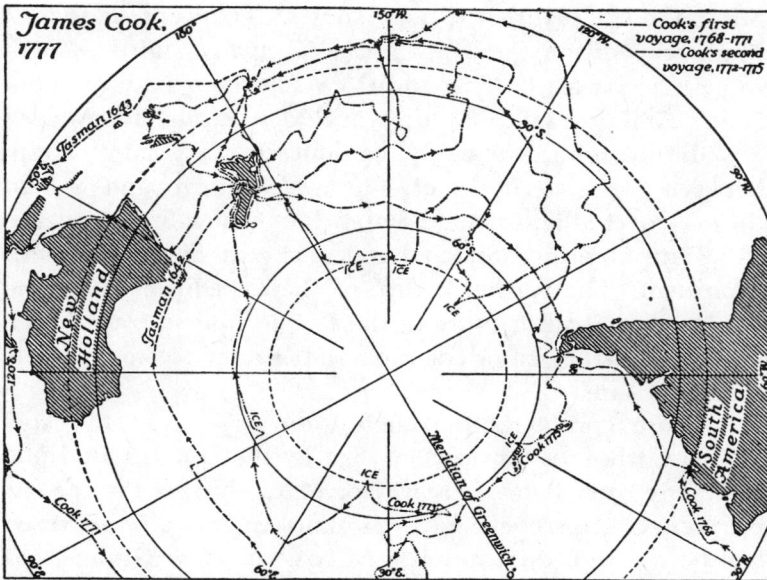

James Cook.
1777

—— Cook's first
voyage. 1768-1771
—— Cook's second
voyage.1772-1775

THE SOUTHERN OCEAN AS EXPLORED BY CAPTAIN JAMES COOK

The area shown in the map on the opposite page as it appears on the map accompanying Captain Cook's account of his second voyage. Tasman's expedition (1642-1643) separated New Holland (Australia) from the Southern Continent but left the relation of New Zealand to other possible lands uncertain. Cook's two voyages removed all doubt that the South Pacific Ocean was open and left room for a large land mass only within the Antarctic Circle. The word "ice" marks the points where Cook was prevented by floating ice from sailing farther southward.

a century of Spanish discovery and conquest had faded. The Columbus of the Southern Land was unable to obtain means for a voyage which perhaps would have cleared from the map the land of his search. The torch of exploration in search of the unknown continent passed into the hands of the Dutch, who went more soberly about the quest, for the sake not of uncounted Austral souls awaiting salvation, but of more tangible worldly goods. We have first to mention the expedition of Jacques Le Maire and Willem Corneliszoon Schouten in 1615-1616, financed by the Dutch West India Company. The Dutch *East* India Company had by its charter the sole right to navigate through Magellan's

Strait for trade with the East Indies. By means of the Le Maire expedition the *West* India Company craftily sought to gain access to the East Indies by finding a passage about South America south of the forbidden Strait. Le Maire's expedition found it easily enough in Le Maire Strait, between the eastern tip of Tierra del Fuego and a land they discovered and named Staten Land, now Staten Island. Le Maire identified Staten Land as a part of the Southern Continent. The further course of the expedition lay along the familiar Spanish route in the Pacific, and so contributed nothing to the knowledge of the Pacific sector of the unknown land.

A more considerable contribution was made by Abel Tasman, who was sent out in 1642 by the Dutch East India Company from Batavia by way of Mauritius for the specific purpose of "discovering the unknown and undiscovered Southern Land, the southeastern coast of New Guinea and the islands lying thereabout." The instructions given to Tasman embody the belief of the functionaries of the East India Company that in "the remaining unknown part of the globe, lying in the south and presumably nearly as large as the Old or the New World," there was good reason to expect "the discovery of many excellent and fruitful lands, in the cold, temperate, and hot zones." Even if no land were discovered, these excellent merchants would be satisfied to find that the South Sea was open and provided a clear passage to Chile, with which country profitable commercial relations might be established.

As is so often the case, Tasman's expedition did not yield a definitive answer to the problems it was expected to solve. Tasman sailed eastward across the Indian Ocean, discovered the island now called Tasmania, and then reached the western coast of New Zealand. From New Zealand he continued northward to the Tonga and Fiji Islands, and returned to Batavia by a route passing north of New Guinea. In his report on the expedition Tasman expressed the opinion that the sea was open to the east of New Zealand, basing his judgment on the fact that he encountered a heavy swell from the southeast when he passed the northern end of that land. This assurance did not satisfy his superiors in Batavia, who in their report to the authorities in Holland grumblingly remarked that if Tasman "had sailed a few degrees farther south, he well might have encountered land, even Staten Land, that ... may extend all the way to Le Maire Strait, perhaps even many miles farther eastward." They, like Queiros, were judging from some such map as the original of that on page 110.

Tasman's route is marked in the map on page 111. It definitely cut off a large area from the possible extent of the Southern Continent, and proved as definitely that New Holland, which later inherited the name of Terra Australis, is not attached to any southern circumpolar land mass. But Tasman found a hitherto unknown coast, that of New Zealand, which was immediately available as a potential shore of the long-sought continent. After following the already long story of the Southern Land, one is not astonished to find another Hollander, Jacob Roggeveen, embarking in 1722 on a search not greatly different from Tasman's, undertaken eighty years earlier. Roggeveen added nothing to the available information concerning the unknown land.

Thus it remained for that prince of navigators, James Cook, to answer the question "whether the unexplored part of the *Southern Hemisphere* be only an immense mass of water, or contain another continent, as speculative geography seemed to suggest." That is Cook's formulation of the problem as it appeared over a hundred years after Tasman. It is written in Cook's account of his second voyage, that of 1772-1775, the object of which he described concisely as "to put an end to all diversity of opinion about a matter so curious and important."

How thoroughly Cook scoured the South Pacific Ocean in order to obtain a final answer to the question he was set to resolve may be seen in the map on page 111, drawn from the map that accompanied the published account of his second voyage. The map shows the routes he followed on his two expeditions into the southern hemisphere. No land mass of any considerable magnitude could escape the net of tracks he threw across the southern ocean. After his farthest thrust southward into the Antarctic, and in view of the ice barrier which prevented his penetration of still higher latitudes, Cook could with good reason write that he was "now well satisfied no continent was to be found in this ocean, but what must lie so far to the south as to be wholly inaccessible on account of ice." We may still be well content with Cook's answer, though we should hesitate to stamp the word "inaccessible" on any part of the map of the earth. The problem of a specifically Antarctic continent was not the one he was asked to solve; he left room for that problem when contemplating the ice barrier that stopped him at his farthest south: "And yet I think there must be some [land] to the South behind this ice."

THE ICE-FREE INTERIOR OF GREENLAND

Too often, in our intellectual arrogance, we speak as if modern modes of scientific thinking were a sufficient preventive of egregious error. Yet it appears that erroneous judgments concerning unvisited parts of the earth have up to the present been corrected by direct observation, not by improvements in methods of speculation about what lies beyond the horizon of known parts. For the purposes of this chapter it is unfortunate that the earth has been too well known, the unexplored parts of it too small, to provide a really searching test of the ability of modern science to predict what would be found when some unexplored area was finally visited. It has thus been impossible to commit any such heroic error as the belief in an Austral continent. One error may be cited, however, that very nearly qualifies for admission into the company of the greater geographic errors of the past. Its qualifications are the magnitude of the country involved and the scientific eminence of the man who committed the error. The country is Greenland,

and the erring scientist was Adolf Erik Nordenskiöld (1832-1901), who won his widest fame by leading the expedition on the ship *Vega* that traversed for the first time the "northeast passage" from the Atlantic into the Pacific. In the latter part of the nineteenth century Nordenskiöld had more Arctic experience than any other living scientist.

In 1883, some three years after his return from the historic voyage of the *Vega*, Nordenskiöld organized an expedition to Greenland, the principal object of which was to investigate the interior of this great island (or continent, as he preferred to call it), which no one had yet penetrated. Some, including himself on an earlier expedition, had ascended to the edge of the ice that everywhere approaches or reaches the coast, and had been unable to see anything toward the interior except an expanse of ice. Some had drawn the conclusion from these observations that Greenland is covered by a glacier of continental dimensions. The other continental glacier of the present-day earth, that of Antarctica, was then even less known than Greenland, so that there was no precedent for believing that so large an accumulation of glacial ice existed. The glaciers that were known were mountain glaciers or such relatively small ice-caps as those in Iceland and Spitzbergen. Nordenskiöld had explored one of the Spitzbergen ice-caps, and so knew by personal observation how puny an accumulation of ice, in comparison with the area of Greenland, the climate permitted there, far up near 80° north, the latitude of the most northerly part of Greenland. He held the conclusion that Greenland is covered with ice, drawn from so scanty observation, to be "wholly unjustified." "On the contrary," he argued, "the following considerations seem to indicate that on the whole it is a *physical impossibility* that the interior of an extensive continent should be completely shrouded in ice under the climatic conditions that obtain south of 80° north latitude on our globe." (He planned to strike inland from the west coast in about latitude 69° north.)

His argument for the possibility—even probability—of the

interior of Greenland's being ice-free is still convincing. It proceeds from a definition of the circumstances under which snow accumulates to form glaciers: over a period of years more snow must fall than can be melted in the summers. This condition must be fulfilled at the source regions of all glaciers, including those which come down to the sea along the Greenland coasts. The essential question was, where are the sources of these glaciers? Is there a great common source in the interior of the land, or is the source in narrow mountainous strips of country that surround a lower interior? From the observations made about the margins of Greenland, either case was within the range of possibility.

The diagram on page 117 is a slight elaboration of a drawing Nordenskiöld published as part of his argument for the possibility that the interior of Greenland is not covered by a permanent sheet of ice. The snow that falls to accumulate in the form of glaciers is precipitated from air that comes from over the sea. The occasion for the greater part of the snowfall is the cooling that results from the ascent of the marine air up the seaward slopes of the land. But if the interior of the land is lower than the highlands about its edge, this air will descend in passing inland from the marginal mountains and be warmed in its descent. Indeed, if the air descends to the same elevation it had before the water-vapor in it was condensed out as snow, it will be actually warmer than before it ascended: for there will have been added to the air the "latent" heat that is released whenever water is condensed. The principle thus appealed to is familiar; it was first developed as an explanation of the warm and dry "Föhn" winds of the northern Alps, the name of which is given to all warm and dry winds of the same origin, wherever they are found.

The argument is unexceptionable. If the glaciers seen along the edges of Greenland are formed from snow that falls on a marginal highland, and if the interior of the land mass is low, the winds that reach this interior must be relatively warm and too dry to contribute enough snow to maintain a permanent cover of snow and ice. It follows that if the opinion of those who believe that all Greenland except a narrow coastal margin is covered with ice is correct, the profile of the land must be different from that shown in *a* below. Rather, the land must be shaped as is shown in *b*—"have the form of a dome with an upper surface that slopes gently and regularly toward the sea." On a land surface having this form, snow may fall all the way into its

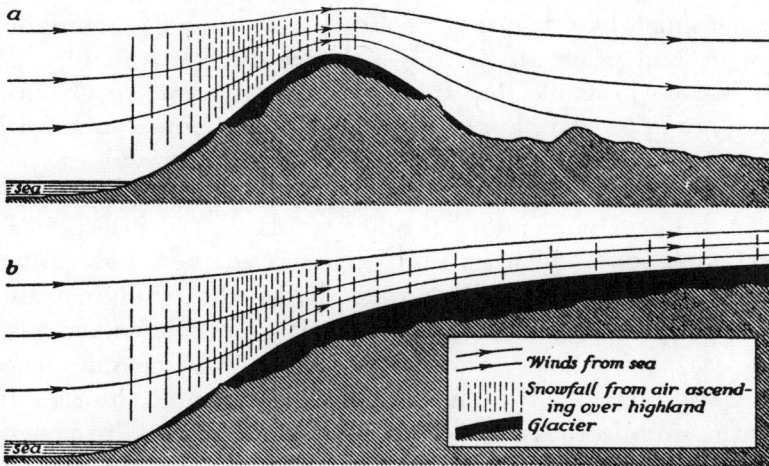

THEORIES OF THE INTERIOR OF GREENLAND

Alternative interpretations of the profile from the coast into the interior of Greenland before the interior was explored: *a*, A. E. Nordenskiöld's interpretation, a mountain range with a lower interior behind it; *b*, a dome-shaped land mass, sloping toward the sea. If *a* were correct, the interior would be free of permanent ice. If *b* were correct, the interior might be covered by an ice-sheet.

highest and most remote interior. The amount that falls in the interior will be slight, but the low temperatures that obtain there will prevent the melting of what little does fall, and thus even more may be accumulated and be consolidated into the ice of glaciers than on the marginal slopes, where higher temperatures melt much of the abundant snow that falls.

Again, the argument is convincing. But why, one asks,

may not Greenland have the form necessary to maintain a cover of ice, such as it appears to have according to the observations made about its margin? The interior of the land being unknown, one judges what it must be from the nature of the surfaces of the lands that are known: "We do not find such a form in any of the continents of the globe whose relief is known, and one can therefore with the utmost probability assume that Greenland does not possess it. On the contrary, the geologic character of Greenland, which in many respects resembles that of Scandinavia, indicates a relief much like that of our country, consisting of mountain ranges and peaks alternating with deep valleys and plains. It is even probable that the mountain backbone in Greenland, as in Sweden, England, and North and South America, runs in the longest dimension of the country along its western coast."

Argument by analogy, in other words. In Nordenskiöld's drawing, from which *a* of our figure is adapted, one easily recognizes a generalized profile across the Scandinavian peninsula from west to east: a glaciated backbone nearest the sea, and a lower, broader area on the eastern side, ice-free and blessed with summers which may be uncomfortably warm, even in the latitude in which it was planned to penetrate the unknown interior of Greenland.

That is not quite all there is to be said. Even more fundamental than the familiar structure of the Scandinavian peninsula, which was clearly in Nordenskiöld's mind when he sketched the original of our figure, was a preconception of the relief appropriate to a land mass of the size of Greenland. There was involved, that is to say, a definition of a continent—not in terms of area or elevation above sea-level, but in terms of a certain general structure expressed in the distribution on it of highland and lowland; a definition abstracted from the detailed contemplation of continents more accessible and better known than the land he planned to explore.

Nordenskiöld's expedition of 1883, planned as the crucial

experiment which should answer once for all the question at issue, was the first that actually reached the interior of Greenland. It found there only the monotonous ice-sheet that later expeditions have made familiar. The observations then made were not, however, utterly conclusive, and the leader of the expedition was not convinced that the inland ice was continuous across the whole width of the land until Nansen crossed it in 1888. Numerous crossings since then make it probable that Greenland, under its mantle of ice, is in fact shaped like the profile drawn in *b* of our figure. Nordenskiöld's error was in the venerable tradition of judgments concerning unvisited parts of the earth, though his deductions were couched in terms of the best earth science of the late nineteenth century. In Elysium its author is certainly not excluded from the honorable company of those who have pursued geographic errors to the remote ends of the earth.

Section II

THE PHYSICAL REALM

Chapter IV

ERROR IN PHYSICS

By W. F. G. Swann

THE PRESENT CHAPTER, *as its title implies, seeks to single out those realms of physics where error has played its part in retarding progress—error, for the most part, in a positive rather than a negative rôle; error active rather than passive; error which saw too vividly that which seemed as though it should be and shunned, ofttimes, that which was; error which gained substance not so much from bad material or from mistakes of the experimenter as from the urge to force the facts available to man into a mould to which they were ill-fitted.*

With this purpose in mind, it has naturally been necessary to omit much in the science that is of interest. The chapter is a psychopathic sketch rather than a normal story of physics; as such, it runs the danger of leading the uninformed reader to the belief that the whole science is sick. It is true that the sickness, in some form or other and to some extent, has threatened every part of physics; for the road to ultimate success is often the road of trial and error. Nevertheless, there are to be found in our science innumerable healthy offspring. Among these are the practically usable things that physics has given to the world. In fact, the offspring, when schooled by the guidance of experience, have shown health and an efficiency far beyond what might have been hoped from the frequently unhealthy parents of which they were born. It was the experiments of Michael Faraday and of Joseph Henry, made in early Victorian times and guided by mental pictures which seem rather bizarre today—it was these experiments, aided by those old-fashioned pictures, that have grown the modern dynamo and motor and all the appliances of modern electrical machinery. It was these experiments that, when viewed through the mathematical mind of J. Clerk Maxwell, gave us the beginnings of wireless telegraphy. It was in days when the mode of operation of the electric battery was but ill understood that the telegraph was born. It was in an age when man had but the vaguest concept of how electricity flowed through a wire that the telephone appeared.

It is a feature characteristic of most of the discoveries that have been of utilitarian value that they came as the result of efforts directed with no utilitarian end in view. Toward the end of the last century, physicists were interested in the processes by which electricity is conducted through gases. It was necessary to invent and improve pumps to produce high vacua, with the result that the pumps and technique associated with them today can produce vacua more than ten thousand times as perfect as were attainable before these investigations started. It is to this development of the art of producing high vacua that we owe the electric house-lamp of today and a multitude of other appliances useful in the affairs of life.

Thirty years ago it was known that wires gave off electricity when heated, and physicists became interested in how and why they did so. At that time it would hardly have been possible to think of a line of investigation having less to do with things of utilitarian value; yet it was these investigations when combined with the modern methods of producing high vacua that provided the essential elements for operation of the modern X-ray tube, the radio tube, and all the hundred and one devices used in wireless telegraphy, wireless telephony, transcontinental cable operation, motion-picture recording, and endless other things.

It seemed a small matter that ultra-violet light was shown to possess the power of ejecting electricity from metals. Yet it is to this fact, pursued and studied for purely scientific reasons, that we owe the photo-electric phenomena, among them the means by which the wireless transmission of pictures has become possible.

The discovery of X-rays, more or less by accident, provided a new tool for medical science. At first that tool was one only for photography of bones and the like through the tissues of the body. Soon, however, it was discovered that the rays had therapeutic properties of untold value in the cure of disease. Physicists became interested in the nature of X-rays. They had a suspicion that they were something like light; if that were so, they should give certain characteristic phenomena when allowed to fall upon a regularly spaced structure, such as a grating upon a metal of fine lines ruled parallel to each other and near together. No such effects were found. It was suspected that the reason was that a closeness of ruling sufficient for the purposes of light was not sufficiently close for X-rays; hence the physicist turned to patterns of regularity more finely grained than those which could be made by man. He turned to those regularities of structure made by Nature which are characteristic of crystal structure. In a flash the X-rays responded to these regularities; when allowed to fall upon crystals, they revealed a set of phenomena which threw an immense light not only upon the nature of X-rays themselves, but also upon the nature of crystal structure.

Almost immediately these new phenomena found a place in the utilitarian field. The power to detect flaws, strains, and other defects in structural materials, such as steel, is of immense value in metallurgy. The new method of X-ray analysis provided a means to look at the matter from the point of view of the crystalline structure of the metals involved. It was as though medical science, having formerly known the nature of injuries and malformation in tissue only by such things as general discoloration of the parts concerned, were suddenly presented with a microscope by means of which the individual cells could be studied. X-ray analysis enabled a more refined and detailed examination to be made of many questions pertaining to the structure and condition of metals than would have seemed possible two or three decades ago.

The discovery of radium and of radioactivity provided us with another set of radiations useful in medical science; and the study of all these things excited our curiosity more and more as to the ultimate nature of matter and as to the structure of atoms and molecules. As these various phenomena of physics became born to our knowledge—X-rays, radioactivity, phenomena concerned with optical-atomic spectra, phenomena produced by ultra-violet light, cosmic rays, and so forth—they all converged in their claims on one central parent to account for their existence and properties. That parent was the atom. Each of these phenomena placed certain demands upon the parent, and so we came to know more and more about its nature.

Theory and experiment went hand in hand. The significance of the relation of the various atoms to one another became clearer and clearer. No longer were hydrogen, chlorine, zinc, iron, gold, etc., simply a heterogeneous conglomeration of entities deposited through accident by Nature in the universe. They now appeared as members of a great family, the relationship of which to one another we began to understand. The age-long dream of the alchemist, the transmutation of the elements, became no longer a far-off dream; for today the conversion of one element into another is becoming a commonplace matter. Not yet can we perform this feat with large quantities of material; but a good beginning has been made, and it is no longer possible to regard as a hopeless maniac him who would turn lead into gold. Though it is true that we think that he who would do this probably misses the essential point in his desires: for we now know that the transformation of one atom to another involves, frequently, release or absorption of large amounts of energy; so that if our alchemist of old could convert his ounce of lead into gold, he would find that the energy released in the process would be enormously more valuable than the gold he had produced. He would be better advised to set up a power-supply plant than a jeweler's shop.

In the development of the tools to dissect the atom, examine its structure, and draw its teeth, strange new devices have been evolved—devices for creating very high voltages, devices for giving particles of matter very high energy; and with these devices many strange things are being accomplished. By bombarding the elements of matter with these high-energy particles, it has been possible to endow many of them with some of the properties of radium. It has been possible to put them into conditions in which they spontaneously emit radiations on their own account, radiations valuable in the cure of disease; so that before long we may be independent of radium for our medical treatments and may rely upon new materials of man's own making.

And now, having made an introduction to show that not all that science has done has been to create error, let us humble ourselves and enter the psychopathic laboratory of physics, that we may see how the imagination of man has led him, by devious paths, with many side-shows, delays, and with indulgence in many philosophical vices, to something which we believe to approach nearer and nearer to the ultimate truth as the ages go by.

SCIENCE AND COMMON SENSE

AMONG the chief falsifiers of the truth in physics stands one who has always been regarded as a most highly respected member of the community of thought. His name is "Common Sense." I fear this statement will shock many who pride themselves on their reverence for this much-lauded authority, who worship him almost as a god, and who see in him the most conservative and guileless of guides in the exploration of the uncharted seas of Nature's mysteries. Now, I shall not deny that "Common Sense" has his merits. In his proper domain he is a counselor of priceless value; and it is because he justly inspires such confidence in that domain that he becomes the most dangerous of deceivers of those who seek his guidance outside it. For "common sense" seeks to pin all thoughts of the new to the fabric of the old; thus ofttimes it distorts the meaning of the new by destroying that form which was inherent in it in its own right, and for no purpose other than to fit it to a pattern with which it has no harmony. The result is a bizarre and shapeless thing, out of harmony with the form into which it has been forced and out of harmony with the form that was its own. Common sense in natural philosophy repatterns itself from age to age. At each stage of its development it seeks to generalize the ideas born of the experience of the immediate past and to weld them into bonds which sometimes restrain the future. Thus the breeders of error in the epoch to come are sometimes the truths of the days that have gone.

In the common sense of ancient and medieval times inanimate things moved in orderly fashion only when they were carried; and so, knowing of no human agency capable of carrying the sun around the earth, men attributed the feat to supernatural agencies, of the same general kind as the natural ones but presumably more potent. Thus, to the

Egyptians the universe was a box. Around the edge of the box ran a river, and on this river traveled a boat which carried the sun. In other schools of ancient thought the sun was carried by an angel. Here we see an example of another fundamental principle which again and again has bred error in science, one active even to this day—the principle of seeking comfort in an explanation that invokes the action of some agency about which we know so little that nobody is able to deny that it may act in the way required. No man could transport the sun; but an angel was a sufficiently mysterious being to enable one to rest content in the belief that he could do anything. His position was rendered still more impregnable by the fact that, being a supernatural person associated with the gods, any doubt as to his powers verged on sacrilege. When the "man in the street" today sees an "explanation" of something that seems mysterious to him, he no longer invokes angels in name; but he often does what amounts to the same thing. He attributes the mysterious phenomenon vaguely to "magnetism" or to some other high-sounding name pertaining to a branch of science about which he knows sufficiently little to be unworried by any complaints from that science itself as to the possibility of its bearing the burden cast upon it.

But to return to common sense: Experience taught a man that when he was moved, he was jolted. Even in reasonably smooth motion he could feel the jolts. But he suffered no jolts as he lay peaceably upon the earth. It was probably common sense that suggested therefore that the earth was at rest and that the heavens moved around it.

And then, as reason developed and the angels and their like were felt to be unsatisfactory, common sense, having by this time accepted into its realm the principle of a stationary sun with planets revolving around it, sought a more sophisticated reason for that motion in the supposition of the existence of an all-pervading medium which whirled the planets around the sun like corks in a whirlpool. Others preferred the supposition that the planets were supported in invisible crystal spheres which revolved in the heavens carrying the planets with them. The sound of these spheres in motion was supposed to be perceptible to a few chosen persons sensitive to it. On the other hand, the great Kepler sought harmonization with reason in supposing that the sun carried spokes which wheeled the planets in their orbits.

Common sense seeks always to interpret the unknown by analogy with the known. This, in itself, is not bad. The great danger of common sense lies in its too frequently seizing upon and magnifying the non-essentials of the situation to the extent of neglecting, or even sacrificing to them, the claims of the essentials. The important thing about the motions of the heavenly bodies was the geometrical regularity of that motion. The important things to emphasize were those principles which now permit us to foretell the state of the heavens a thousand years hence in terms of its state now. The angels gave nothing of that story. The boat of the Egyptians, Kepler's spokes, and Descartes' whirlpools were equally silent on that question. Their silence would have mattered much less were it not that what little they had to say, if said, would have denied the actual facts had the logical consequences of their story been demanded of them.

SCIENCE AND PRECONCEPTION

But while man has erred in seeking in the new too close an analogy with the old, while he has erred in seeking too materialistic an explanation of the new as materialism was understood in his day, he has occasionally gone to the other extreme—the extreme of abstractness. Fired with enthusiasm over some scheme of regularity of principles in some branch of arithmetic or geometry or in some creed, he has sought to force all the phenomena of Nature into the form that has tickled his fancy. Thus, to prove to his own satisfaction that the world is perfect, a philosopher of ancient times argued thus: "The world is composed of solids. Now, solids have *three* dimensions. But *three* is the most perfect number.

We do not speak of *one* as a number; of *two* we say *both;* but *three* is the first number of which we say *all.* Moreover, it has a beginning, a middle, and an end." The Pythagorean school placed much weight on the properties of numbers as symbolic of the structure of the universe. Much elated by the discovery that the length of strings that gave any note, its fifth, and its octave were in the ratio of six to four to three, they affirmed that the distances of the planets from the earth must be representative of a musical sequence, which was incorporated in the "music of the spheres."

At a much later date Kepler sought to force the frame of the universe into harmony with a certain geometrical design. If a cube be inscribed in the sphere containing Saturn's orbit, the sphere containing Jupiter's orbit will just fit within the cube; if a tetrahedron be now inscribed in Jupiter's sphere, the sphere of Mars' orbit will just fit within it, and so on for all the five regular solids.[1] To Kepler this fact, now known to be only an approximation, was probably significant of a design made intentionally by the Creator, as an artist might paint a picture. The Greeks, given much to philosophy, had sought to decide how the universe must be fashioned by arguing from certain preconceived notions founded ultimately upon beliefs which were regarded presumably as obvious or suggested by the supposed habits and characteristics of the gods. Thus, the Aristotelians argued that Nature abhorred a vacuum, a belief which persisted until the seventeenth century. Torricelli (1608-1647) showed that if a tube closed at one end be filled with mercury and inverted, the mercury will stand up to the top of the tube only if the tube is less than 30 inches long; evidently Nature's abhorrence of a vacuum was limited. The fundamental postulate that the universe, having been created by God, must be perfect, combined with the idea that the circle was the perfection of symmetry, led to

---

[1] The *regular* solid is one having all of its edges, faces, and face-angles equal. The five regular solids are the cube, the tetrahedron, the octahedron, the dodecahedron, the icosahedron.

the belief that the planets must necessarily move in circles, or if not in circles, then in curves compounded out of circles. This postulate, originally formulated by Hipparchus (second century B.C.), led finally to complexity upon complexity until the Sovereign of Castille, when the system was explained to him, was driven to remark that had he been present when the universe was constructed, he could have given the Deity some good advice.

In the school of Aristotle the falling of bodies to the earth was regarded rather as a desire on the part of the body to unite with the earth than a result of a continued pull on it such as was suggested later by mechanical considerations. The feeling of Aristotle was to the effect that the greater the size of the body, the stronger its determination to get to its natural goal, and the greater, therefore, its speed of fall. The curious thing is that such a belief should have persisted for two thousand years without its truth being tested by experiment. The basic reason for the belief, flimsy as it appears to us today, seems to have been stronger in its appeal than the direct evidence of experiment. For even when experiment showed the falsity of the arguments, there were those who sought to save the principle by seeking in the experiment disturbing causes which resulted in its truth being apparently vitiated.

Toward the beginning of the seventeenth century, in a world reeking with the dogmas of an ancient past and firmly in the grip of the philosophy of an all-powerful Church, we find Galileo, a young enthusiast enjoying the princely stipend of fifteen cents a day as professor of mathematics in the University of Pisa. He has done a very wicked thing. He has invented a telescope, by which man can see in the heavens more than his unaided eye could show him. Such a device must most assuredly have been inspired by the Devil; for had man been intended to see these things, his eyes would have been made so that he *could* see them. The young upstart, trying to improve upon the creations of the Almighty, has only himself to blame if the Devil is allowed to bewitch his impudent device so as to cause it to reveal to him bizarre monstrosities which cannot exist in Nature. The young man has asserted that Jupiter has moons attendant upon him. This is likely to upset everything; for the Florentine astronomer Francesco Sizzi has cogitated upon the mysteries of the universe as revealed to him with the senses that God gave him. He has observed that there are seven days in the week, seven openings in the head, seven metals (as he thought), and I believe he found a few more sevens. The number seven obviously had a mysterious significance. There should be seven planetary bodies, and no more. Moreover, according to the doctrines of the common sense of the day, all of these things were made for the benefit of man, the crowning achievement of creation; so common sense would obviously decree as nonsensical the creation of anything so useless as satellites which the eye could not see, which consequently could send him no light, and which would therefore be utterly useless to him. It was unthinkable that the Deity should create such things. But young Galileo has done even worse. He has turned his obnoxious appliance upon the sun and has found spots upon its surface. Is it for a young fifteen-cent-a-day upstart to find blemishes upon this most perfect of all heavenly bodies, the sun? No! The young man has had the effrontery to question the perfection of Nature, and Nature has hurled back at him through his telescope a mass of falsehoods, that he may see Nature as a madman sees it, and as he deserves to see it.

But the worst is yet to come. For Galileo has busied himself with the infant science of motion and has dared to dispute the great Aristotle. On the basis of certain experiments in which he has actually observed the fall of bodies from the top of the Leaning Tower of Pisa, he has affirmed that light and heavy bodies fall at the same rate.

Today the ideas born of Galileo's researches on motion form the foundation of our most conservative thinking. When we can see phenomena in a new realm acting in a manner explicable in terms of motions that can be thought

of in the same sense as Galileo and Newton pictured
the motions of the heavenly bodies, we feel that we understand them. That which was nonsense to the contemporaries
of Galileo has become the common sense of today. And so
firmly has it become engraved in our consciousness that
only with the greatest strain of the imagination has the
physics of the last few years gained courage to advance still
further, to a realm of thought where once again one must
seek a new basis of common sense, a basis in which even the
laws of Galileo stand shorn of any ultimate claim to a fundamental obviousness apart from the fact of their experimental truth over the restricted realm of phenomena that
they covered.

In criticizing the attitude of mind of the critics of
Galileo and of the Greek philosophers, we must be careful
lest we throw stones from glass houses. It is not because
Francesco Sizzi sought in the properties of numbers the
correlation of the facts of the universe that we should laugh
at him; for a critical examination of most modern theories
of atomic structure will show that, in the last analysis, our
theories are based upon the properties of numbers in a
manner which, from the standpoint of a naïve concept of
reality, is as artificial as the mental meanderings of Sizzi.
No, it is not on the basis of artificiality that we must criticize
these men of bygone days. The hypotheses and dogmas of
modern physics differ from those of the ancients not so
much in the matter of artificiality as in the fact that modern
hypotheses are *chosen* with Nature as a guide. They are
chosen so as to fit Nature, whereas the hypotheses of the
ancients were chosen ofttimes from principles having no
immediate connection with Nature or with the branch of it
under discussion, and this choice was then followed by an
attempt to force Nature into them. It is as though two
persons should view a picture or design which, at first glance,
seems difficult to comprehend. The first, symbolic of the
ancient point of view, says: "I will try to understand this
picture by showing how its lines can be constructed by com-

pounding them out of a series of circles of different sizes and positions, because I believe the circle is the perfection of symmetry, that the artist was a good one, and that therefore he must have designed the picture on the basis of circles." The second student of the picture, symbolic of the physicist of today, says: "No! I see no reason for starting with circles. I will form *no* preconceived ideas. I will leave my mind open as to the fundamental forms that have guided the artist until I find those forms. Then, when I think I have found them, the simplicity with which I can reconstruct the picture out of them shall be my guide as to their worth. I care not if I have never thought of these forms before; their power to correlate the features of the design in the picture shall give them all the prestige they need for me." Possibly our student may find that a figure S or a figure 8 may be the form that he chooses ultimately. The object of modern physics is to seek *in the facts themselves* some starting-point which may be taken as the key-note or notes of all the other facts, so that those other facts may follow as a natural consequence from them.

Perhaps the first great and typical example of the departure from these older paths of thought which had never led to success is to be found in Newton's formulation of the Law of Gravitation. From a consideration of the actual motion of the planets, as experiment revealed that motion, Newton concluded that a planet moves at each instant in such a manner that its acceleration towards the sun is inversely proportional to the square of its distance from the sun. When, in its orbit, it is far away from the sun, it moves so that its acceleration towards the sun is small. When it is near to the sun, the corresponding acceleration is large. Now it matters very little whether the reader understands the exact meaning of this statement of Newton's law. The essential thing to observe is that in it *one* statement is made about the motion of the planets; and from that one relatively simple statement accepted as a truth all the complexities of the planetary motions follow as a natural and accurately calculable consequence. From this law of Newton one can calculate the time of recurrence of eclipses, the time of return of comets, the shapes of the orbits of the planets, the speed of revolution of the planets around the sun, and a hundred other things. But, following this great generalization of Newton, it was difficult for man to avoid being seduced by the beauty and simplicity of the law to seek a *cause* for it. Once again the voices of the ancient philosophers rang out from Valhalla, and common sense joined in the dictum, "To every effect there must correspond a cause," and demanded an answer to the question, "What is the cause?" And ever since Newton, even to this day, many of us have puzzled our brains to find the "cause."

THE SEARCH FOR CAUSES

Now when we seek a cause for anything, it is perhaps incumbent upon us to seek the basis of our contentment in the *status quo* that we believe would exist in the absence of the cause. A stone moves along the floor. If the floor is rough, the stone soon comes to rest. If it is level and smooth, the stone moves a long way with sensibly undiminished speed. If the stone is thrown on level ice, it moves still farther and retains its speed still longer. When we look at the rough floor, we see the little projections that obviously played a part in stopping the stone. It is easy to see how we can be led to think that the stone's natural desire is to go on forever with undiminished velocity, that it is only because of the participation of extraneous phenomena—the roughness of the floor—that it comes to rest at all. Thus we form the concept of uniform motion in a straight line with undiminished velocity as the state natural to the body when uninfluenced; and any departure from that state seems to call for a "cause."

Now the ancients were not so content in the thought of even uniform motion being a thing satisfying in itself and requiring no further thought. Incidentally, the conventional school-boy has much in common with the ancient philosophers in this respect. He usually starts out with the concept that what he regards as a force is necessary to keep a body moving even with constant linear velocity. He is surprised if told that, but for the friction of the water, the *Europa,* without her engines working, could be pulled through the ocean at 40 knots tied to one end of a hair—or even without the hair. The school-boy, if left to himself, feels some necessity of seeking a cause for the ship's motion. However, guided by such experiences as those gained from the stone and the rough floor as compared with the smooth floor and the ice, the more sophisticated investigator of the post-Galilean period has come to an attitude of mind in which he is content if a body moves in a straight line with constant velocity. He seeks no cause. Straight line with constant velocity represents the condition of nothingness as regards anything happening that the mind takes as its origin from which to elaborate its thoughts. In the contemplation of this state the mind is happy—no questions are asked.

But the existence of any other state of motion has come to be regarded as calling for a cause. In the last analysis, this attitude of the mind is purely arbitrary. It is an attitude which has become stimulated doubtless by observation, unconscious or otherwise, that when any motion other than uniform linear motion of a body is contemplated, there will usually be found in the vicinity other bodies in terms of whose positions and activities the departure from the simple type of motion can be simply expressed. Thus, Newton's Law of Gravitation expresses in simple form the relation between the departure of a planet's motion from that of uniform linear velocity and the positions of the sun and the other planets. However, the mind has usually not been content with a mere statement of the motion, but has felt that here at least a "cause" is demanded. When we seek satisfaction in the discovery of a cause in a situation of this kind, we must be on our guard to avoid seeking contentment in illusions.

I may perhaps be pardoned for using here an illustration

that I have used elsewhere.[2] It concerns one who seeks contentment in the thought that perhaps the pull of gravitation may be regarded as similar to the pull of a piece of elastic upon a stone when the stone is tied to one end of the elastic and swung around in a circle, with the other end of the elastic held in the hand. Our seeker of a cause may feel content until I ask him why a piece of elastic pulls. To this he will probably reply: "Oh, of course, we do not know all about that; but in a general way we believe that the elastic is composed of a lot of little molecules, and when these are separated from each other by the pull, they tend to come together again." "But," I shall ask him, "why do they tend to come together again since they are separated?" "Well," he will answer, "of course, when I said separated, I really did not mean entirely separated. We believe that between the molecules there is an all-pervading medium of some kind, a medium with elastic properties, and this draws the molecules together." "But," I shall persist, "what do you mean by the medium having elastic properties?" "That it acts like a piece of elastic," he will reply. "But why does the elastic pull?" is my rejoinder. "Well," says he, "because it is composed of a lot of little molecules, and when these are separated ... etc., etc.," and so on *ad infinitum*. You see, all that even a successful appeal to the elastic idea about gravitation could do would be to show that this thing "gravitation" which we do not understand acts in the same sort of way as that other thing about the elastic which we also do not understand, but think we do. There is a sort of unification of ignorance in the matter. But this unifying of ignorance must not be despised. It constitutes much of the plan and purpose of physics. Thus, we like to correlate the workings of the atom in our minds in such a way that it acts like a little solar system, or an electrical machine, or the ripples in a bowl of water—something that has become familiar to us in our everyday life and with whose behavior

[2] In "The Trend of Thought in Physics," *Science*, Vol. 61 (1925), No. 1582, pp. 425-435; also in *The Architecture of the Universe* (New York, 1934).

we have come to be satisfied. We were happy about the elastic before we had grown to the age when we were compelled to think about it; and by the time we had become old enough to think about it, it had become so familiar to us that we felt no necessity to think about it any further. The elastic was a god—the origin of all things—who himself needs no certificate as to his own origin. And so physics, like everything else, has its starting-points, its postulates. And the postulates of physics are its gods.

In the search for causes of phenomena in physics there are two goals—the first a thing of worth, and the second an illusion. The first goal is one in which the phenomenon in question is shown to be the direct outcome of some starting-point that can serve also as the starting-point for the deduction of other phenomena of nature. No cause is sought for the starting-point in itself. It exists in its own right, and its value as a sort of grandfather of the various phenomena it predicts is in pointing out the relationship of these phenomena to one another. If two phenomena, at first sight distinct, can be traced back to a common starting-point, the number of postulates that one must make in order that he may understand Nature in terms of them is reduced to that extent. The maximum of satisfaction is to be found in ultimately understanding Nature in terms of the acceptance of the minimum number of hypotheses.

The second goal, often an illusion, is one which offers that kind of contentment of understanding of the reason for the phenomenon symbolized by the person who, in the illustration above cited, sought comfort in the thought that the pull of gravity was like that of a piece of elastic. The chief danger in the search for this goal lies not in the *contentment* of illusion; for if the searcher for a reason is happy in his reason, why disturb him? The danger lies rather in the effort to adorn the mechanism that has given satisfaction to the mind with irrelevant appendages to enhance its reality, appendages which ofttimes cause trouble by demanding on their own account certain requirements which would be inconsistent with the fundamental requirements of the main problem.

IRRELEVANCE IN PHYSICAL THEORIES

Many are the errors of physics and philosophy that have arisen through catering too closely to the demands of things fundamentally irrelevant to the discussion in hand. In most of our pictures of phenomena, and this is true also of physical phenomena, the irrelevant parts of the picture tend to be the superficially more obvious parts. An electron is the fundamental unit of negative electricity, a thing so small in size that millions of millions of them would be required to take up the space of one inch. Sometimes I have been asked what its color is. To one who is familiar with the properties of electrons, the asking of such a question is something like asking what political views are held by the majority of those unicellular organisms called amœbae. The latter question is meaningless because we do not believe that in the amœba there is anything to correspond to the kind of intelligence that would permit the amœba to have any political views at all. Now if I should meet someone who said, "I cannot understand or think about any animal or organism that has no political views," then it might be an advantage to tell him that most amœbae are Republicans. He would then be able to think about the amœba and investigate it. He would make no use of his information to the effect that the amœbae were Republicans, but the information would make him happy (or otherwise, as the case might be). In this state of mental exhilaration, stimulated by the Republicanism of the amœbae, he would proceed to investigate their digestive processes, their methods of cell growth, and so forth. All would be well, and no harm would be done unless some day he should proceed to draw upon the qualities of Republicanism in the amœba. If he should say to himself: "Now Republicans are for the most part stanch capitalists, and capitalism goes with banks, stock-market, and the like; therefore I will proceed to look for banks and stock-markets among the amœbae," he would of course land himself in difficulty. The difficulty, moreover, would be one which never had any reason for its existence. It would be something which existed as an appendage to the Republicanism of the amœbae, and this thought of Republicanism was introduced for no purpose other than that of stimulating the brains of the investigator to think about the amœba. So, with the electron, if one attributes color to it, no harm will be done unless he comes to look for some results caused by the color itself. A characteristic example of the part played by irrelevance in physical thought is afforded by the many attempts that have been made to understand the propagation of light, heat, etc., through space through the agency of an all-pervading medium, an ether.

The ancients believed that sight came about through the agency of feelers which extended out from the eye and "felt" the object looked at. This seems a very natural and sensible idea, strongly suggested by the analogous means by which we locate objects through the sense of touch in our fingers. However, several obvious difficulties arising from such a view finally resulted in a realization that sight was to be attributed to something emanating from the body seen and traveling to the eye that it finally influenced. The most obvious "something" to think of was a particle; and so at the time of Newton there was a strong school of natural philosophers, headed by Newton himself, who adopted the view that light consisted of streams of charged particles, or *corpuscles*, as they are often called. The rival view was to the effect that light consisted of *waves* in some all-pervading medium. The corpuscular view gave a natural interpretation of the existence of shadows, a phenomenon not so easily explained by the wave theory, since waves bend around corners. However, it was realized later that regular trains of waves would bend around corners but little if the lengths of the waves were short enough, and the little that they would bend was just such as to explain several optical phenomena which became revealed by careful observation—

phenomena such as the colored fringes seen when we look through our eyelashes, the multiple images seen of a lamp when viewed through wire gauze, and so forth.

What was regarded as the most conclusive evidence in favor of the wave theory and against the corpuscular theory, however, was the fact that light was found to travel faster in air than in water. It was well known that when a ray of light penetrated the surface of water, its direction was bent towards the normal (the perpendicular) to the surface. Newton explained this by supposing that the light particle was attracted towards the surface as it approached. The result of such an attraction would be to increase the velocity in the direction perpendicular to the surface and leave unaltered the velocity in the direction parallel to the surface. The net result would be to bend the ray towards the normal as the facts required, but at the same time to increase the *resultant* velocity of the particle. On the other hand, the behavior of a train of *waves* on entering the water would be such that their direction of propagation would be altered towards the direction normal to the medium provided that the velocity of the waves was *less* in water than in air. Here, then, a crucial test seemed to suggest itself. If the velocity of light were found to be less in water than in air, the fact would decide conclusively in favor of the wave theory. The experiment was tried. The velocity was greater in air than in water; and apparently the corpuscular idea was eliminated for all time.

I suppose that one could hardly think of a more typical example of an apparently crucial experiment to decide between two theories than this experiment on the velocities of light in water and air which stood for two hundred years as a barrier against all further thought of a corpuscular theory of light. And yet, I think, we must recognize in the attitude of natural philosophers toward the conclusiveness of this experiment an example of one of the errors of thinking in physics. For modern research has drawn us from the old wave theory in the sense in which it was understood a hundred years ago. It has driven us from the old corpuscular theory in the sense in which it was understood two hundred odd years ago. We now have a theory which, while too abstract for exposition in detail here, harmonizes the concepts of corpuscles and waves. It permits light to be thought of, in a certain sense, as composed of corpuscles; but the laws that these corpuscles obey are such that there is, of course, no inconsistency in the matter of the relative velocities of light in air and water. The point I wish to make is this: It was not the supposition that light was a corpuscle that really contained the source of trouble in Newton's day. It was not the supposition, characteristic of a corpuscle, that light struck only here and there instead of everywhere, as a wave would strike, that led to difficulty. It was not the corpuscularness of the corpuscle that was at fault. It was something else, something irrelevant to that corpuscularness. It was the hypothesis as to how the corpuscle was attracted to, or otherwise affected by, the water surface at approach that was the real seat of the trouble; and this hypothesis, natural in itself in terms of the common-sense criteria of the day, is not something required by those features of "corpuscularness" which are really valuable in the modern theories or which, indeed, were the significant features in the story of Newtonian days. Frequently these irrelevant encumbrances of which we have just seen an example arise as the offspring of some attempt to endow the hypotheses of physics with a semblance of what is called reality. We can find few better examples of the effects of these irrelevant appendages than is to be found in the story of the "ether."

THE ETHER

When the wave theory of light had established itself securely in the minds of physicists, common sense demanded a medium, an "ether," in which the waves could travel. The analogy of sound was before us. Sound waves could travel in water, or in air, or in solids. Now it seemed that a solid must be ruled out because the ether had responsibilities other than the transmission of waves. It had to allow the planets to move freely through it, and how could a solid do that? And yet, a liquid or a gas would not do for another reason. In the case of sound waves in a gas or liquid, the particles of liquid swing back and forth in a direction parallel to the direction of propagation of the wave; but light shows certain characteristics associated with what is called polarization, which demand that if we are to think of it as composed of waves, the back-and-forth motions of the particles of the medium transmitting a wave must take place in a plane perpendicular to the direction of propagation of the wave. Such a state of affairs could exist in waves transmitted by a solid, but not in those transmitted by a liquid or gas. Thus we were confronted by a dilemma. For light waves the ether had to behave like the most rigid of solids, but in permitting the passage of the planets through it, it had to behave as the most limpid of fluids.

The question now was whether it was possible to conceive of something which was neither a solid nor a liquid nor a gas and which was capable of fulfilling the desired purposes. Well, the mathematician James MacCullagh (1809-1847) was able to devise such a medium. When I say devise, I mean that he was able to specify, in a self-consistent way, a set of properties for a medium which would cause it to behave in the manner desired. The trouble was that there seemed to be nothing tangible on the face of the earth that looked anything like what this cantankerous ether had to be. It was as though we desired to understand certain occurrences by attributing them to a particular kind of criminal whose characteristics as judged by his deeds were that he was meticulously honest in some things and a thief in others, that he was an intolerant infidel yet nevertheless gave all his ill-gotten gains to the church, and so on. If we had set out to understand the doings of this individual by thinking of him as a man, there is a question as to how much our primitive instincts as to what a man would do would help us in predicting his actions. We should be driven to say, "Well! he is a strange fellow, unlike anybody else; and all that can be said in describing him is that he is the kind of person who acts in such and such ways on such and such occasions. Those intuitions which serve us as a 'reason' for the actions of other beings are totally inapplicable in his case." We were in some such state as this with regard to the ether when the mathematicians had succeeded in formulating its properties for us.

Then, about the time when men were exercising themselves on these matters, came the great discoveries of Michael Faraday (1791-1867) and Joseph Henry (1797-1878) on the nature of the relationship between magnetism and electricity, followed by their culmination in Clerk Maxwell's (1831-1879) masterly correlation of these discoveries in a unified picture, a picture which demanded that electromagnetic effects should be capable of propagation through space with a finite velocity in the form of waves. Moreover, the nature of these waves was such that if there was to be a medium, that medium had to have characteristics the same as those which served for light. Indeed, Maxwell was able to find from purely electromagnetic measurements what the velocity of propagation of electrical effect should be; and the velocity turned out to be equal to that of light, suggesting that light itself was really an electromagnetic phenomenon. This gave the ether a great boost in prestige, so that after a time we were able to swallow our prejudice concerned with his eccentric behavior as compared with known media such as solids, liquids, and gases, and we were happy to admit him as a new but respectable member in the brotherhood of "media." He had at least one normal trait in company with his brothers: any little piece of him could occupy a definite position in space. We could form a sort of picture of waves going through this ether, and we could think of ourselves as forming a picture of something going on in the medium to correspond to these waves, even though we did not know what that something was. And then, as though itching to free itself from all respon-

sibilities to a medium of the so-called material type, Nature
proceeded to reveal a class of phenomena which rendered
nonsensical and meaningless this one remaining claim of
the ether to reality, in the sense of the reality of the past—
the claim that each of its little pieces could be thought of
as occupying a definite position in space. For the picture of
this fixed ether suggested that we should seek to ascertain
how we on our earth were moving in relation to it. An
epoch-making experiment which was designed to investigate
this question may be illustrated by the following analogy:

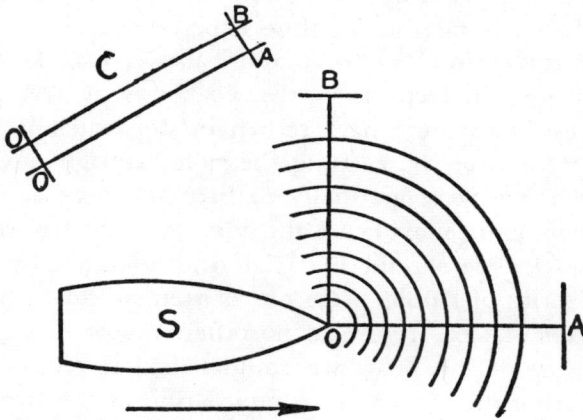

Suppose that S represents a boat which can sail along a
river in the direction of the arrow. Let us attach to the
prow of the boat two mutually perpendicular rods OA and
OB, one parallel to and one perpendicular to the length of
the boat. If we start some ripples at the point O by dropping
a stone into the water, these ripples will travel out in circles
in relation to the water. If at A and B we attach boards
which dip into the water, the ripples will be reflected from
these boards and will return to the point O. If the boat
is at rest in the water and the poles OA and OB are of equal
length, the ripples will return in step. If, however, the boat
is in motion in the direction of the arrow, the time taken
for a wavelet to travel from O to the board at A will be

increased by the motion of the boat, because the board at
A runs away from the ripples. On the other hand, the return journey will be shortened in time because the prow
of the boat runs to meet the ripples. It is a comparatively
easy matter to show that the loss of time on the outward
journey of the ripples is greater than the gain of time on
the return journey, so that, on the whole, the time taken
for a ripple to go out and return is increased by the motion
of the boat. While not so obvious, it can readily be shown
that the time for a ripple to travel from *O* to *B* and return
will also be increased by the motion of the boat, but to a *less*
extent than the increase in time experienced by the wave-
lets that travel from *O* to *A* and back. Thus, if the ripples returned in step when the boat was at rest in the
river, they would no longer return in step when it was in
motion. Moreover, if, keeping the poles at right angles to
each other, we were gradually to turn the boat across the
river, keeping it moving at the same velocity in relation
to the water, we should find that the amounts by which
the two series of ripples were out of step would alter; and,
as a matter of fact, if we had no other way of determining
the velocity of the boat, we could calculate that velocity
by measuring the *alteration* in retardation of the two trains
of ripples as the boat was turned through a right angle.

Now, using mirrors and light waves, Professors Albert A.
Michelson and Edward W. Morley performed an experiment exactly analogous to this in an attempt to measure
the earth's velocity through the ether. The result of this
experiment was regarded as just what might have been anticipated if the earth were at rest in the ether. But if we
are to adopt the picture of an earth embedded in a surrounding medium, it cannot be at rest in relation to that
medium continually—since it moves around the sun—unless
the earth drags the ether along with it. This hypothesis was
attended with difficulties concerned with the passage of light
to us from the stars, difficulties so great as to rule the supposition out of court. If in the corresponding experiment

with the boat a result similar to that obtained by Michelson and Morley had been found, and if we had been unwilling to admit that the water was in motion down stream with the boat just floating along with it, but had insisted on a relative motion between the boat and the water, there would have been only one way out of the dilemma. We should have had to say that the lengths of such rods as OA and OB changed according as those lengths ran parallel to or perpendicular to the direction of their motions, and that these changes took place as the boat was rotated and were of such amount as to annul, completely, the results that would otherwise have been expected.

Drastic as such a hypothesis seems, it came to be looked upon with favor. The development of its consequences and the logical extensions into the realm of time measurements necessitated by these consequences led us into a state of sophistication intermediate between that of the Victorian era and that which we have today. In that state of sophistication our picture was one of a stationary ether, with measuring-rods and clocks for measuring position and time. To cement the picture as definitely as possible, we can imagine the whole universe as filled with these scales and clocks, all the clocks going simultaneously, all being "right" in relation to the standard. But having established this happy state of affairs for ourselves, we had to picture anybody who resided on a planet moving relatively to us as in a curious state of confusion of which he was all unconscious. As he was moving in relation to us, all of his scales were erroneous, but he happened not to know it because any other scale he used to test them was wrong in the same way. All his clocks were wrong, but he did not know it because any other clocks he used to test them were wrong in the same way and conspired to conceal the error. If it so happened that this other observer was moving relatively to us with a velocity nearly equal to that of light, it would be necessary for us to suppose that his clocks went so slowly in relation to ours that the passage of a thousand years as recorded by our clocks would on his clocks appear only as an hour. Moreover, we would have to suppose that he would be all unconscious of these things because, his animal metabolism taking place so enormously more slowly than ours, it would take him 3,000 years to get hungry again after a meal, so that he would be quite happy about the indications of his clock. He would take his meals by it with comfort.

RELATIVITY

Now we might have continued to enjoy this picture in which we and our measurements were "right" while any other being moving with high velocity relatively to us was hopelessly wrong—though unconsciously and unharmfully so —were it not that, having figured out from the relation between our measures and his how we must look to him, it became evident that he would think exactly the same things about us as we were thinking about him. He would believe that all his measurements were right and ours wrong. Indeed, it turned out that there was nothing that we could do to prove that we were right and that he was wrong that he could not do to prove that he was right and we were wrong. The fact is that both are right. The apparent obstacle in the way of both being right was the fact that our intuitive concepts of absolute length, and absolute equality of the lengths of two rods in different places, together with our intuitive consciousness of absolute simultaneity of events taking place in different places, carried with them the belief that there was only one meaning to these concepts. Having been forced by the foregoing considerations to analyze a little more carefully the nature of our concepts, we had to admit that these concepts of absolute length, absolute time, absolute simultaneity, etc., were not as impregnable as had been supposed. We were forced, in fact, to the conclusion that not only were they not impregnable, but that they had no meaning. I am well aware that this attitude is apt to startle the layman. Some people feel that they have a deep-seated consciousness of the meaning of absolute length and time which is impregnable, and they are even apt to conclude that anyone who says otherwise is lacking in some feeling which they possess. Let me assure any such devotee of the absolute that the physicist is not unconscious of those feelings to which he refers. The physicist has had those feelings himself. He has held on to them with the greatest tenacity for years and years. He has molded his theories and viewpoints again and again in order to find a natural home for them. Only when the cold facts of Nature drove him to abandon them was he willing, reluctantly, to let them go.

But when these concepts of absolute space and time have gone, what becomes of the ether? The ether is now a medium which is satisfying, intuitively, only to an observer in one state of motion. Such comfort as he gets out of it is derived from certain thoughts about it which become totally meaningless to an observer in another state of motion. Thus, in the logical extension of the ideas above cited, we realize that, in the language of the electromagnetic theory of Maxwell, the observer sees the ether playing its part through the interaction of magnetic forces which he thinks of in one kind of way and of electrical forces which he thinks of in another kind of way. The clear-cut distinction between these electric and magnetic forces is at the root of that intuitive satisfaction which he gets from contemplation of the ether. And now he has to realize that these forces which were quite distinct in his own mind have to be all mixed up to provide the equally intuitive picture adopted by some other observer. What A calls a magnetic phenomenon is to B a mixture of magnetic and electric phenomena; and what to B is a pure magnetic phenomenon is to A a mixture of magnetic and electric phenomena. All that the ether gave in the way of an intuitive picture with an absolute meaning to the concepts in it becomes spoiled. This ether and its intuitive concepts now cause nothing but trouble; and the pity of it is that it is just the concepts that cause all the trouble that really contribute nothing at all to the working feaures of the mathematical theory. The satisfaction derived from the intuitive picture of an ether and of the mechanism of its propagation of effects from one place to another is illusory; and it is for this reason that I have ventured, in another place, to define the "ether" as a "medium invented by man for the purpose of propagating his misconceptions from one place to another." Though of all subtle fluids invented for the stimulation of the imagination it is the only one that has not been prohibited, it has been revealed as a sort of drug which gives rise to a mental hallucination of contentment which has no justification in Nature.

This chapter, which is devoted primarily to "error in physics," is no place for an attempt to expound the theory of relativity. It must suffice to say that in the various experimental investigations and theoretical discussions that have led up to the theory of relativity, we have come to realize that one of the errors of physics which has persisted for the greatest period of time has been the belief that there is an absolute meaning to such things as the real length of a body regardless of what means the observer takes to measure that length, and that there is an absolute meaning to equal intervals of time, and to the simultaneity of events taking place at different places, apart from some arbitrary experimental procedure on the part of the observer to define such a meaning.

THE ATOM

The ancient Greeks had been driven to surmise that matter had an atomic basis. However, having no experimental data of the type that we have today, they were unable to enrich that hypothesis by attaching properties to the atoms such as would give them some working content. To say, with the degree of meaning implied by the Greeks, that matter is composed of atoms is something like saying that Heaven is populated by angels. The statement does not tell us much unless we know something about the characteristics of angels. We might infer a few consequences, such as that two parts of the population of Heaven could exist in the same place since there would be plenty of room between the individual angels; but our means of inferring more would be very limited.

A hundred years ago physics dared no more than to think of the molecules as hard, elastic spheres whose distances apart in the case of a gas for example, were large compared with the size of each. Even this crude concept of the atom was sufficient to harmonize a number of the phenomena of matter. We could understand how the pressure exerted by a gas comes into existence provided that we believed these molecules to be in rapid motion; for their continual bombardment of the walls of the containing vessel would give rise to the equivalent of a steady pressure, just as a stream of sand directed at the hand gives the sensation of such a steady pressure. We could see how it was possible to reduce the volume of a gas to an amount small compared with its original volume. We could see how such diminution in volume would be accompanied by an increase in pressure, since it would result in an increase of the number of impacts per second on the walls of the containing vessel. These, and many other things about the behavior of gases in relation to change of pressure, conduction of heat, and so forth, could be understood in terms of the simple picture to which I have referred. With the increase of our knowledge of chemical phenomena, and especially of the complexity of the light emitted by glowing gases, it became necessary to conceive of the atom as an object more complex than a hard ball in order that such phenomena could be harmonized. The discovery of radioactivity and of a multitude of phenomena concerned with the conduction of electricity in gases increased further the demand for more detailed study of the atom itself; and then some new phenomena, revealing as they did new elements in matter's structure—elements such as the electron, gave suggestions as to the fundamental bricks out of which it was reasonable to suppose that the atoms might be built.

And so, about the end of the last century, physicists had molded their clay, made their bricks, and set out to erect the scaffolding for that building, the atom, in which they hoped to find a place for the phenomena of Nature. The bricks—the electron and so forth—were new, but the atomic architects set out to make with them a building of conventional design. Now the trouble was that the people whom it was intended to house in that building—the newly discovered phenomena of X-rays, light, electrons, etc.—were very modernistic people. They could find no use for much that was in the building and found total lack of provision for what to them were their most important needs. It was as though we had built a medieval castle to house a group of modernistic musicians, painters, poets, and débutantes. The building had been controlled much in its design by the desire to provide a chapel, but the modernistic tenants have no use for a chapel. Moreover, the chapel gets in the way of something which means much to them, the provision of a great, long, narrow hall going right through the building which by its great uninterrupted length facilitates the performance of some feat of gymnastics very fundamental to the doings of the modernists.

To leave the language of parables, we may say that the physicists of the late Victorian era hoped to see the atom work as a watch works, or as a dynamo works, or as the ripples on a bowl of water work. They desired to see the atoms composed of parts held together by forces resulting from some state of stress in the medium around them. The kind of picture that would have contented them, intuitively, would have been one, for example, where the atom was likened to a lot of marbles embedded in a jelly, so that if any one of them was given a knock, they would all vibrate under the elastic forces of the jelly. To go back to something like a jelly is very satisfying. If anyone were asked just why a jelly behaves as it does, why it exerts elastic forces when strained, and so forth, he would be hard put to it for an answer. But he would probably seek content-

ment in the fact that "everybody knows that a jelly wob-
bles." And so it was hoped to be able to tell the story of
the atom by an appeal to such activities as take place in
springs, jellies, etc. Then, alas, it was discovered that the
kind of behavior to be expected from such things was
entirely different from that found. That kind of picture
told us that the light emitted by atoms should be fundamentally different from what we found it to be. It told us
that the distribution of energy among the different lengths
of waves in the sun's heat radiation should be drastically
different from what it is. It told us that if light, in the
form of waves, should fall upon matter, whatever effects
were produced would be produced in all of the atoms alike,
and that these effects would be very small, a mere tickling
of all the atoms. Yet the facts showed that the effect of light
was only to do something here and there in the matter
upon which it fell. Where it did act, however, it acted with
a degree of drasticness out of all proportion to what could
readily be accounted for on conventional ideas. It was as
though, when a wave-front advanced upon matter, the in-
finitely many parts of the wave-front called a council of
war and said: "We are all so weak that if we advance upon
that fortress, none of us will produce any noticeable effect.
Let us concentrate all our strength into one of our members,
and let him use it to attack one place." But while we can
understand how an army can call such a council of war
and put its decrees into action, it is not so easy to see how
the light waves can do a similar thing.

And so we had to admit into physics concepts which were
strange to our sense of "reality." If I were asked to ferret
out the greatest error that physics made in the period of
growth and development of the theory of atomic structure
to its present form, I should have to seek it in the attempt
to achieve harmony with Nature by forcing the laws of
atomic structure into a form which satisfied a certain sense
of reality without realization that those criteria of reality
were in the last analysis perfectly arbitrary. We have been

to some extent in the position of the person, cited earlier in this chapter, who sought contentment in the thought that gravity acted like a piece of elastic. In seeking to explain the elastic he conceived of it as composed of a lot of little molecules, but instead of admitting for these molecules some properties unfamiliar to him which by their combined action produce in matter in bulk the phenomenon of elasticity, he was content with no account of their behavior other than one founded upon the same thought of elasticity that he set out to explain. To explain the actions of matter in bulk we have been driven to the thought of molecules and atoms. Then we have felt the necessity to go further and formulate a set of laws for these atoms that would account for the behavior expected of them. By a strange kink in the psychology of mankind we have held on to a hope that those laws would be the same as for the matter in bulk. They were not, and so for a time they seemed artificial to many, and to some they seem artificial today. Bound up with our sense of reality was the thought of the predetermination of the universe, the idea that everything that happens is the result of some state or happening which occurred before, so that if we only knew enough about the universe at the present instant, we could predict its course for all time. This doctrine was always a stumbling-block to the man of science. If it were not true, his whole scheme of physical law seemed to break down, his whole subject evaporated. And yet he was frightened when he thought of free will; so he went to church on Sundays and believed in free will, and during the remainder of the week he sought to show by his experiments and theories that everything was predetermined. While my own view as to this question of predetermination versus free will, etc., holds that it is rather a tempest in a teapot and maintains that, in the last analysis, there is not as much difference between them as is usually supposed, the limited space of a chapter will not suffice to give the reasons for this attitude. The point that is to be emphasized here, however, is that we have come to realize in physics that an expression of the laws of Nature in such a form that we say such and such a thing has only a certain chance of happening, rather than a certainty of happening, is not such an expression of incompleteness of knowledge as we formerly thought. We can have the laws in this form as a practical basis for unifying our experiments; and the old-time desire for something which seemed more "complete" is rather analogous to the desire of one who asks the question, "If an elephant could solve mathematical equations, would he also be a good musician?"

While the desire to see the atom conform to certain arbitrary principles of reality impeded the development of the scheme of atomic structure to some extent, the delay has been enormously shorter than would have been caused a hundred years ago by far less revolutionary revisions of our ideas. Physicists have learned to change quickly in adapting their concepts to Nature. The "new deal" in physics has progressed with a knowledge that the "constitution" of physical laws—the conventional view as to what should be and what should not be: the criteria of common sense—was determined in a day when the phenomena presented to the physicist by Nature were very different and very much less complex than they are today. The laws of the new contain the laws of the old where those older laws have a right to be applied, but the old does not contain the new.

Sometimes physics in going through the throes of a change of vision adopts a sort of half-way picture. It was the hope of the physicist of fifty years ago that some day we should be able to draw a picture of the atom, and even make a working model of it. To ask for the picture of an atom today is something like asking for a portrait of an angel taken subject to the conditions that the being portrayed has to be represented as weightless, capable of flight with infinite velocity through a vacuum, temperamental, and invisible. In passage from one of these extremes to the other we

stopped to take breath; and the stopping-place was the
Bohr theory of atomic structure. Space will not permit me
to say more than this concerning it. It laid hold of the
significant elements of the true laws of the atom and traced
their consequences to the point of harmonizing in terms of
them some of the most characteristic and beautiful phenomena concerned with the light emitted by atoms. In
doing this it presented a model of the atom which was
something like the solar system on a small scale, the electrons filling the rôle of the planets and a central nucleus
the rôle of the sun. Here at any rate was something that
we could draw and look at; but the electrons had to be
robbed of some of the properties they would inevitably
have were they really electrons. Hence the comfort that
we derived from the contemplation of the model was largely
illusory. It was as though, in trying to represent the characteristics of a certain animal, I hit on the idea that he was
an elephant, but found that in order to explain his activities it was necessary to assume that his trunk was used only
for purposes of wagging while the purpose of his tail was
to carry things about, that his ears were not used for hearing but for the absorption of food. It is doubtful whether
I should be justified in deriving much comfort from the
thought that this animal was an elephant. And so we have
here an example of one of the errors of physics, the error
of seeking comfort in a situation where we conceive of
some sort of model composed of things which are more or
less familiar to us and then find it necessary to discard most
of the natural properties of these things in order to make
the model work. Of course, it may be said in favor of the
procedure that it is valuable provided that it stimulates
the mind to think. It is, indeed, difficult for the human
mind to think at all unless it has something to think about.
The danger lies in the temptation to think too much about
those features of the model which are really irrelevant, the
feaures that cause trouble because they really have no legitimate place in the picture.

When we speak of the qualities that dominate the characters of individuals, we speak of their temperaments, their capacities for mathematics or for music; we speak of the extent to which certain circumstances affect them, and so forth. Sometimes the psychologist seeks a deeper reason for these things, but for practical people who have to live with the individuals these characteristics form our fundamental starting-points. The modern theory of the atom is one in which its fundamentals are expressed more after the manner of the characteristics of individuals than in terms of springs, mechanisms, and so forth. It is as though the different atoms had different temperaments; and to try to correlate their activities by thinking of them in terms of forces and ordinary mechanics is something like trying to correlate the actions of an operatic prima donna by an appeal to laws of springs, weights, and machinery, or the so-called laws of common sense.

SCIENCE AND THE LAWS OF NATURE

It has always been the ideal of the natural philosopher to seek some scheme of laws that will harmonize all the phenomena of Nature, regarding as unfortunate makeshift any scheme of law that applies only in a limited field. It is doubtful whether this ideal and attitude are as impregnable as might appear at first sight. There has been a tendency to presuppose, moreover, that there is one and only one theory of the universe that is right, one and only one system of rules in terms of which the phenomena of Nature may be correlated. Now it is quite possible that this view is an illusion and that there may be many theories satisfying the criteria demanded, the determining fact being not which is right, but which is the most convenient. On such a view the ideal of harmonizing all Nature under one scheme loses much of its dignity; for that dignity had its origin largely in the thought that if only we could find this scheme of laws that harmonized everything—vital phenomena, physics, chemistry, economics

—we should have got down to bed-rock and learned to view Nature as she actually is and to see the real truth in all its glory. If, however, there are many theories which would harmonize the facts, the reason for the ideal evaporates, and it is incumbent upon us to count the cost of an attempt to attain it and weigh that cost against the value of the ideal. If spectacle-makers had refused to make spectacles until the theory of diffraction in glass had been completely solved according to modern atomic theories, many spectacle-makers would have gone out of business. Moreover, they would probably have gone out of business had they attempted to use the theory when it had been formulated. The spectacle-makers thought in terms of geometrical optics. They thought of light as Newton thought of it, and the scheme was wide enough for their purposes.

As we widen the scope of our theories to embrace phenomena which are of a vastly different kind, the theories take forms in which they are not very simple in statement for any of the individual limited realms of phenomena which they are called upon to include. Different theories are like different languages for describing the same phenomena. A nation of bankers who made a language for their own purposes would probably make one quite different from that created by a nation of painters or musicians. A language suitable to cover the interests of both would probably be more complicated than the language of either the bankers or the painters considered separately. When the bankers had to include in their interests artistic values as well as the values of bonds and the like, their language would have to take a more general form which would cause it to lose some of the specificness of appeal that it had when aimed directly at the phenomena of finance.

Perhaps the greatest error inherent in the ambitions of the theorists of the past lay in the hope that when the right theory was found, the universe and its laws would appear not merely understandable but obvious. There seems to have been a feeling that if only we could get far enough down to bed-rock, the activities of "Nature" would *appear* "natural." But, alas, in the vague sense in which the word "natural" was used, Nature refused to be "natural." If we may claim an advance in sophistication today, we must claim it in the fact that our theories seek not to "explain" in the old sense of the word, but to correlate the phenomena of Nature. They form a shorthand notation for remembering all the facts of Nature. We seek some way in which, by saying few things, we may deduce as their consequences many things which are true in fact. We no longer concern ourselves with the "reasons" for the few things we started with. It is not that we have become disheartened and given up the task of finding such reasons, but rather that we have come to realize that there is no meaning in the quest for them.

In addition to correlating the known facts in the above sense, our theories frequently suggest that certain other phenomena should occur, phenomena not as yet known. Search for these phenomena has frequently been rewarded with success. Thus, about ninety years ago, Newton's theory of gravitation when applied to the observations on the planets suggested the existence of another planet then unknown. It was sought, and thus Neptune was found. Such suggestions made by our theories are of great value, but they carry with them a great danger. An error made frequently in the past by those who pursued knowledge in science was to suppose that discovery had come to an end. Again and again physics has reached a stage where the horizon of discovery seemed also its boundary. It reached that stage when the immediate consequences of Newton's gravitation theory had all been worked out. It reached that stage when Maxwell's great electromagnetic theory seemed to have reached the final goal that man might hope to reach in knowledge. It almost reached that stage when the consequences of the theories of Niels Bohr and Arnold Sommerfeld had been traced as far as it seemed practicable to trace them; and were it not for the confidence gained by physics in having peered successfully through the apparently impregnable barriers of the past, it probably would have reached that stage when the modern atomic theory had conquered the first line of defense of the atom against man's prying eyes.

In these periods of depression, when discovery seems to have ended and science seems dead, Nature seems to divide her phenomena into two catagories, those concerning which we seem to know everything, and those of which it appears we can never know anything. I believe the origin of this state of affairs is not far to seek. When a new set of phenomena have been successfully correlated into a theory and the theory has done its service of predicting other phenomena which are subsequently found, that theory exercises a sort of censorship over everything. Having done good service in predicting what should be so, it predicts just as emphatically what should not be so, and so closes the door of intuitive reason to further search. If we have come to feel the spirit of the theory in our bones, it seems a waste of time to seek for anything that seems nonsense in the light of it. And then some unsuppressible radical does one of those "foolish" experiments and gets a disturbing result, a result which seems like nonsense in terms of the theory. If, after criticizing his experiment and browbeating the impudent experimenter, we are forced by the facts of repetition of his experiment to believe he was right, it becomes necessary to do something about the theory. We remold it to a greater or less extent as required by the exigencies of the situation imposed by the results of the new experiment. Then, when we have done this, the theory, as though in the hands of the torturers, says: "If you make me admit this new experiment, I might as well admit many other things which in my older guise I should have denied." And so the flow of the blood of discovery is started once more, and the factories of knowledge hum with activity until in due time order is restored, with the new theory dominant over all and claiming, as did its predecessors, the right to censor all phenomena not included in its own particular code.

Perhaps one of the most characteristic features of the discoveries of the last thirty years has been the increasing speed with which we have learned to adapt ourselves to new lines of thinking. Common sense has not had time to establish itself at any stage of progress in the hectic development of our mental scenery, but is left trying to wade its way out of the sticky ether of the nineteenth century. There are many who do not like the mad panorama of thought presented to them. They long for the conventionality of the past. They feel that Nature could never have been so radical and bizarre, and that, if we only knew how, we could understand her in terms of the criteria of "good horse-sense." And yet, how unkind it seems to have to remark that, in the last analysis, "horse-sense" is, in all verity, but the kind of sense that a horse has.

Supplementary Note on Superstition in Mathematics

By Eric T. Bell

IN ANY STORY of human error, mathematics is likely to play a less conspicuous part than some of the natural sciences. The subject-matter of mathematics cannot be confronted with fact in the scientific sense. Apart from trivial slips, the errors of mathematicians have usually been concealed in the foundations of their vast edifice. Before the major error that has infected mathematical thought is pointed out, it will be well to state emphatically that this error is still a cardinal dogma in the creed of many professional mathematicians, particularly those whose environments incline them to reactionary orthodoxy in religious and social matters. Thus what some mathematicians consider a blunder of the first magnitude is in fact the corner-stone of the working creed of many other mathematicians. However, not all mathematicians by any means are superstitious about their subject.

The early mistakes in mathematical outlook, seen most clearly in the number-mysticism of the Pythagoreans, can be ignored. Only the mentally incompetent are impressed by any of this meaningless mysticism today, and likewise for the rank growth of the sacred theory of numbers in the Middle Ages. A more important mathematical superstition, however, grew naturally out of the crude mysticism of the Pythagoreans. The doctrine that *everything is number*—the italicized phrase being taken in the most literal sense imaginable—has survived in not too subtly refined forms to the present day. For example, we find an eminent popularizer of astronomy declaring that the Great Architect of the Universe (whatever that may mean) begins to appear as a pure mathematician. This survival of Pythagoreanism is more likely to find favor with physicists, astronomers, and other scientists who are not mathematicians than with professional mathematicians.

Nevertheless, the "God ever geometrizes" of Plato or the "God ever arithmetizes" of Jacobi has induced sympathetic vibrations in the devout bosoms of scores of reputable mathematicians in the present century. Even some who treat the subject of these famous epigrams as a hypothesis, in the manner of Laplace, subscribe to the modernized form of the ancient faith which the sayings express. The truths, they assert, of mathematics are necessary and eternal, *a priori* and, in the philosophic sense, synthetic. This is the faith; part of it is ascribed to Kant. Newton, according to the faithful, did not *invent* the binomial theorem, as the behaviorists might have him do; he *discovered* it as a pale reflection of an unimaginable, eternally existing and timeless truth in the everlasting realm of Platonic ideas. And Newton was enabled to make his great *discovery* because, in the beginning, part of that ideal binomial theorem had been woven into the warp and woof of his own soul as it revolved clockwise with the Platonic soul of the universe in a perfect circle. If the last sounds absurd to some mathematicians, the soberer assertion that the binomial theorem is synthetic and *a priori*, in the philosophic sense, sounds definitely false.

The "truths" of mathematics, it is now believed by many, are analytic—again in the philosophic sense. They are tautologies. This is the conclusion that is accepted as more likely than the synthetic, eternal-truth theory to be true by the majority of those who have made it their business to examine the nature of mathematics during the past twenty years. The creed that mathematics has a superhuman, necessary truth not shared by other constructions of human beings was the great error inherited from Pythagoras, Plato, and Kant—if, of course, it *was* an error. Anyhow, many reputable mathematicians today agree that it was of the same stuff that wishful dreams are made of.

The concomitant belief in the absoluteness of space still hangs on. A century and a half ago there was but one space and but one geometry, and Euclid was the Mahomet of both. Lobatchewsky, Riemann, and their successors by their creation of non-Euclidean geometries in shoals disposed of this error. "Space" however still maintains its ontological mystery over and above all the swarms of geometries that have bred like flies on the decaying remains of its ghostly carcass; but mathematics is content to leave "space" as a thing-in-itself to philosophy. "Time" in the absolute Newtonian sense outlasted "space." But "time" never played the part that "space" did in contaminating mathematics with superhumanism. When Einstein took liberties with "time" in 1905 and again in 1915, the mathematical repercussions were negligible.

To sum up: The errors of mathematics have not been markedly different from those of the physical sciences. In both fields an atavistic yearning toward supernaturalism has misled imaginative mathematicians into occasional profound investigations of empty mares' nests; but no great harm has come to the investigators themselves from their fruitless researches. The innocent bystander, the so-called layman or man in the street, alone has seen the signs and wonders to which common mathematical sense is congenitally blind.

CHAPTER V

ERROR IN CHEMISTRY

By Charles A. Browne

IN THE IDEAL *exact sense of the word, chemistry may be defined as that branch of natural science which gives a perfectly truthful account of the composition and properties of matter and of the changes undergone by matter under different states and conditions. Obviously, in such a sense there can be no errors in a science which is a repository of perfect truth; the errors are not in chemistry but in man's interpretation of it.*

The history of chemistry is largely an account of the mistakes that man has made in his effort to arrive at a satisfactory explanation of the constitution, properties, and transformations of matter. It is a history of trial and error, many explanations of one era being rejected as erroneous by the next in the light of more perfect knowledge. The chemical texts of a hundred years ago are now rejected because of certain errors of statement. A century hence the chemistry books of our time will meet with a similar rejection. The beginning of chemistry as a science is referred by some to Boyle in the seventeenth century, while others would postpone the dawn of the science until the time of Lavoisier—a full century later. Actually we can fix no such line of demarcation, for the boundary of time that separates what men accept as the truth and what as error is a continually shifting one.

We must, therefore, for our present purpose reject the ideal definition of chemistry as one which is humanly impossible of attainment. To coin a definition that will apply to any era of history we may say that chemistry is a contemporaneous record of man's attempts to explain the constitution, properties, and transformations of matter. The errors of chemistry are thus the explanations that succeeding generations of men have found to be untrue as a result of more careful observation and experiment.

The science has been divided into the two branches of applied and theoretical chemistry. Applied chemistry, as indicated by the name, considers the employment of chemistry for useful practical ends, whereas theoretical chemistry deals solely with a consideration of the laws that govern the composition of matter and the changes that it undergoes. Of these two branches of the science, applied chemistry is much the older, dating back to the primitive arts of prehistoric man such as making fire, preparing food, tanning hides, baking brick, dyeing wool, fermenting grape-juice, crystallizing salt, and smelting ores. The explanation of these processes did not concern him, and it was only after man's rise to a higher degree of civilization that he began to speculate about the nature of freezing, melting, burning, evaporating, breathing, fermenting, and other chemical phenomena of daily life. Crude conjectures began to be formed; opposing forces were personified as the operations of love and hate, and mythological conceptions were introduced.

As man's powers of observation became more sharpened, a classification of substances according to their properties—hard, soft, dry, wet, warm, cold, etc.—began to be made. Resemblances and differences were noted, and then came at last the first great step that marked the birth of theoretical chemistry—the attempt to explain the changes in the states of matter by the action of some unifying principle or law. Such were the attempts of the old Ionic Nature philosophers, who reduced all the operations of Nature to the action or interaction of the four elementary principles earth, water, air, and fire. Such, also, were the attempts of the early atomic philosophers, who, penetrating deeper into the constitution of matter, attributed the differences in properties of substances to differences in the nature of the minute indivisible particles of which they were supposed to consist. Rival schools of philosophy were built around these opposing views, but the art of verifying hypotheses by carefully controlled experiments had not yet been developed. Consequently chemistry for the next two thousand years consisted largely of a mass of erroneous philosophic speculations overlaid with astrological, alchemical, and other pseudoscientific ideas. It was only in the sixteenth and seventeenth centuries, when Francis Bacon (1561-1626), Van Helmont (1577-1644), and other inquirers began to make use of experiment for the discovery of truth, that the errors

*handed down from antiquity began slowly to be cleared away.
But in removing these old errors the devotees of the new experimental philosophy committed new mistakes in explaining the
results of the experiments they had devised. Later experiments
by other inquirers with improved appliances gradually eliminated the fallacies of their predecessors; thus the science of chemistry has been brought slowly and laboriously to its present state.
Selected and significant examples from the history of this development are interpreted in the following chapter.*

THE FOUR ELEMENTS

THE CHEMICAL ARTS of dyeing, ceramics, salt-making,
metallurgy, fermentation, tanning, and many other
primitive manufactures were all well developed before man
began to speculate about the constitution and properties of
matter. Though we find crude conceptions about the nature
of inorganic and organic substances to have existed among
the early Egyptian, Chaldean, Indian, and Chinese philosophers, it is to the Greeks that we owe the development of
those first theories of chemistry which controlled the progress
of the science down to the beginning of the modern era.

In order to understand the nature and extent of Greek
influences in chemistry we must go back twenty-five centuries
to the time of Thales (640-546 B.C.), who, early in the sixth
century B.C., sought to reduce the manifold phenomena of
Nature to the operations of one common principle which
he believed to be water. This conception is noteworthy as
it marks the beginning of attempts to unify the explanations
of natural phenomena. The idea of water being the original
"materia prima" or *"Urstoff"* found adherents even down to
modern times. The Flemish chemist Van Helmont (1577-
1644), one of the founders of modern experimental chemistry, demonstrated to his own satisfaction that the conception of Thales was correct.

A second Ionic philosopher, Anaximenes (sixth century
B.C.), proposed air as the original element from which the
universe was derived; a third, Heraclitus, considered fire to

be the primal basis of all material existence, the suggestion for this belief coming perhaps from Zoroastrian influences. Fire, according to Heraclitus (about 500 B.C.), underwent successive degrees of condensation through air and water until the so-called downward way was reached in earth. Everything was thus in a state of flux, or, as Heraclitus expressed it, πάντα ῥεῖ, "all things flow." This evolution of all material substances from one element gave rise to another saying of Heraclitus, ἓν τὸ πάν, "the all is one," which many centuries later became the motto and cardinal principle of Greek alchemy.

Instead of regarding one particular principle as the original basic element of all matter, the Sicilian philosopher Empedocles (about 455-395 B.C.) recognized four such elements, fire, air, water, and earth, as the first roots of all existing things. The combination of these elements in different proportions, according to Empedocles, gives rise to the multiplicity of natural substances, a condition of love, or attraction, and of hate, or repulsion, between the elements being the activating force that controlled these combinations. The theory of four elements, proposed by Empedocles and adopted by Plato and Aristotle, became the central doctrine of chemistry for the next two thousand years.

The Greek philosopher who exercised the greatest influence upon modern chemistry was Democritus (about 400-357 B.C.). According to his conception, all matter was composed of exceedingly small indivisible atoms, indestructible and unchangeable, but differing greatly in form, weight, and size. The basic material of all atoms was the same; it was only their variation in shape and magnitude that caused the differences in the properties of substances. Thus, the atoms of water and iron are of the same substance; those of the former, being smooth and spherical, do not cohere, but, slipping easily over one another, give water its fluidity; the atoms of iron, on the other hand, being rough and jagged, cohere, thus giving the metal its rigidity. Democritus regarded the space between the atoms as a void and attributed the variations in density of different substances to the degree of compactness with which the atoms filled the given space. Matter itself he held to be indestructible, although its forms were continually changing because of the constant disintegration of older atomic combinations and the building therefrom of new rearrangements.

Had the atomic theory of Democritus met with immediate acceptance, the rise of modern chemistry might have been advanced two thousand years. But his explanations of natural phenomena without recourse to any ideas of purpose or design were repugnant to the Greeks and especially to Aristotle (384-322 B.C.), who blamed Democritus because he rejected the doctrine of final causes and explained everything by the action of necessity. He also criticized the atomic school because the reciprocal convertibility of the four elements was denied. The evaporation of water was explained by Aristotle and his school as an actual conversion of the liquid into air; the atomic philosophers denied any such conversion and attributed the disappearance of the water to the separation of its invisible atoms by wider intervals of space. When these spaces were reduced, the atoms came closer together again and the liquid phase of water was restored. As we now know, the explanation of Democritus, with its recognition of the existence of the same substance in different states, was much nearer the truth. We may indeed date the commencement of modern chemistry from the seventeenth century when, under the influence of such leaders as Bacon, Galileo, Gassendi, Boyle, and Newton, the atomic conception of Democritus was revived and the physical theories of Aristotle, which had hampered the progress of science for two thousand years, were overthrown.

THE FOUR QUALITIES

Closely related to the doctrine of the four elements was that of the four qualities—warm, cold, moist, and dry—upon which Greek thinkers from earliest times based all the relationships of man to the exterior world. Health, temperament, diet, agriculture, and many activities of daily life were regulated according to the proper balance of these hypothetical basic qualities. We may cite as an example the Hippocratic doctrine of the four humors—blood (warm and moist), phlegm (cold and moist), yellow bile (warm and dry), and black bile (cold and dry)—upon the perfect balance of which health was supposed to depend. Fevers were attributed to an excess of yellow bile and colds to an excess of phlegm. People having fevers or a hot, choleric temper should partake of cooling food such as melons; people having colds or a phlegmatic disposition should receive warming food such as mustard. In agriculture cold soils were considered necessary for the best growth of cooling foods and warm soils for the best production of warming foods. If the premises were accepted, the deductions seemed logical and consistent. The doctrine of the four qualities remained practically unchanged as a cardinal principle of physiological and agricultural chemistry from the time of Hippocrates (about 460-357 B.C.) until after Boyle. Louis Lemery (1677-1743), Regent-Doctor of the Faculty of Physic at Paris, thus wrote of cucumbers: "They contain much phlegm, moisten and cool much, quench thirst, allay the sharpness of humors and too great a fermentation of the blood. Cucumbers in hot weather are proper for young persons of a hot, bilious constitution; but weak and old people of a phlegmatic temper should avoid them." [1] The expression "cool as a cucumber" is a modern survival of the old Greek doctrine of qualities. Lemery, although twenty-one centuries distant from Hippocrates in point of time, was scarcely removed from him as regards his ideas upon the chemistry of nutrition.

ERRORS OF DEDUCTION BY ANCIENT EXPERIMENTERS

It is interesting to note the interpretations that the Greeks gave to some of their experimental observations, such as those in regard to combustion, which has always been a

[1] *Treatise of Foods . . . According to the Principles of Chemistry and Mechanism* (Paris, 1702).

pivotal problem in chemistry. Philo of Byzantium, who lived
in the second century B.C., described the familiar experiment
of burning a candle under an inverted vessel over water.
After the candle goes out, water will rise a short distance
into the neck of the vessel. According to Philo, the corpuscles of air in the vessel are converted by the flame into
the finer particles of fire, which escape through the pores
of the glass, as is evident by the warmth imparted to the
hand. A partial vacuum is thus created which causes the
water to ascend into the neck of the vessel. This highly in-

PHILO'S EXPERIMENT OF THE BURNING CANDLE
From Wilhelm Schmidt's edition of the *Pneumatics* of Hero of Alexandria
(Leipzig, 1899).

genious but erroneous explanation satisfied the minds of
men for many centuries to come.

Hero of Alexandria (second century B.C.?), one of the
most skilful experimenters of antiquity, also described combustion in the introduction of his work upon *Pneumatics*.
"Bodies are consumed," he wrote, "by fire which transforms
them into finer elementary particles—namely, water, air, and
earth. That they are actually consumed is evident from the
carbonaceous residue, which, although occupying the same
or a little less space than before combustion, nevertheless

differs greatly with respect to the weight the material had at the beginning."

We have in this passage with its allusions to weight an early hint of the application of quantitative methods to the analysis of organic substances. Philo and Hero described highly developed and elaborate forms of apparatus; without

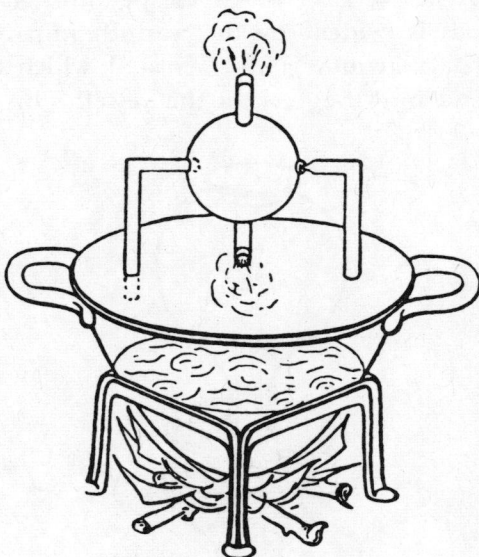

HERO'S EOLOPILE

The precursor of the modern steam turbine. Steam from the boiler passes through the side tube on the left into the sphere from which it escapes through the upper and lower tubes with ends bent in opposite directions. The recoil of the escaping steam causes the sphere to revolve. The Greek mechanician Hero devised numerous pieces of automatic apparatus which were operated by the pressure of air or water. From Schmidt's edition of Hero's *Pneumatics* (Leipzig. 1899).

knowing it they had at their disposal every device necessary for establishing the true composition of the gaseous products of combustion. The only obstacle to their accomplishing this was the general acceptance of a false system of scientific inquiry which, working by the unproductive *a priori* method, employed experiments to illustrate preconceived theories rather than to discover new truths.

ALCHEMY

It was because of this fallacious *a priori* method of reasoning that the Greeks of the later Alexandrian and Byzantine period wasted so much time and energy on efforts to transmute the baser metals into silver and gold. Since the conversion of the four elements into one another was taught by Plato (427-347 B.C.) and Aristotle, it was argued that the transmutation of copper or lead into silver and gold was a possible accomplishment. All fusible substances such as wax, glass, and metals were held by Plato to consist of earth with a certain amount of water to give them fluidity. The rust or corrosion of the baser metals was regarded as an exudation of this earthy ingredient. Of all the metals, gold was considered to contain the least amount of earth, this residual impurity being of so fine a quality that there was no exudation of rust or tarnish. The removal of the excess of the supposed earthy constituent of base metals by a process of derusting (ἐξίωσις) was thus the first step of transmutation.

The belief in transmutation was enhanced by observing the gold- and silver-colored alloys that the skilled Egyptian technologist prepared from base metals—alloys which in many cases were used for counterfeiting. A number of their workshop receipts are contained in old Greek papyri in Leyden and Upsala; their fraudulent character is clearly indicated by the following translations:

Clean white soft tin four times; melt six parts of the same with one mina of white Galatian copper. It becomes prime silver that will deceive even skilled workmen who will not suppose it to be made by such a treatment.

Add six parts of purified tin and seven parts of Galatian copper to four parts of silver and the resulting product will pass unnoticed for silver bullion.

These quotations are good examples of the chemical receipts of the Greco-Egyptian period before they were ob-

ΤΩ ΥΒΑΛΛΟΜΕΝΟ,ΥΕΙC ΤΗΝ ΚΡΑCΙΝ ΤΟΥ ΑCΗΜΟΥ
ΚΑCCΙΤΕΡΟΝ ΚΑΘΑΡΟΝ ΑΠΟ ΠΑΝΤΩΝ ΧΩ
ΝΕΥΕ ΚΑΙ ΕΑCΟΝ †ΥΓΗΝΑΙ ΚΑΙ ΜΕΙΤΑC ΕΛΘΟΝ
ΚΑΙ ΔΙΕΙC ΠΑΛΙΝ ΧΩΝΕΥΕ ΕΙΤΑ ΤΡΙΤΑCΕΛΑΙΟΝ
ΚΑΙ ΑCΦΑΛΤΟΝ ΚΑΙ ΑΛΑC ΑΛΕΙΤΩΝ ΚΑΙ ΕΚΤΡΙΤΟΥ ΧΩ
ΝΕΥΕ ΚΑΙ ΕΑΝ ΧΩΝΕΥΘΗ ΑΠΟ ΒΟΥ ΚΑΘΑΡΙΟC ΙΤΑΥΝ
ΑC ΕΙΤΑΙ ΓΑ ΩC ΑΡΓΥΡΟC CΚΛΗΡΟC ΟΤΑΝ ΑΕΑΝ ΤΙ
ΑΡΓΥΡΩ ΜΑΤΩΝ ΕΡΓΑΖΕCΘΑΙ ΘΕΛΗC ΙΝΑ ΜΑ ΘΗ ΚΙ
ΕΧΗ ΤΗΝ ΤΟΥ ΑΡΓΥΡΟΥC ΚΛΗΡΙΑΝ ΠΡΟC ΜΙΚ ΓΕ ΤΟΙC
ΤΕΤΡΑCΙΝ ΜΕΡΕCΙΝ ΤΟΥ ΑΡΓΥΡΟΥ Ε Γ ΚΑΙ ΓΕΝΗ
CΕΤΑΙ ΤΟ ΠΡΟΚΕΙΜΕΝΟΝ ΩC ΑΡΓΥΡΩ ΜΑ
        ΚΑCCΙΤΕΡΟΥ ΚΑΘΑΡCΙC
ΠΙCCΑΝ ΥΓΡΑΝ ΚΑΙ ... C ΦΑΛΤΟΝ ΕΝ ΚΑΘΕΝ ΒΑΛΩΝ
ΧΩΝΕΥΕ ΚΑΙ ΚΗ ΝΕΙΟΙ ΔΕ ΠΙCCΗC ΙΝ ΡΑC ΙΕ
ΑC ΦΛΤ ΟΥ Λ ΙΒ
        ΑCΗΜΟΥ ΠΟΙΗCΕΙC
ΚΑCCΙΤΕΡΟΥ Λ ΙΒ Υ ΑΡΓΥΡΟΥ Λ Δ ΓΗ ΕΧΕΙΑC
Λ Β ΧΕΝ ΕΥCΑC ΤΟΝ ΚΑCΙΤΕΡΟΝ ΕΠΕΜΒΑΛΕ
ΤΗΝ ΤΓΙΝ ΤΕ ΤΡΙΜΜΕΝΗΝ ΕΙ ΤΑ ΤΗΝ ΥΔΑΡΓ
ΥΡΟΝ ΚΑΙ ΧΕΙΝ ΕΙC ΕΙΤΗΡΩ ΚΑΙ ΧΡΩ
        ΑCΗΜΟΥ ΔΙΠΛΩCΙC
Η ΔΕ ΔΙΠΛΩCΙC ΤΟΥ ΑCΗΜΟΥ ΓΗΝΕΤΑΙ ΟΥ ΤΟC
ΤΟΥ ΕΧΙ ΩΜΕΝΟΥ ΧΑΛΚΟΤ Λ Μ ΚΑΙ ΑCΗΜΟΥ Λ Η
ΚΑΙ ΚΑCCΙΤΕΡΟΥ ΒΟΥΝΗC Λ Π ΧΩΝ ΩΥΕΤΑΙ ΑC
ΓΙ ΡΩ ΤΟC Ο ΧΑΛΚΟC ΚΑΙ ΜΕΤΑ ΔΥΟ ΠΥΡΩCΕ ΙC Ο ΚΑ
CCΙΤΕΡΟC ΕΙ ΤΑ Ο ΑCΗΜΟC ΕΙ ΤΑ Ο ΤΑ ΝΑ ΜΦΟ ΤΕΡΑ
ΜΑΛΑΚΓΕΝΗ ΤΑΙΑΝ Ω ΧΩΝΕΥΕ CΠΟΝΑΚΕΚ ΚΑΙ
ΚΑΤΑΤΥΓΕ ΤΩ ΠΡΟ ΕΙΡΗ ΜΜΕΝΩC ΚΕΥΑCΙΑΤΙ
ΕΙ ΤΑ ΠΛΑΤΗΝ ΑΕ ΤΑΙC ΑΥ ΤΑΙC ΟΙΚΟΝΟΜΕΙΑΙC ΑΝ
ΑC ΛΗΧΕΤΩ ΚΟΥ ΦΟΝΘΩ ΚΑΙ ΤΡΙ ΠΧΩ CΙC ΙΕ ΙΝΕ
ΤΑΙ ΤΑΙC ΑΥ ΤΑΙC ΟΙΚΟΝΟΜΕΙΑΙC ΚΑΤΑ ΜΕΡΙ ΖΟΜΕΝ
ΑΙC ΤΑΙC ΟΝ ΚΑΙ C ΙC ΩC ΕΙC ΙC ΠΡΟ ΕΙΡΗ ΤΑΙ

PORTION OF A PAGE FROM THE LEYDEN PAPYRUS

The earliest known Greek alchemical manuscript, dating from about the close of the third century A.D. This page gives recipes for the purification of tin, making of asem (an alloy of silver and gold), duplication of asem, etc. The metals mentioned are lead, tin, silver, copper, mercury, and asem.

scured by the allegorical and mystical interpretations of the
later alchemists, who, under the influence of philosophical,
gnostic, astrological, and religious ideas, believed a simula-
tion of the colors of silver and gold by means of counter-
feit alloys to be an actual transmutation of the base into
precious metals.

Alchemy, as a pseudoscience, began to attract the attention of Greek writers during the third century A.D. Poets,

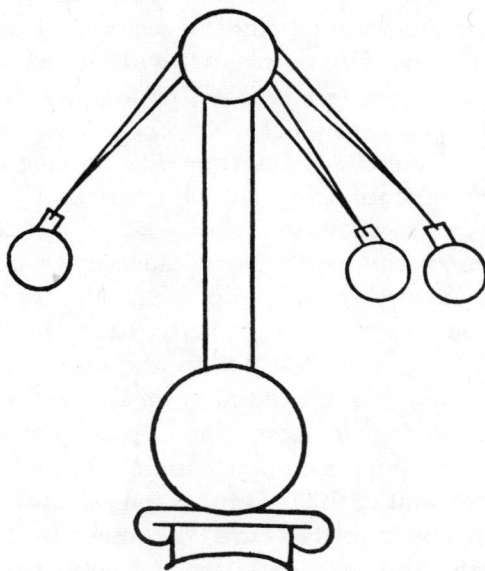

A GREEK TRIBICOS OR ALEMBIC WITH THREE RECIPIENTS
The alchemist's apparatus for making distillations. From Berthelot's *Al-
chimistes Grecs*, Vol. I, after a drawing in a tenth-century Greek alchemistic
manuscript in the Library of St. Mark, Venice.

priests, philosophers, and physicians, as well as laymen, became interested in this so-called "sacred art." In the five
centuries between 300 and 800 A.D. there was a vast output
of manuscripts upon the subject. The oldest collection of
these remains of Greek alchemy is the famous Codex 299, in
the Library of St. Mark at Venice, written in a hand of the
eleventh century. Some of these manuscripts bear the names

of mythical authors, as Hermes and Isis; others are pseudo-graphs falsely attributed to Moses, Democritus, and other ancient sages; others are genuine productions of Zosimus (fifth century A.D.), Synesius (about 370-430 A.D.), Olympio-dorus (fifth century A.D.), Stephanus (seventh century A.D.), and other writers of the later Alexandrian and Byzantine periods.[2]

The literary remains of the Greek alchemists form a mixed collection of philosophical treatises, dialogues, poems, enigmas, prayers, invocations, oaths of secrecy, allegories, visions, and magical incantations, together with numerous laboratory receipts. Astrology, mythology, philosophy, and religious ideas of Egyptian, Chaldean, Greek, Jewish, and Christian origin are all hopelessly intermingled, in language which is deliberately obscure, in a vain effort to explain some theory or process of transmutation.

But even in this period of decadence we note the same desire of the early Greek philosophers to arrive at a broad, unifying conception of the universe that would include man and all the works of Nature. In the same way that man, the microcosm, was viewed as a small counterpart of the great universe or macrocosm, the Greek alchemists regarded the metals as having a similar cosmic relation to man upon the one hand and to the universe upon the other. Each of the seven alchemistic metals was supposed to be generated in the earth by emanations from the seven heavenly bodies —gold by emanations from the sun, silver from the moon, quicksilver from Mercury, copper from Venus, iron from Mars, tin from Jupiter, and lead from Saturn. Remains of these old cosmic ideas in Greek alchemy still linger in such chemical terms as "microcosmic salt" and "lunar caustic."

The ancient technologists imparted the color of gold to base metals by means of yellow-colored oxides and other mineral substances. In this way originated the alchemistic idea of the philosopher's stone, or powder of projection—a

---

[2] Selections from these works have been edited and translated under the direction of the French chemist Berthelot (1827-1907).

miraculous tinctorial substance of infinite potency, of which, it was said, a single grain could convert oceans of base metal into gold. This transmuting substance was held to be a universal medicine, an elixir which would cure disease, prolong life, and restore youth. The quest of this fabulous stone or elixir became eventually the obsession of all classes of men, from slaves to emperors.

No better illustration of the fatuity of alchemy can be found than that during its revival at the time of the Renaissance the writers of the West could only reëcho what the Greeks had written ten centuries before. As in the alchemy of the Greeks, so in that of a thousand years later we find the same symbols, allegories, prayers, oaths, invocations, poems, and obscurities of language. The Western writers could only retrace the landmarks of the ancient Greeks.

We may summarize the accomplishments of the Greeks in chemistry by saying that their work was distinguished on the one hand by many brilliant conceptions, some of which had a great influence on the future history of the science, and that it was characterized on the other by a great lack of experimental ability, such as was needed for subjecting their speculations to practical tests. They failed to make proper use of such experimental knowledge as they already possessed, for had they applied the densimetric method of Archimedes (287-212 B.C.) to testing the gold and silver that they thought to obtain by the transmutation of base metals, they would have demonstrated the fallacy of alchemy at the outset and thus ·saved the world from one of the greatest illusions of the human mind.

Nicephorus Blemmydes, one of the later Greek writers of the thirteenth century, once stated that the four requirements for successful operations in chemistry were "a balance, a furnace, a supply of fuel, and a mind that was subtle and unlimited." The subtle and unlimited mind the Greeks always possessed, but their recognition of the value of the balance came too late. The opportunity of drawing fundamental deductions from experimental work with this instrument was reserved for future time to the younger civilizations of the West.

THE THREE PRINCIPLES

The alchemistic ideas of the Alexandrian Greeks were assimilated in the seventh century by their Arab conquerors, who for the next five centuries were the only race to make advancement in chemistry. The science is indebted to the Arabs for new knowledge and also for new errors. Arab influences are indicated by such words as *elixir, alkali, alcohol,* and the word *alchemy* itself. Avicenna (980-1037), Rhases (852-932?), the more or less mythical Geber (eighth century A.D.), and other writers composed works upon alchemical subjects; some of the works appearing under their names, however, are pseudographs. It was from the Arabs that scholars of the Western nations derived their first knowledge of alchemy; many of the Arab works were translated into Latin.

A conception introduced into alchemy and promulgated by Geber was that all metals consist of sulphur and mercury. This dualistic idea, however, did not satisfy the Christian alchemists, who had to give everything a Trinitarian conception. As the Deity consisted of three persons and man himself was endowed with a body, soul, and spirit, so the metals by analogy must have a triad of components. Thus arose the doctrine of the three principles—mercury, sulphur, and salt. This doctrine, which was adopted by Paracelsus (1493-1541) and the iatrochemists, although shorter-lived than the other fallacious doctrines of the four elements and the four qualities, was just as pernicious as these other errors in its effects upon the future of chemistry. Both schools of alchemists held that metals owed their varied properties to differences in the proportions of their components and that the art of transmutation consisted simply in abstracting from a base metal more of one ingredient, such as earth or salt, and adding more of another, such as fire or mercury. The difficulty that none of these hypothetical components could be found in the metals was met in both schools of alchemists by resorting to Aristotle's differentiation of *actual* and *potential* qualities. The earth, water, air, fire, or mercury, sulphur, and salt of metals were held to be not the actual substances thus designated, but potential or abstract entities. To prevent the confusion that might thus arise, these abstractions were sometimes referred to as "the mercury of the philosophers," "the sulphur of the philosophers," etc. By means of such false assumptions the contradictions between theory and practice were easily dismissed, and the alchemists became only the more deeply involved in a network of ignorance and error.

IATROCHEMISTRY AND THE REFORMS OF ROBERT BOYLE

In the fifteenth century, following the invention of printing, when books upon science began to be widely circulated, men were stimulated to think for themselves upon the problems of the constitution and transformations of matter. The long-accepted doctrines handed down from antiquity began to be questioned; from this time forward there was a growing wave of scepticism concerning the purposes and aims of alchemy. One of the earliest leaders in this chemical revolution was Paracelsus (1493-1541), who asserted that "the object of chemistry is not to make gold but to prepare medicines." This statement, although a great advance over the aims of alchemy, did not, however, involve a repudiation by its author of the principles of the pseudoscience. Paracelsus was a firm believer in the actuality of transmutation. He cited, as ocular proof, the apparent conversion of iron into copper when a nail is placed in a solution of blue vitriol— an experiment which the alchemists repeatedly performed for the purpose of silencing their critics. He believed in the doctrine of the four elements and declared, "Man derives his animal body from earth and water, but his animal life from fire and air." He appropriated the doctrine of the three principles and made it the basis of his system of iatrochemistry. "If anyone wishes to read and understand my philosophy," he proclaimed, "let him know that sulphur, mercury, and salt are the best and surest guides of the physician." Astrology, necromancy, magic, and superstition fill the pages of his works. He discourses upon the microcosm, the quintessence, the alkahest, homunculi, amulets, and salamanders. Digestion is explained by the action within the body of an intelligent *archeus* or demon which separates from the food the ingredients that are useful for growth. Each animal has its special *archeus* or alchemist. "The peacock eats snakes and lizards, the salamander fire, and the pig dung. The alchemist of the pig is much more subtle than that of man, for it separates food from dung, which the alchemist of man is unable to do. Wherefore pig excrement is not eaten by any animal. For no alchemist is so skilful as the pig's alchemist in the separation of food." According to Paracelsus, too much sulphur in the diet gives rise to fever, too little to gout. An excess of mercury causes paralysis, and an increase of salt produces diarrhea and dropsy. It is thus seen that in shunning the vain practices of operative alchemy Paracelsus rushed to other extremes which were equally fantastic and unintelligible.

But notwithstanding his extravagances and crudities, Paracelsus by making the preparation of medicines the main object of chemistry infused new life into a science which was being threatened with extinction by the fruitless endeavors of alchemy. He broke effectively with tradition; as a result of his work chemistry soon became an indispensable study in the education of every physician. He introduced new preparations of mercury, antimony, iron, copper, and other metals into medicine, and the work he thus initiated was carried forward enthusiastically in the succeeding century by Libavius (1540-1616), Van Helmont (1577-1644), Glauber (1604-1668), Kunckel (1630-1715), and other followers of the iatrochemical school. As a result of their labors many new chemical compounds and processes were discovered, while at the same time many of the extravagant conceptions of Paracelsus himself were rejected. Pharmaceutical chemistry was created in this period; and by the new discoveries the way was paved for the great reforms of Robert Boyle (1627-1691), who in his *Sceptical Chemist* (1661) effectively demolished the arguments of the alchemists for the Aristotelian doctrine of four elements and of the iatrochemists for the Paracelsian theory of three principles.

By announcing that the true aim of chemistry was to discover the composition of substances and not to make gold or medicines, Boyle was the first to direct the science in its proper path. He helped to revive the atomic theory of Democritus and predicted that the time would come when many more chemical elements would be known than were assumed in his day. He enunciated the fundamental axiom that any substance that cannot be decomposed into simpler constituents is an element. Yet Boyle himself was a firm believer in the doctrine of transmutation and even claimed to have demonstrated it experimentally by converting gold into silver by means of his *menstruum peracutum,* an experiment which we now know to have been erroneous because of a contamination of his gold with silver. Although an opponent of the doctrine of the three principles, Boyle included in the list of his remedies not only all the mercurial, antimonial, and other preparations of the iatrochemical school, but such disgusting ingredients as powders of dung, dead men's skulls, earthworms, wood-lice, and vipers. The existence of these and many other blemishes in the publications of Boyle do not detract, however, from the value of his reforms. He builded better than he knew, and largely as a result of his work most of the errors of chemistry resulting from the mistakes of alchemy and iatrochemistry were eliminated in the following half-century. He deserves to rank as the first of the great founders of modern chemistry.

PHLOGISTON AND THE REFORMS OF LAVOISIER

The eventful period in the history of chemistry between the death of Boyle (1691) and the death of Priestley (1804) has been called the period of phlogiston. It was an era when more brilliant chemical discoveries were made than in all previous time, yet many of the achievements during this period were accomplished by men who strove to vindicate a wholly erroneous theory. Chemists, working under the influence of Boyle's reforms, began to forsake the false paths of alchemy and iatrochemistry in order to turn their attention to solving the problems of the constitution of matter—more particularly the basic problem of combustion, which had continually intrigued the attention of inquirers since the time of the old Greek fire-philosopher Heraclitus. Combustion, or calcination, according to the ancient philosophers and the later alchemists, was simply a conversion of heavy earthy substances into the lighter elements of water, air, and fire, the remaining ash being simply an unconsumed residue of earth.

For the genesis of the phlogiston theory we must go back to one of Boyle's contemporaries, Johann Becher (1635-1682), who, at first a dabbler in alchemy, became dissatisfied with the vain pretensions of the gold-makers and chemical healers. He sought to reform the old Paracelsian doctrine of three principles by substituting in their place his so-called three fundamental earths—the *terra pinguis* or fatty earth in place of sulphur, the *terra mercurialis* or mercurial earth in place of mercury, and the *terra lapidia* or stony earth in place of salt. Upon the proportions in which these three earths were combined the nature of any material was held to depend. According to Becher, the combustion or calcination of substances was due to the escape of the *terra pinguis* or fatty earth.

Becher's revolutionary conception of combustion was taken up zealously by his pupil Georg Stahl (1660-1734), an ardent searcher for truth, who elaborated the theory of his teacher and gave the name *phlogiston* (from the Greek word φλογιστός, meaning inflammable) to the principle of inflammability or *terra pinguis* of his teacher. According to Stahl, all combustible substances and all metals susceptible to calcination contain phlogiston, which escapes when the substance is ignited. The more inflammable the substance, the more phlogiston was supposed to be present. In the case of a metal, its phlogiston was thought to exist in combination with its calx, or ash, which was left behind when the phlogiston was expelled. To regenerate a metal from its calx the phlogiston must be added again; this was done by heating the calx in contact with any substance rich in phlogiston, such as coal. In the same way Stahl proved to his own satisfaction that sulphur was a compound of phlogiston and sulphuric acid; for when the latter was heated with coal, sulphur was again regenerated.

While Stahl's theory of phlogiston seemed to offer a satisfactory explanation for many phenomena of combustion, there was one great stumbling-block it failed to remove: that was the increase in weight that metals underwent upon calcination. If phlogiston were expelled upon heating, the calx that was left behind should weigh less than the amount of metal that was taken. Actually the calx or ash always weighed more, which would indicate that something was added to the metal instead of some constituent being removed. In fact, Boyle, in his essay "Fire and Flame Weighed," had already shown that copper, silver, tin, lead, iron, and zinc when calcined all increased in weight, a phenomenon which he wrongly attributed to the absorption by the metals in their pores of small particles of flame. The old idea of the corpuscular nature of fire still persisted in the mind of this skilled experimenter.

To overcome the objection of the increase in weight by metals during calcination, the defenders of Stahl's theory took refuge in the explanation that their hypothetical phlogiston had a negative weight—in other words, that it was repelled and not attracted by gravity. Its removal from a metal would, therefore, make the residual calx weigh more than the metal itself. This explanation, however illogical it may appear, seemed to satisfy the minds of many eminent chemists; Marggraf (1709-1782), Macquer (1718-1784), Black (1728-1799), Cavendish (1731-1810), Priestley (1733-1804),

Bergman (1735-1784), Scheele (1742-1786), and many other noted chemists who made discoveries of epoch-making importance were all advocates of the erroneous theory of phlogiston. Indeed, Cavendish, by his discovery of hydrogen, thought he had isolated phlogiston, the inflammable gas that was evolved when metals were treated with acids being thought to have come from the metals themselves and not from the acid.

On every issue the phlogistonists took the opposite of the correct explanation. Yet their system in many ways was consistent, and although absolutely erroneous in principle, it helped greatly to the advancement of chemistry. The very discoveries of the phlogistonists, however, proved in the end to be the demolition of their theory. The discovery of oxygen by Priestley and Scheele was the key that in the skilful hands of Lavoisier unlocked the door that opened up the immense vistas of modern chemistry. Lavoisier demonstrated by exact quantitative experiments that the increase in weight after calcination was due to the combination of the metal, not with the hypothetical fiery effluvium that Boyle postulated, but with the dephlogisticated air that Priestley himself first brought to Lavoisier's attention and which the latter afterwards renamed oxygen (from the Greek word ὀξύς, acid, and γενής, producing). The mystery of combustion was thus finally solved, and with it those other problems of reduction, respiration, corrosion, regeneration, and decay which had puzzled the minds of men since the time of the far-off announcement of Heraclitus that "all things flow."

But the old phlogistonists were slow to give way before the advancement of the new knowledge. Dr. Joseph Priestley, the last of the phlogistonists, remained faithful to Stahl's theory and died still believing in its correctness. It is sad to think that the last scientific work from the pen of this great chemist was a vain attempt to establish the truth of the already discredited doctrine of phlogiston. The last shot in the memorable conflict between phlogistonism and the new chemistry was fired by Priestley from the stronghold of his laboratory in Northumberland, Pennsylvania.

THE ATOMIC THEORY

The epoch-making career of Antoine Lavoisier (1743-1794) was cut short in the midst of his brilliant discoveries by his unfortunate execution during the stormy period of the French Revolution. In addition to solving the problem of combustion, he had assisted in replacing the old chaotic nomenclature of alchemy and iatrochemistry with our present rational system of classification. Following Boyle, he defined an element as a substance which cannot be further decomposed, and upon this basis he constructed the first table of simple substances or elements. In this table were twenty-three of our present elements and eight others which were known in combination but had not yet been isolated. The inclusion of light and heat in Lavoisier's table of elements is a relic of the ancient Heraclitean idea that fire or caloric is a material substance—an error which was not eliminated for several decades yet to come. We know that Lavoisier at the time of his untimely death had vast projects of research in view; his execution by the guillotine may be regarded as the greatest tragedy in the history of chemistry. There were great unsolved problems which awaited his attention, the greatest of these being the explanation of the laws that governed the combination of the elementary substances in his table. Had the life of Lavoisier been spared, his brilliant intellect would no doubt have speedily solved this problem also, and chemistry might thus have been spared the long succession of errors during the next sixty years before the unraveling of the problem of atoms and molecules was finally attained.

The old atomic theory of Democritus, revived by Boyle and Newton, was the weapon with which John Dalton (1766-1844) first attacked this problem. Reasoning from a demonstration of Newton that a gas consists of minute atoms which repel each other by a force which increases in proportion

as the distance between the atoms diminishes, Dalton concluded that the diffusion of mixed gases could only be
explained by assuming that "gases of different kinds have a
difference in the size of their atoms." Knowing the density
of oxygen, hydrogen, and nitrogen, Dalton conceived that
he could calculate the relative sizes of his hypothetical atoms
if he knew the relation of the weights in which these
elements combined. His study of these relations led him to
the discovery of the law of multiple proportions, which
states that when two elements unite in more than one combination, these stand to each other in a simple arithmetical
ratio. When two elements unite in but one proportion, Dalton assumed that but one atom of each entered into combination. He accordingly adopted the formula ○⊙ for water
and ⊕○ for ammonia, ○, ⊙, and ⊕ being Dalton's symbols
for hydrogen, oxygen, and nitrogen respectively. Dalton's
analyses showed that one part by weight of hydrogen combined with 6.5 parts by weight of oxygen and with 5 parts
by weight of nitrogen, these combining weights, as they are
now called, being designated by Dalton "relative atomic
weights." Because of imperfections in his analytical methods
Dalton's combining weights are somewhat different from
those accepted at present. These relations are seen more
clearly in the following table:

Element	Symbol Dalton	Modern	Relative atomic weight, Dalton	Combining weight, Modern
Hydrogen	○	H	1	1
Oxygen	⊙	O	6.5	7.94
Nitrogen	⊕	N	5.0	4.63
Carbon	●	C	5.4	5.95

When two elements united in more than one proportion,
Dalton selected the simplest ratios that would satisfy the
results of his analytical determinations. Thus, for the three
oxides of nitrogen Dalton postulated the formulas ⊕⊙⊕
(N_2O) for nitrous oxide, ⊕⊙ (NO) for nitric oxide, and

⊙①⊙ (NO_2) for nitrogen dioxide—the modern symbols of these compounds being inserted in parentheses. Dalton's general conclusions may be summarized as follows: (1) Every element is composed of homogeneous atoms whose weight is constant; (2) chemical compounds are formed by the union of atoms of different elements in the simplest numerical proportions.

Dalton's atomic theory was received most favorably by chemists, as it clarified considerably the problem of chemical combination. It was soon confirmed in many respects by Gay-Lussac's discovery (1805) of the law of volumes. Gay-Lussac (1778-1850) found that, no matter which gas was in excess, 100 volumes of oxygen always combined with 200 volumes of hydrogen to form water; that 100 volumes of ammonia always reacted with 100 volumes of hydrochloric-acid gas to form ammonium chloride; and that 100 volumes of nitrogen always united with 50 volumes of oxygen to form nitrous oxide, with 100 volumes of oxygen to form nitric oxide, and with 200 volumes of oxygen to form nitrogen peroxide. From these data the general law was deduced that chemical combination between gases takes place in simple volume ratios, and that when contraction occurs in such combinations, the diminution in volume bears a simple ratio to the volumes of the original gases. The existence of simple ratios for combining weights, as announced by Dalton, and for combining volumes of gases, as announced by Gay-Lussac, seemed at once to indicate that equal volumes of gases must contain the same number of atoms and, therefore, that the weights of equal volumes of gases were exactly proportional to the atomic weights. This conclusion, however, was not acceptable to Dalton, who in 1807 raised the following objection: "*Query.* Are there the same number of particles of any elastic fluid in a given volume and under a given pressure? No, azotic [that is, nitrogen] and oxygen gases mixed equal measures give half the number of particles of nitrous gas [that is, nitric oxide, NO] nearly in the same volume."

Dalton's view can be understood more clearly by the following diagram, in which the atoms of the two gases are represented by the modern symbols:

$$\boxed{\begin{array}{c} O \\ \hline N \end{array}} \quad - \quad \boxed{ NO }$$

2 vols. with 2 particles — 2 vols. with 1 particle of NO gas which agrees with experiment but is contrary to the theory of equal numbers of particles in equal volumes.

If the theory that the ultimate particles of all gases occupy the same volume is correct, then, according to Dalton, the volume of the NO gas should be only one-half that of the two combining gases or

$$\boxed{\begin{array}{c} N \\ \hline O \end{array}} \quad - \quad \boxed{ NO }$$

2 vols. with 2 particles — 1 vol. with 1 particle which is contrary to experiment.

A simple solution of this difficulty, which perplexed chemists for many years, was proposed in 1811 by the Italian physicist Amadeo Avogadro (1776-1856), who stated that the contradictions in this and all other similar cases could be harmonized if we regard the ultimate particles of all elementary gases as compounds, so that instead of N and O we should substitute N-N and O-O as the reacting units. Our diagram would then show

$$\boxed{\begin{array}{c} NN \\ \hline OO \end{array}} \quad - \quad \boxed{\begin{array}{c} NO \\ \hline NO \end{array}}$$

2 vols. with 2 molecules — 2 vols. with 2 molecules of NO gas which agrees with both theory and experiment.

Avogadro termed his compound particles "integral molecules" (for which we now retain the word *molecule*) and his simple particles "elementary molecules" (for which we now substitute the word *atom*). His hypothesis may then be defined: "Equal volumes of gases at the same temperature and pressure contain equal numbers of molecules."

It is a singular fact that this clarifying announcement of Avogadro fell upon deaf ears; chemists by postponing their recognition of his fundamental differentiation between atoms and molecules were lost in a quagmire of confusion for the next fifty years. Dalton, in the query previously quoted, had already propounded Avogadro's hypothesis and at once rejected it; the idea, if it occurred to him, of doubling the ultimate particle of an elementary gas probably appeared irrational. Such was no doubt the attitude of other chemists. It was not until 1860, at the International Chemical Congress of Karlsruhe, when chemists from all European countries convened to bring order out of the chaos of misunderstanding about the meaning of the terms atomic weight, combining equivalent, atom, molecule, basicity, etc., and about the methods of expressing chemical formulae, that Stanislao Cannizzaro (1826-1910), by his insistence upon the correctness and importance of Avogadro's hypothesis, indicated the path that led chemists out of the wilderness to the establishment of the present system of chemical notation. The atomic weight of oxygen was then established as 16 and not as 8, and the formula for water rewritten as H_2O and not as HO as first proposed by Dalton. It was twenty years, however, after the Karlsruhe congress before this and the other reforms resulting from Avogadro's hypothesis were generally adopted.

The question might be asked, "Why did chemists persist in error for two generations when the truth of the matter had already been announced?" Many other instances of the delayed acceptance of truth can be cited in the history of chemistry, a more recent example being the important phase rule of Willard Gibbs (1839-1903), which explained so many phenomena relating to the states of matter. This is probably in part the result of a natural conservatism with many scientific men to accept hypotheses unless they are accompanied by experimental proofs. Chemistry is rightly regarded as an experimental science; had Avogadro and Gibbs accompanied their announcements with practical demonstrations and suggestive applications, the acceptance of their theories would no doubt have been greatly hastened. It must also be remembered that in Avogadro's time chemical science was so young that the actual significance of his suggestion in the solution of the problems that were pending was not understood.

VITALISM

From the earliest times in the history of chemistry there have existed two schools of thought which held opposite views with regard to the nature of the chemical processes of plant and animal life. The view of the ancient school of Democritus was the purely materialistic one that the operations within a living body are controlled by the same mechanical laws of atomic action as prevail elsewhere in the universe. Contrary to this opinion is the belief of the Hippocratic school that there is within the body an innate vital faculty or force which governs the operations of digestion and the like. The conflict between these two rival schools is reflected in the pages of Galen (131-201 A.D.) and other ancient writers. The vitalistic theory in various forms was passed on through the Middle Ages to modern times. Paracelsus, and, following him, Van Helmont and other iatrochemists, as we have noted, held that there was an intelligent demon or *archeus* which regulated digestion and other bodily functions. Stahl, the founder of the phlogiston theory, adopted a somewhat similar mystical explanation, holding that it was the soul that directed the functions of the body; other phlogistonists held a similar belief. This idea of a directing vital force, which was neither chemical nor mechanical in action and operated in living cells only, pre-

vailed after the time of Lavoisier. It was held by the great
Swedish chemist J. J. Berzelius (1779-1848), who declared
that it was impossible to synthesize in the laboratory the
organic compounds that were produced in plants and animals. This belief that organic compounds were formed solely
by a vital force was overthrown in 1828 by Fredrich Wöhler
(1800-1882), who synthesized the organic substance urea,
which is secreted in the urine, by heating ammonium isocyanate which can be prepared from inorganic materials:

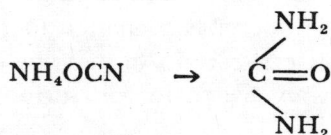

$$NH_4OCN \rightarrow \underset{NH_2}{\overset{NH_2}{C}}=O$$

Ammonium isocyanate   →   Urea

The old vitalistic belief, however, was slow to give way,
and it was only after many other compounds of plant and
animal origin had been synthesized in the laboratory from
inorganic substances that it became generally recognized
that the laws of chemistry and of the conservation of energy
hold as rigidly in the living as in the inanimate world.

THE THEORY OF FERMENTATION

Next to combustion probably no other subject has aroused
so much speculation and controversy in the history of chemistry as fermentation. The action of a minute quantity of
leaven in fermenting a large mass of dough was cited by the
ancient alchemists as an evidence of the immense converting
power of a single grain of the philosopher's stone. They
regarded the transmutation of metals as a fermentation. The
ideas of the iatrochemists about fermentation were similarly
vague. Stahl, the phlogistonist, was the first to suggest a
mechanical explanation of the phenomenon by assuming a
transfer of the motion of a fermenting particle to other particles. A similar view was held by Sir Isaac Newton (1642-
1727) who stated: "In fermentation the particles of bodies

which almost rest are put into new motions by a very potent principle which acts upon them only when they approach one another and causes them to meet and clash with great violence and grow hot with the motion and clash one another into pieces and vanish into air and vapors and flame." Lavoisier, in his thoroughly scientific way, left speculation alone and demonstrated that the amount of alcohol and carbonic acid produced in fermentation had the same weight as that of the sugar fermented, which helped him to make the important broad generalization that matter is never destroyed but only undergoes a change in form. While Lavoisier failed to detect certain minor by-products of the alcoholic fermentation, his main conclusion was fundamentally correct.

L. J. Thenard (1777-1857) focused his attention upon yeast as the cause of alcoholic fermentation and taught that the latter must not be confused with other fermentations, such as that of vinegar in which the alcohol of wine is oxidized to acetic acid. A vitalistic conception of a living ferment then entered into the discussion; but it was soon shown that finely divided platinum could by mere contact with alcohol oxidize it to acetic acid. This seemed to strengthen the mechanistic explanation of fermentation by contact without the intervention of life. Berzelius adopted this contact theory and introduced the term "catalytic action" to explain the action of yeast upon sugar. Justus von Liebig (1803-1873), the celebrated founder of modern organic chemistry, took a somewhat similar view: that yeast was simply an unstable albuminous body which underwent spontaneous decomposition, the internal vibrations thereby produced being communicated to the sugar and bringing about its decomposition.

Louis Pasteur (1822-1895), whose classic researches demonstrated that each type of fermentation—alcoholic, acetic, lactic, butyric, etc.—was due to a specific organism, completely demolished this theory of Liebig. Pasteur's aphorism, "No fermentation without life," seemed to confirm the vitalistic view, until Eduard Buchner (1860-1917) in 1897 isolated from yeast a lifeless ferment (i.e., enzyme) which converted sugar into alcohol and carbon dioxide. This work disproved the vitalistic theory as expressed by Pasteur; the view of fermentation held at present is that yeasts, bacteria, and other microörganisms produce catalytic chemical changes by means of enzymes which are present within their cells. A reconciliation was thus effected between the opposing views of the different workers, each one of whom, although partially in error, contributed his share to the final solution of the problem.

Space is lacking to discuss other "errors in chemistry," such as the error of Lavoisier's oxygen theory of acids, the error of Berzelius' dualistic theory of chemical compounds, the error of Mulder's humus theory of plant nutrition, the error of Liebig's theory of base exchange in soils, the errors in the discovery of spurious elements, the errors in atomic-weight determination, and many others which might be mentioned. Enough has been shown, however, to illustrate the fact that in chemistry, as in other sciences, the quest of truth is always accompanied by the commitment of errors. Errors are but the waste and débris that have accumulated in the erection of the beautiful structure of modern chemistry which is still far from complete. Our modern chemical literature is filled with errors which future research will gradually eliminate. The mere search for errors, however, is a fruitless endeavor if we lose sight of the great truths that have been gleaned by the men whose mistakes have been described in the present chapter.

It should not be forgotten, moreover, that many of the so-called errors of chemistry that were long ago rejected have been found in the end to be true. The breaking down of uranium into radium and of radium into lead are natural confirmations of the old doctrine of transmutation; many other conversions of the elements recently achieved in the laboratory are but modern vindications of the ancient car-

dinal doctrine of alchemy. The old rejected doctrine of
Heraclitus that "the all is one" has again come into its own
with the discovery that the atoms of all elements are composed of the same *materia prima*—protons and electrons. The
once discarded hypothesis of Prout (1785-1850) that the
atoms of all the elements are but condensations of atoms
of hydrogen is now revived by the modern theory that the
mass of all atoms resides chiefly in the protons and that the
number of protons in the atoms of the different elements
are even multiples of the single proton of hydrogen. All of
which is but a confirmation of the old adage that "truth
crushed to earth shall rise again."

Section III

THE LIVING REALM

Chapter VI

ERROR IN ZOÖLOGY

By Howard M. Parshley

ZOÖLOGY *is the inclusive science that deals with animal life in all its phases. Its field includes the structure, functioning, habits, distribution, classification, genetics, and evolution of animals. All animal species, from the shapeless primordial speck of living jelly called* Amœba *to the magnificent ruler of creation who calls himself* Homo sapiens, *are included within its scope. Before the dawn of history, man's curiosity and observation—foundations of all science—were directed toward the animals that were useful or injurious to human interests; while man's fear and imagination—chief sources of all error—played with their false light upon whatever prowled beyond the narrow circle of the camp-fire and upon whatever happened without due cause apparent to the dim intelligence and faulty observation of the time. For long, man himself was held to be entirely unique, a creature apart from the animals that helped or hurt him; and even now the common failure to recognize that such studies as psychology and medicine are really branches of zoölogy tends to perpetuate this first and foremost of all zoölogical errors. The newly emerging science of human biology is finally dispelling this ancient misconception; and we are beginning to reap in all fields the advantages—partially realized in the development of physiology and medicine—growing out of the application of zoölogical or, more broadly, biological principles to human affairs.*

In the following pages we shall trace the history of zoölogical science, with our interest centered upon the errors that have marked its course. We shall find that errors are not all of a kind. They differ not only in magnitude and in their effects upon progress, but even in their very nature. Indeed, there is a type of error quite inseparable from the process—called the

*scientific method—through which alone it is possible to gain
access to the truths of nature. We shall begin, therefore, with
some general remarks upon truth and error, by way of necessary
introduction to what follows.*

TRUTH AND SOME FORMS OF ERROR

THE DISTINCTION and the importance of genuine
observational science depend upon the fact that such
science alone affords a general method of approach to true
and useful knowledge. Through observation and experimentation, tentative suggestion of hypotheses, testing of such
hypotheses, and verification, we may get at the truth in particular instances, as near to absolute truth as it is given
man to see. Yet every science embraces errors and unsolved
as well as unrecognized problems, along with the accumulated body of truth which by its continual growth makes
the science a reliable guide to an ever-widening field and
justifies our constantly increasing trust in its findings. That
philosophy and religion assert their own allegedly prior
claims as ways to truth simply means, at bottom, that
thought, opinion, and *belief*—the "creative intelligence" as
opposed to disciplined *curiosity* and *observation*—got into
action before science developed, and established pretensions
that remain unrelinquished to this day. The errors of infallibility, whether it be logical or theological, are permanent.

The delusions and errors of science, on the contrary, last
only as long as the ignorance that gave them birth. The
zoölogist derives his professional satisfaction as much from
exposing an old mistake as from discovering a new fact or
principle of Nature. Indeed, the scientific method itself,
founded deeply upon the basic behavior-pattern of animal
life, proceeds by the way of trial and error and thus makes
fruitful use of the false assumptions, faulty deductions, and
wild surmises that spring luxuriantly from "creative"
thought. The processes of thought are certainly necessary in

arriving at a comprehensive theory; but if the theory is to be reliable and true, it must be freed from incorrect assumptions, deductions, and surmises—to say nothing of more or less conscious wishes—by the refining and error-eliminating powers of the scientific method. A wrong hypothesis, employed temporarily, is soon revealed for what it is when subjected to crucial experiment, further *ad hoc* observation, or corroborative efforts by other qualified investigators; but in this process it may frequently serve to suggest other and more promising lines of research, among which sooner or later the right one will ultimately be found. Thus, under the method of modern investigation, zoölogy, like other sciences, advances to the accompaniment of innumerable errors; but these errors are harmless when duly regarded as tentative notions and finally repudiated as they fail to pass the test of verification and prediction.

If we could know the complete history of zoölogy, therefore, we should have an endless list of errors to deal with: "the history of error," says Sarton,[1] "is, of its very nature, infinite." But the fact is that a vast majority of these, so to speak, routine errors of the scientific method never come to light, save sometimes in the original detailed reports of technical experimentation. They constitute the slag and dross which is rejected and forgotten—and quite properly—while general attention is centered upon the purified items of truth that are obtained from successful investigations. Only those who have themselves worked in the research laboratory can realize how many false starts, how many ruefully if promptly abandoned schemes and hypotheses, how many prematurely published and quickly repudiated conclusions, mark the esoteric history of even productive lines of research—not to mention the countless abortive efforts, including not a few that were the work of a lifetime, in which error was soon or late perceived to be supreme.

These routine errors of the scientific method constitute

1 George Sarton, *Introduction to the History of Science* (Washington, 1927).

the first of the three main categories that we may distinguish in our effort to understand the place of error in zoölogical science, namely: (1) *tentative*—errors inherent in the scientific method of trial and error, which are consciously regarded as on trial before they are unmasked and discarded; (2) *imaginative*—false information and beliefs held by laymen and zoölogists in common simply because the questions involved have not been subjected to any or to adequate technical investigation; (3) *scientific*—mistaken views held by zoölogists more or less generally and for a longer or shorter period of time, which are based on unreliable, misinterpreted, or fraudulent data. Like all biological classifications, this summary disposal of all zoölogical errors in three groups is somewhat arbitrary, in that some mistakes no doubt exist that do not fit readily into any of the categories, while others are of such character that they might be placed with equal reason in one or another of them; yet this grouping has a distinct value, because it distinguishes at least the most important kinds of zoölogical error and, with few exceptions, fixes the status of the several errors as encountered.

It may be noted further that these three types of error bear a general relation to the historical development of zoölogy and to its practical values. Imaginative errors mark especially the earlier periods, when science as we understand it was scarcely existent. In those days even the ablest minds, however keenly objective by nature, had weak defenses against fanciful beliefs, because of their restricted experience and lack of both a critical attitude and technical equipment. Ancient zoölogy, even at its best, was infiltrated with imaginative mistakes and thus had little of practical value to offer either to medicine or to agriculture. That it had any scientific character at all is a magnificent tribute to the intellectual power of a few individual observers, of Aristotle particularly. Later, during Roman and medieval times, this light failed, or rather existed as a pale reflection, incapable of promoting scientific progress and powerless against the surrounding shadows of imagination, until at last the modern period of rapid scientific advance was made possible by the development of technical equipment and improved methods of research. With these came the conscious use of tentative error and the tremendous growth and success of practical application, only occasionally disturbed by the waxing and waning of such scientific errors as in the nature of things were bound to arise from time to time, as they still do.

ANCIENT ZOÖLOGY

Primitive man was ignorant and correspondingly superstitious. His zoölogical knowledge consisted simply of practical facts about wild animals that were dangerous or useful, domesticated species which shared his communal life, and the more obvious biological processes of life, birth, and death that he shared with them. All else was imaginative error. It appears that the human species, once established as such, possessed then as now a brain adapted for thinking, wishing, and believing; and the result was that, apart from the immediately practical and extremely narrow round of daily affairs, man built up for himself a vast body of fanciful lore concerning the animals, which, instead of picturing the truth of Nature, reflected man's wishes and fears and satisfied his inner need to explain everything. Thus it was that he imagined the existence of a god outside Nature—how else could Nature have been created and continue to exist?—and a multitude of spirits dominating unseen the activities of animals no less than the manifestations of the weather, the flow of streams, and the growth of trees.

From this beginning developed the mythology of antique civilization, with its array of imaginary animals—the unicorn, the phœnix, the centaur, Pegasus—which were too fabulous for any but the most popular and ignorant belief even in the times of Aristotle and Pliny and which therefore can hardly be regarded as representing zoölogical errors at all. From the same origins magic developed also, at one time a general doctrine or view of the universe and later a mere collection of rites and feats believed in by the credulous. Aristotle, while almost free from such belief, retained some vestiges of it, as is indicated by occasional passages in his works which occur incongruously among his records of extraordinarily numerous and exact unquestioned facts. For example, he says that "the salamander shows that it is possible for some animal substances to exist in the fire, for they say that fire is extinguished when this animal walks over it"; regarding bees he relates that while they make wax from flowers, the honey is simply gathered by the insects after "it falls from the air, principally about the rising of the stars and when the rainbow rests upon the earth," generally after the rising of the Pleiades. By way of support for this astrologic belief he notes that "in the autumn there are flowers enough, but the bees make no honey."

But more often the errors of Aristotle were caused simply by the ignorance which his technical resources were inadequate to overcome, while the element of magic and superstition was far stronger in his follower Pliny, as we shall see. Regarding the breeding habits of the eel, which are so singular and obscure that their discovery constitutes one of the more recent triumphs of modern zoölogical research, employing every resource of marine exploration and oceanography, Aristotle says:

Eels are not produced from sexual intercourse, nor are they oviparous, nor have they ever been detected with semen or ova. ... After rain, they have been reproduced in some marshy ponds from which all the water was drawn and the mud cleaned out; ... they originate in what are called the entrails of the earth, which are found spontaneously in mud and moist earth. They have been observed making their escape from them ... and they originate both in the sea and in rivers wherein putrid matter is abundant; in those places in the sea which are full of fuci, and near the banks of rivers and ponds.

Even this passage is notable not so much for the error of ascribing spontaneous generation to the eel, as for the sev-

eral statements, some of them dubious enough on the surface, that fall truly into place in the modern account of
the matter. Eels do reappear when drained ponds are filled
again with water; they can be observed making their way
over moist and marshy earth in their now familiar migrations; they really live both in the sea and in fresh water;
their young can be seen among thick seaweed at the sea,
shore and also near the banks of rivers in the shallows!
Quite naturally, again, Aristotle went astray in regard to
certain phenomena that could hardly be seen and understood without optical magnification. For instance, he failed
to observe the eggs of the smaller insects and accounted for
their origin in various ways. Some, like fleas and mosquitoes,
appeared by spontaneous generation from decaying matter
or, like bedbugs and lice, from the bodies of their hosts;
others, like flies and moths, were produced by sexual intercourse in the form of maggots or caterpillars. Yet Aristotle
saw the nits of lice and speaks of oviposition and the white
eggs inside the body of the female grasshopper, apparently
without quite getting a clear picture of the uniform rôle of
eggs whether large or small.

But the errors of Aristotle were trifling and highly ex-
cusable in comparison with those of Pliny (first century A.D.),
who, a brave and honest soldier, had an uncritical passion
for collecting all sorts of information about animals from
all available sources, however unreliable. This information
he recorded with omnivorous credulity in his vast *Natural
History*. Even though it included much of Aristotle's true
science, Pliny's work was so stuffed with the errors of the
time that it undoubtedly takes the palm as the greatest
single repository of misinformation known to man. This compilation became enormously popular throughout the medieval world but did little or nothing to lighten those dark
ages; its effect was rather to reinforce the power of superstition. It did, however, serve as a useful basis when science
was at length resurrected, after fifteen hundred years.

It is not remarkable that there should have existed an

endless capacity for belief in marvels known only from hearsay among men who could credit the wildest errors concerning the commonest and most familiar features of human biology. For example, Pliny discusses menstruation, and after noting that "among the whole range of animated beings, the human female is the only one that has the monthly discharge" (an error), he goes on to describe the "marvellous effects" (all erroneous) of the menstrual blood:

On the approach of a woman in this state, milk will become sour, seeds which are touched by her become sterile, grafts wither away, garden plants are parched up, and the fruit will fall from the tree beneath which she sits. Her very look, even, will dim the brightness of mirrors, blunt the edge of steel, and take away the polish from ivory. A swarm of bees, if looked upon by her, will die immediately; . . while dogs which may have tasted of the matter so discharged are seized with madness, and their bite is venomous and incurable.

The delicacy of these bees is matched by the remarkable friendliness of the dolphins, "which recognize in a most surprising manner the name of Simo, and prefer to be called by that rather than any other," whether this taste originates in the applicability of the name (*simus*, flat-nose) or in the mere sound of the word. These sportive creatures of the waves can leap over ships, according to both Aristotle and Pliny; but, best of all, they are music-lovers, always ready to be charmed *symphoniae cantu!* Both these authors, however, knew the species to be a mammal—"they suckle their young"—and thus were superior to the widespread imaginative error of the present day through which these animals are thought to be fishes.

In Pliny we find well displayed the points of contact that existed between zoölogy and magic, that most colossal and enduring of human errors. Though he firmly disclaimed belief in magical performances—like the free-born American who jokingly but none the less carefully avoids walking under the ladder or omitting his religious duties—Pliny tells in all seriousness of charms and remedies that evidently he believed in, while scorning some of the beliefs of others. For example, he reports that eating the flesh of the stag or wearing the longest tooth of a fish can ward off or cure fevers, and that the wearing of the carcass of a frog deprived of its "claws" has therapeutic virtue. It may be noted that this attribution of magical values helped decidedly in maintaining the untenable doctrine, dear to Pliny and to the ignorant even now, that every animal, as well as all other natural objects, have some use—"what were they made for?"

During the Middle Ages, the fifteen hundred years of scientific stagnation that followed the decay of the Roman Empire and the invasions of uncivilized hordes, zoölogy hardly existed as a field of productive study. In accordance with the doctrine of Thomas Aquinas, theological interests connected with the "kingdoms of grace and blessedness"— sinful men on earth and saved souls in heaven—took precedence of secular interest in the "kingdom of Nature," for which the works of Aristotle (and in lesser degree those of Pliny and the authors of the bestiary known as *Physiologus*) were held to be quite adequate. Edification through animal stories was the chief aim of such Nature writing as was done. In the absence of any serious zoölogy it would hardly be worth while to single out particular errors from the preposterous mass of superstition and foolishness. Among the Arabians true science was cultivated more or less extensively during this time; they knew Aristotle and Hippocrates, and even in the thirteenth century they published commentaries which included new observations. One of them, Sakarja ben Muhammed, in *The Wonders of Nature*, advanced an erroneous theory of fossils, to the effect that they were animals petrified by steam emerging from the ground.[2]

EARLY MODERN ZOÖLOGY

As the Middle Ages merged into modern times (*circa* 1500 A.D.), two main characteristics of the Renaissance, re-

[2] Erik Nordenskiöld, *History of Biology* (New York, 1928).

covery and wide dispersal of classical writings and the new
freedom of interest in natural phenomena, led to a revival
of zoölogical science and to a rapid development of the
subject which has continued without interruption to the
present day. At first, quite naturally, works were produced
which followed Aristotle and Pliny in general, retaining
some of the anecdotal and marvelous features of the latter
and all of the inadequate classification and physiology of
the former. The chief aim was to identify the animals
mentioned by the ancients and to record along with them
the large numbers of newly discovered species continually
coming to light as a result of the world-wide explorations
then in progress. The new arts of printing and illustration
furthered the publication and wide distribution of these
works, of which the finest example was the *Historia Ani-
malium* of Konrad von Gesner (1516-1565), commonly regarded as the starting-point of modern zoölogy. It is, indeed,
one of the greatest of all biological works, comprising about
thirty-five hundred pages in five enormous volumes and
containing many illustrations that could hardly be improved
upon today. Its chief failings were the encyclopedic inclusion
of folklore and speculation and the failure to trace similarities of form and function among the animals described.
From this time on, the works of zoölogists were progressively
marked by the elimination of superstitious and imaginative
errancy, and by the increasing emphasis placed upon relationships—two movements which were to receive their full
justification in the work of Darwin.

From the time of Aristotle to the middle of the seventeenth century an abundant source of error lay in the small
or microscopic size of many animals and their parts. This
caused great confusion in connection with the reproduction
of animals that do not display conspicuous eggs like those
of birds and reptiles. It led also to the mistaken belief,
universal up to the time of Marcello Malpighi's dissection
of the silkworm (about 1675), that smaller forms of life
contain no internal organs. Using the recently developed

microscope, this marvelous technician, much to his own surprise, discovered in insects a fine anatomy no less complicated than that of larger animals. Such microscopic work was necessary not only for tracing the hitherto neglected relationships among zoölogical groups, but also for the final correction of long-standing errors in such major aspects of physiology as the circulation of the blood [3] and for the elucidation of such unsuspected fundamentals of biology as the cellular structure of animals and plants, the ignorance of which led to all sorts of mistaken views and fantastic speculations.

But the early microscopic work, carried out with the aid of very imperfect apparatus, contributed new errors of its own, of which we may note two examples. As late as 1840, Christian Gottfried Ehrenberg, a great student of microscopic forms of life, clinging to the erroneous belief that all animals must possess one common type of organization, vitiated his otherwise valuable work by insisting that he was right in his belief that he could discern a digestive system, reproductive organs with eggs, and many other familiar structures in the minute bodies of animals that have nothing of the kind. Five years later Karl T. E. von Siebold showed their true nature as single cells of protoplasm without true organs and called them the Protozoa. Much earlier, in 1677, Antonius van Leeuwenhoek and his student Stephen Hamm discovered the sperm, an active element swimming in the seminal fluid; whereupon others, especially Niklaas Hartsoeker (1656-1725), straining their eyes to see the minute bodies through their imperfect microscopes and straining their imaginations to fit what they *saw* with what they *believed* about the process of reproduction, persuaded themselves that each sperm contained a little image, complete in every part, needing for full development only to be embedded in the nourishing substance of the egg. One member of this school of thought said that he saw in the human sperm "two naked thighs, the legs, the breast, both

[3] See Chapter VII.

arms, etc., the skin being pulled up somewhat higher did cover the head like a cap." Another group believed they could discern the necessary parts all preformed in the egg.

This erroneous *preformation* theory led logically to the fantastic idea of Charles Bonnet (about 1750) that every female contains the germs of all her descendants, enclosed one within another. He actually did see a new generation forming within and emerging from the body of the plant-louse and felt justified in formulating his idea as a general principle by such additional observed facts as the existence of the young new plant as a rudiment within the seed and the presence of the adult insect with all its parts within the shell of the pupa. Although orthodox in religious belief, Bonnet was keen-minded enough to combat the erroneous scientific application of the vitalistic philosophy—a zoölogical error which persists in some quarters to this day—by pointing out that the bodily functions work on mechanistic principles and that the use of the "soul" as an explanation is merely a facile begging of the question. The crude preformation theory is a good example of how a luxuriant growth of mistaken theorizing can spring from a small root of observational error, or at least be nourished by it. It was only toward the close of the nineteenth century that the advance in microscopic technique established the very different truth regarding germinal preformation or determination by disclosing the chromosomal apparatus and the organization of the egg.

RECENT AND CONTEMPORARY DELUSIONS

Since 1850, zoölogy, like other sciences, has made extraordinary advances through the conscious and intense application in research of what we have called the scientific method. This, of course, has involved a vast multiplication of *tentative* errors, of which a great majority have never been heard of outside the laboratories, though a few have had their brief day of more or less general credence. *Imaginative* error—the inevitable outcome of ignorance and superstition—has largely disappeared with the growth of knowledge, being represented currently among scientists almost entirely by religious and spiritualistic beliefs.[4] A few still maintain that they find in scientific knowledge convincing evidence of purpose in Nature and a personal Intelligence behind it all, and thus continue (however feebly) in such works as *The Great Design*[5] the tradition of William Paley and his *Evidences of Christianity* (1785), which, according to Singer (1931), "hypnotized until our own time the University where Newton had taught." But for the most part, among zoölogists, this tradition is all but extinguished and finds its last and most attenuated representation in the reconcilers who hold that we can go "through science to God," or who maintain that science and religion are entirely separate provinces because "there are two great revelations to mankind," or who occupy some position between these two extremes.[6] Historians in the future may well associate the end of all this with the famous "monkey trial" at Dayton in Tennessee, where the reconcilers held their last great rally in the attempt to mediate between Bryan, the anti-evolutionist, and his victim, Scopes.

A number of *scientific* errors in nineteenth- and twentieth-century zoölogy remain to be considered. These are mistaken views which have been held by zoölogists more or less generally, by reason of faulty data. Many of them involve simply questions of fact that have been more or less recently settled by scientific research, though the erroneous beliefs may persist as popular fallacies.[7] Others have more general im-

[4] See, for example, Alexis Carrel, *Man, The Unknown* (New York, 1935).
[5] A volume of essays by fifteen scientists edited by Frances B. Mason (New York, 1934).
[6] Howard M. Parshley, *Science and Good Behavior* (Indianapolis, 1928).
[7] This type of error is illustrated by the following list of beliefs, each of which has at one time or another received support from zoölogists: that inbreeding causes degeneration; that ants and bees are intelligent; that certain fly larvae, living as guests in ants' nests, are mollusks; that selection can gradually change the nature of a hereditary factor, or gene; that toads at the breeding-season find ponds through an ability to sense water from a distance; and that fishes can sense a current as such and swim against it when unable to see or feel the bank or bottom of the stream.

portance. Modern research in the behavior and reproduction of animals, for example, has disposed of so vast a body of error that in many respects these fields of biological study have been completely revolutionized, with resulting gains for psychological and physiological science, as detailed elsewhere in this volume. But there are other misconceptions of even wider significance, both theoretical and practical.

Among these scientific errors which bear alike on zoölogical theory and on human affairs, there are two that merit detailed consideration, namely, the doctrine of special creation and the belief in the inheritance of acquired characteristics. Both of these ancient ideas were widely held by people in general and were quite naturally accepted for a time by zoölogists, uncritically, even unthinkingly, as self-evident facts. A time came, however, when both were questioned by certain bold and lonely spirits; then others more or less belatedly began to look over the data. Enough mistaken, misunderstood, or actually fraudulent evidence was available in support of these erroneous beliefs to elicit considerable scientific support for them for a time. The special-creation error, moreover, had the sanction of Christian doctrine, to say nothing of well-nigh universal primitive belief. Thus the struggle to replace it by the theory of evolution produced the most bitter and dramatic controversy of recent times. Belief in the inheritance of acquired characteristics, in contrast, had no such theological implications; and so its fall was attended with less public uproar, although its significance for human welfare can hardly be overestimated. Since these two doctrines are closely interrelated in certain respects, they will be considered together in the following discussion.

SPECIAL CREATION OR EVOLUTION

Aristotle and some other ancients, as well as Saint Augustine, had a vague conception that, as the Stagirite says, "Nature can only rise by degrees from lower to higher types"; but this was metaphysically more or less confused with

Aristotle's idea of the scale of Nature—the erroneous belief
that there is in contemporary Nature a continuously graded
advance along a single line from the inanimate, through
plants, and through the simpler and more complex animals,
to the higher mammals and man. And, in the case of the
Saint, the conception that all things "grew up in such manner as they are now known to us, in due time, and after
long delays," was similarly involved with his theological
exegesis of the Biblical story of creation. But in the absence
of scientific knowledge adequate for supporting a convincing theory of evolution, the Christian world—including biologists—failed to appreciate the insight of these two men
and fell back into a simple and literal acceptance of the
words of *Genesis*. That each species was created as it is now
and always has been since the beginning was the belief
universally held up to the middle of the eighteenth century.

For another hundred years, until Charles Darwin published *The Origin of Species* in 1859, the error continued
as an almost undisputed principle. The poorly founded
though ingenious and prophetic evolutionary hypotheses of
Buffon, Erasmus Darwin, Jean Baptiste Lamarck, and Geof-
froy Saint-Hilaire (all writing shortly before or after the
year 1800), seemed to be inconsequential aberrations from
the straight and narrow path. Apart from the basic error
of belief in the inheritance of acquired characteristics, which
Lamarck (1744-1829) was the first to formulate clearly
(though it reached back to Aristotle's self-questioned
"maimed parents produce maimed children"), there was a
fatal lack of factual evidence supporting the ideas of these
pioneer evolutionists; and this, together with the absurdity
of the examples (representing the supposed action of the
environment) presented by Lamarck and Saint-Hilaire
(1772-1844), made them an easy prey to such powerful opponents as Cuvier and his followers, who upheld the orthodox doctrine of special creation. How alone the evolutionist
of 1809 really was is indicated by Lamarck's own words.
In stating his case he referred to the special-creation theory

as "the conclusion adopted up to this day," and to the idea of gradual evolution with environmental modifications as "my personal conclusion."

Cuvier (1769-1832), a truly great zoölogist especially ac-plished in comparative anatomy, was the most influential and politically powerful figure of contemporary biology. His influence, together with the deficiencies of the proponents, prevented the spread and acceptance of evolutionary ideas at the time. He opposed Lamarck until the latter lost his eyesight and retired from the field, and then he kept up the fight with Lamarck's one and only disciple, Saint-Hilaire. But both sides labored alike under a hopeless load of confused error and rightness. The evolutionary principle was right; but the effort of Saint-Hilaire, for example, to maintain the homology of the cephalopod with a vertebrate bent in the middle so that its anal region lay close to the head was absurd and easy prey for Cuvier's refutation. Cuvier's principle of types, that is, that there were four main types of animal structure not derivable from one another, was correct in general and even proved, later, to be assimilable with modern evolutionary theory, which regards these types simply as diverse end-products of divergent lines of descent; but Cuvier's basic belief in the fixity of animal species and his theory of successive geologic catastrophes to account for fossils were wrong. He taught that the Biblical flood and other such legendary events wiped out all life over wide areas and thus made the fossils; while different species, which had been in existence all the time in out-of-the-way places spared by the catastrophes, later came in to repopulate the devastated areas and thus to provide the succession of different forms demonstrated by geology and paleontology.

In the interval before Darwin's work was published, most biologists were special creationists, but here and there one appeared who had more or less tenable evolutionary views. Richard Owen (1804-1892), famous English anatomist who lived through the whole Darwinian period, was—long before 1859—a believer in a limited scheme of evolutionary change, chiefly by degeneration from earlier types. But he elaborated the erroneous notion, following Goethe, that the vertebrate skull represents a series of modified vertebrae; he believed in the former existence of perfect type-forms, had no doubt of the continuous operation of supernatural forces on Nature, and ranged himself among the opponents of Darwin. He suggested that the larval forms of animals might separate and become new species and even believed in the daily spontaneous generation of microscopic creatures. More typical was Agassiz, the Swiss who in 1846 became a Harvard professor and a leader of conservative zoölogy. Despite his great ability and sound achievements, he held certain erroneous views which led him to oppose Darwin to the death. He regarded the categories of taxonomy— species, genera, families, etc.—as immutable modes of thought of the Supreme Intelligence, and he interpreted vestigial organs as created to perfect the design. In general Agassiz posed as a supporter of the Biblical creation story—insincerely and to avoid trouble, according to Ward [8]; but he certainly was a firm believer in the fixity of species and later became the chief enemy of the evolutionary idea in the United States. So nearly universal was the acceptance of special creation among zoölogists at this time that Weismann says, "In my own student days in the fifties I never heard a theory of descent referred to."

Darwin's work effected an extraordinarily prompt and well-nigh universal change in the views of biologists and wrought a revolution in the character of philosophy, history, economics, and anthropology. The evolutionary conception, neglected though clearly enunciated before, now gained almost universal acceptance. Darwin's exhaustive and masterly marshaling of the factual evidence was the new and decisive factor. Yet in the matter of the causes of evolution there was room for error. Darwin placed his chief reliance upon the slight variations that are ever occurring

[8] Henshaw Ward, *Charles Darwin* (Indianapolis, 1927).

among individuals; but he could not distinguish those which were really hereditary and those which resulted from environmental influences. He was forced to admit cautiously and without enthusiasm the principle of the inheritance of acquired characters, though he never accepted the excesses of Lamarckism. Thus there developed, especially among Darwin's more extreme supporters, the erroneous notion that natural selection was an all-sufficient explanation of the evolutionary process. Zoölogists of the period lacked the knowledge of Mendelian inheritance and of the process of mutation, which was to clear up these mistakes after 1900. It may be said that Darwin's chief error in this connection was his unavoidable failure to recognize the significance of those striking and hereditary changes called "sports" by practical breeders. He thought they were too rare to be of importance in evolution, and he could not realize the fact that they were essentially identical with the far more numerous small hereditary variations, not distinguishable in his time from non-inheritable environmental modifications. These errors began to lose their hold when August Weismann (1834-1914) advanced his conception of the continuing germ-plasm as distinct from the somatoplasm and thus did away with a universal misconception of the relation between the bodies of animals and their eggs and sperms. He made it clear that the germ-cells form a continuous line, not only between generations but also through the body of the individual in each generation, and thus supplied not only a firm basis for the development of a science of heredity—now called genetics—but also a necessary element, hidden from Darwin, in evolutionary theory.

INHERITANCE OF ACQUIRED CHARACTERS

But one of the implications of Weismann's work—following upon this discrimination of temporary body and continuing germinal material—was not universally agreed to by zoölogists. There thus persisted well into the twentieth century a zoölogical error of the greatest importance, significant not only for biological theory but also for practical human affairs. This was the Lamarckian—and common —belief in the inheritance of acquired characteristics, the supposition that bodily or somatic changes experienced by the individual as a result of either environmental influences or his own activities would somehow be passed on to his offspring and reappear in some measure in the absence of similar influences and activities. Lamarck based his theory of evolution on this belief. He suggested, for example, that the long neck of the giraffe—to our insight obviously hereditary—was produced in the course of time by the slight stretching effort made by each generation in attempts to reach the tender leaves at the tops of bushes. Most zoölogists, in fact, while avoiding the more extreme views of Lamarck and some of his followers, took it for granted that bodily changes induced by exercise and by the environment, if continued over long series of generations, would become hereditary and thus contribute to evolutionary advance. Under the influence of Weismann's work, however, almost all zoölogists abandoned this belief, realizing that there was no conceivable way by which changes in the soma or body could affect the germ-plasm (eggs and sperms) so specifically as to cause the reappearance of the same bodily traits in the next generation. It was recognized that highly deleterious agents might weaken or destroy the germ-cells, but that is a very different matter. Nevertheless, many scientists and leaders of thought in other fields still held to the exploded belief; educationists, philanthropists, and politicians assumed its truth, and in many cases still do, relying upon it as the basis of social progress.

Many zoölogists have carried out extensive experimentation in recent years in the effort to obtain definite objective evidence upon this question, with the result that in all save one or two doubtful cases the error of the Lamarckian principle has been clearly revealed. Rats living in constantly rotated cages acquire an inability to walk straight, and in certain experiments carried out by Detlefsen their young showed the same new characteristic although not themselves rotated. This result created widespread interest among biologists, since it appeared to be in accordance with the Lamarckian theory and contrary to accepted modern ideas. But it was soon discovered that some of the rats were suffering from an infectious disease of the inner ear and that the staggering of their young was caused by this disease, communicated in the ordinary way, and not by any inherited trait that had been acquired by the parental animals.

This case neatly illustrates the readiness with which experimental results may be falsely interpreted. There is another instance which has received far wider currency, that of Ivan Petrovich Pavlov and the mice that he trained to come for food at the ringing of a bell. According to the report of the great Russian physiologist, successive generations of these mice required fewer and fewer repetitions of the conditioning stimulus. It appeared that the ability to respond in this manner—an admittedly acquired character—was becoming fixed in the race by the inheritance of the acquirement. This erroneous view is quoted in many books as proof of the Lamarckian principle. The truth is, though few seem to know it, that Pavlov himself, at the 13th International Physiological Congress (Boston, 1929), withdrew his former interpretation and stated that the experimenters were improving in their teaching methods, the inherited traits of the mice meanwhile remaining quite unchanged.

The chief zoölogist of standing who upheld the Lamarckian doctrine of the inheritance of acquired characters in recent years was Dr. Paul Kammerer, of the Institute for Experimental Biology, University of Vienna. He published a book on the subject in 1924. In this work he brought together a great amount of data bearing on both sides of the question, including reports of his own very extensive experimentation, and endeavored to show that the evidence favored his very decidedly pro-Lamarckian opinion. This book illustrates clearly the dangers and difficulties of the problem. A superficial reading conveys a strong impression that the author is in the right, but a closer study shows that none of Kammerer's own research and none of the many instances he cites from the work of others will bear careful analysis. It is found throughout that the data are quite inadequate for the standard statistical treatment which all modern geneticists always apply to their experimental materials; in every instance the facts he records may be accounted for by preferable and alternative interpretations, in accordance with the evidence supplied by statistically significant researches.

One or two zoölogists were impressed by Kammerer's lectures in England, even though he failed to present satisfactory examples of his experimental products and even though others were unable to repeat and verify his work. He was invited to occupy an important position in Russia. At this juncture an American zoölogist, Dr. G. Kingsley Noble of the American Museum of Natural History, examined some of Kammerer's preserved material in Vienna and found that the supposed inherited character had been artificially and fraudulently fabricated, either by the investigator or by one of his assistants. This exposure ruined Kammerer's prospects and cast doubt on his work in general, strengthening the effect of previous scientific criticism. Kammerer died shortly after by his own hand, and his death may well mark the end of the zoölogical error on which he wasted his life work.

This survey of zoölogical error shows how the human liability to err is manifested among those who devote themselves to the scientific study of animals. It suggests the conclusion that true knowledge has always been reached through observation and verification, which afford the only safe pathway through the infinite possibilities of error, while superstition and preconceived beliefs serve only to beset that pathway with obstacles. Error can never be wholly eliminated from our view of living Nature; but there is sound satisfaction in the thought that *imaginative* and *scientific* error tend to disappear as knowledge increases, while the error that will always remain a part of the scientific method—that error which we have called *tentative*—has lost its power to mislead, has become, in fact, an inevitable and useful step in the search for truth.

Chapter VII

ERROR IN PHYSIOLOGY

By Homer W. Smith

THE EMERGENCE *of physiology from primitive superstition has been complicated, as much as or more than any other subject, by the belief that certain phenomena are explicable only in terms of supernatural forces. The Ionian atomists, and Leucippus and Democritus, laid down a plan of inquiry along naturalistic lines, but the metaphysical reaction of Socrates, and more particularly of Plato, defeated its development. When Plato's idealistic, mystical philosophy, deduced a priori from human needs and predilections with no consideration for fact, was incorporated with Hebraic sacred traditions to form the essential substance of Christian theology and faith, inquiry into the operation of the body was stifled for nearly two thousand years. Spirit, soul, and more recently an imponderable entity called mind, all discoverable by faith alone, have throughout that period obscured real problems, deterred men from undertaking exploration, or set them upon the wrong track. It has been said by Andrew D. White, who has considered the matter most carefully, that "exactly as the world approached the ages of faith it receded from ascertained truth, and in proportion as the world receded from the ages of faith it has approached ascertained truth." There might be some philosophical quibble about what, in the last analysis, constitutes "ascertained truth"; but for our purposes it is enough to exemplify truth by the instances that the blood circulates in the body, that the body burns fuelstuffs to derive energy for growth, heat, and muscular action, and that copulation leads by fertilization to the reproduction of human beings. It is scarcely necessary to enlarge on the reasons why faith, Christian or otherwise, has invariably proved to be an effective barrier to intellectual advancement, but there is one aspect of the problem that warrants consideration here.*

The human organism is so constituted as to achieve what we may call almost perfect egocentricity. All immediate experience relating to the external world, all subjective visceral sensation and emotion, as well as vague and imperfect memories of both, are brought to a more or less sharp focus to produce an integrated conscious personality. So perfectly is this fusion effected from moment to moment that, in the normal man, experiences widely separated in time appear to "happen to" an immutable ego; and on the other hand this ego in turn beholds the entire universe, from the tip of the nose to the remotest stellar nebula, as converging upon itself. It is a natural consequence of this fact that men fell into the naïve error of believing themselves at the center of the universe. The two most outstanding revolutions in human history were concerned with the demolition of false systems of thought engendered by this egocentricity; the Copernican revolution revealed that the universe was not geocentric, and the Darwinian revolution revealed (the revolution is far from being over!) that even this little earth was not homocentric.

There is not the slightest doubt that man is the most wonderful of all natural productions. But because he is so egoistic, it is only within the past hundred years, strictly speaking, that he has been free to see himself as anything short of supernatural. He has repeatedly let spirits, soul, mind, or other metaphysical figments obscure his vision of himself. It is not surprising, therefore, that his climb out of physiological superstition has scarcely begun. If the views of Leucippus and Democritus had not been vitiated by Plato and Augustine, what an understanding he might have of himself by this time! What a command he now might have over his health, his powers, his thoughts and even over his alleged "spirituality"! But two thousand years and more of metaphysical speculation cannot be undone.

Every man must, at some time or other, and however vaguely, ask himself questions like these: How does my body operate? What is the "life" that is in me? What significance is to be attached to my existence, to my sensations and fear, love, and hope? No man would venture to speak for another in answering these questions, and in the following pages I have merely expressed a personal view. If it be contended against me that proof that the material body operates by naturalistic means can never constitute disproof of an immaterial soul, spirit, or mind, I know no better answer than that given by Thomas Huxley to a similar argument fifty years ago: "It is wholly impossible to prove that any phenomenon whatsoever is not produced by the interposition of some unknown cause. But philosophy has prospered exactly as it has disregarded such possibilities, and has endeavored to resolve every event by ordinary reasoning." And I would urge the contender, before he assumes the responsibility for any metaphysical thesis whatever, to review the history of human thought, society, and welfare from the beginning of historic times up to the present day. Or he might concentrate on the fifteenth and sixteenth centuries A.D., when faith in the supernatural and burning of women for witchcraft simultaneously reached their peak.

BEGINNINGS OF PHYSIOLOGICAL INVESTIGATION

A S SOON as the savage has satisfied his primal needs of food, warmth, and safety and can relax in an idle moment of contemplation, it is revealed to him that he is different from the rest of the world. The warmth of his body, the directed motions he can make, the strange sounds he can emit, are a source of pride; he is forthwith induced to make more careful comparisons between himself and inanimate Nature, to relate himself to the phenomena of day and night, the seasons, the weather, and to interpret happenings external to himself in terms of his internal sensations. By the time he has reached, in this egocentric survey of his universe, the intellectual stage of classifying things, his speculations are apt to fall into two categories, one centering about life and the other centering about death. Since speculations concerning death can be neither proved nor disproved, it is not surprising that in the gradual progression of human thought they got off to an earlier start and long held the advantage over speculations about life.

One of the earliest questions the intellectual Dawn Man asked himself, no doubt, was how he happened to come into

the world. Just as he early discovered the causal relation
between the ingestion of food and the keeping of the breath
of life in his body, so he must have discovered the relation
between sexual intercourse and reproduction. The discovery
was perhaps long in coming because the connection is by
no means self-evident. There are savage peoples in the world
today who scarcely recognize that reproduction has its beginnings in mating. In many primitive societies promiscuity
begins with adolescence and does not end with marriage,
even where this is monogamous. Moreover, intercourse and
gestation are so widely separated in time that the chances
of associating them are very small indeed. Even where some
fortuitous circumstance may suggest that the one is the
cause of the other, the odds are against crediting such a gro-
tesque notion. It is easier for primitive man to ascribe preg-
nancy to the influence of something mystical or miraculous:
a meteor that flashed through the night, some extraordinary
aspect of the sun, wind, or rain, peculiar food one ate,
strange animals, or even unusual dreams; for to him one
cause in Nature is just as reasonable as any other.

Whoever first puzzled the matter out we do not know,
but a rough idea of the reproductive process was possessed
by the older civilizations and was transmitted by the Egyptians to the Greeks. The Greeks were not only inquirers
of the first order, but also anatomists, that is, men who looked
to the structure of the body to explain its functions. Their
theories were necessarily approximate and vague, but none
the less distinctly in advance of what had gone before. Empedocles taught that there were four elements, earth, air,
fire, and water; the human body, as all else, was made up
of these primordial substances which could neither be created nor destroyed. All phenomena were but transformations, a flowing of one form or combination into another.
Human generation was an elaboration of the body out of
the dissimilar elements; death was a decay, a return of fire
to fire, air to air, earth to earth, and water to water. This
view of generation was not an entirely abstract, philosophic

one, however, for Empedocles taught that the embryo is formed from the union of the male and female semen, and the child resembles the parent who has contributed the more material for its formation. Though misleading in the latter respect, his description was such as to place the reproductive process on a rational plane, to coördinate it in a valuable manner with the relations of man and woman, and thus to exert a profound influence upon the structure of society.

One would think that with a start in theory such as this, men would quickly make significant discoveries from their ordinary experience, but not so; a certain type or attitude of mind is necessary. Hippocrates (early fourth century B.C.), founder of rational medicine, and particularly of the art of diagnosis, followed Empedocles closely in time and thought. He was a man of great genius for making careful observations and for distinguishing generalities from particulars and accidentals from essentials; but he contributed little or nothing to the problem of generation, primarily because, with classic Greek instinct, he directed his inquiry rather more toward the externals of the body than toward its less prepossessing internal organs. It was not until two generations later that Aristotle broke this esthetic handicap by engaging in extensive studies in comparative anatomy and embryology. His curiosity led him to dissect all manner of animals and to make (so it is reported) what was perhaps the first scientific experiment. The experiment was a simple one, and particularly noteworthy for having been performed on a human being. A current theory had it that the sex of the child was determined by which testis supplied the reproductive fluid. Aristotle bound the right testis of a man and observed that the subject subsequently generated children of both sexes. That the result of the experiment was in a sense negative does not detract from either its scientific or social significance. Apart from this, if indeed the story is a fact, Aristotle did not improve upon the concepts of his time. He taught that the embryo was formed from the union of male semen with female menstrual blood, the former contributing form, animation, and soul, and the latter, matter, substance, and body. He held that life might originate from foam, as Aphrodite sprang from the foam of the sea; in this spontaneous generation soul was taken from the air and heat from the sun. The heart was the principal organ of the body, being the seat of the soul, the source of the blood and its innate heat. He postulated that an intermediary agency, the entelechy, united the soul with the body and enabled it to form and control the latter, and by so doing to realize or actualize itself. Thus, while advancing the problem of reproduction by his experiment, Aristotle made a backward step by speculating about his entelechy. It is not unlikely that, by the weight of his authority, his derivation of the soul from the male semen did much to fortify the social differentiation of the sexes in subsequent centuries.

THE AUTHORITY OF GALEN

Aristotle, no doubt the greatest intellect of pre-Christian times, was destined to have less influence upon human thought than Galen (second century A.D.), who was in many respects a lesser man. A successful physician of the Roman Court, versatile, imaginative, and endowed with tremendous energy, Galen was the most prolific writer among the ancients. He thought, investigated, talked, and wrote about everything pertaining to the body. In large measure he perfected the experimental method used by Aristotle, whereby a question is put to Nature under such conditions that Nature can answer it. He was a true experimenter, but he was at the same time dogmatic and given to unjustified theorizing; he had an answer for any question, explaining everything with astonishing facility in the light of his elaborate theories. He was convinced that man was the center of the universe and that his welfare was its most important concern, and he believed that all Nature bespoke and was planned in accordance with the essential goodness of its Creator. Not only did he start investigation upon the wrong track in several specific matters—by his description of the function of the heart and the circulation of the blood, by his doctrine that wounds must suppurate to heal, and by his theory of vital spirits—but, more significantly, he gave men to believe that when he spoke, the Final Authority had spoken and the last word had been said.

Galen's works were so complete that (like many Gargantuan monographs of today) the inquirer found it easier to consult them than to seek an answer directly of Nature. The sheer weight of the massive tomes no doubt fortified the innate feeling of every inquirer that what is written is permanent; it cannot be wrong. His cock-sureness gave his theories the tone of an oracle, and men have ever sought oracles, in science as in all else. But, above all, it happened that after Galen's time intellectual activity became largely a province of the Church; Galen's piety, his homocentric philosophy, his doctrine of the essential goodness and the design of Nature, all recommended his works to ecclesiastic sanction. The Church was pleased to place its imprimatur upon his books. From that time on, things being as they were, no man dared doubt. For fourteen centuries, an unthinkably long stretch of barren years, the written word took the place of experience in most fields of endeavor; argument about the human spirit supplanted observation of the human body; and men turned from the study of life as it is to the study of what the masters had taught. The fault was not all Galen's, of course, for much of the stagnation of medieval thought must be ascribed to the concatenation of other forces throughout that dreary waste of time; but his name will always be associated with the dual errors of authority and egoistic philosophy, both of which direct men from inquiry to belief. The fact remains that after him men slipped into the most dangerous and the most widespread error ever made. They lost self-confidence, they lost the agnostic's faith that they must take the universe apart if they would learn what makes it tick.

For the renascence from the Dark Ages of biology, the world is indebted to one Andreas Vesalius (1514-1564), born

in Brussels, schooled at Louvain, trained in medicine at
Paris, and in 1543 ocupying a chair of surgery and anatomy
in the University of Padua. In that year Vesalius published
a book entitled *The Structure of the Human Body*, a book
noteworthy not because of any revolutionary discoveries described therein, but because it dared to question the authority of Galen, if only in a modest way. There appeared
in its pages the spirit of inquiry, for the first time, so far
as the Christian world was concerned, in fourteen hundred
years. It is necessary to embed oneself in medieval dogma
and tradition, to imagine oneself living in a world of venerable custom and belief and threatened constantly by the
fearful authority of the Church—in short, it is necessary to
eliminate from one's point of view everything that is significant in modern thought—to appreciate the intellectual effort,
the courage, required to say, "I doubt...." Vesalius had
many influential and capable students, and somehow he impressed upon them the value of appeal to observation. It is
significant that he never made a second Galen of himself; his
pupils, although they thought him greater than Galen or
Hippocrates, held him, together with these authorities, subject always to confirmation.

It so happened that when, in the seventeenth century, men
returned to the path of inquiry, their interest was centered
around the circulation and respiration rather than around
the problem of generation. A long road was to be traversed
towards a fuller knowledge of the breath of life before the
subject of life's origin was to be substantially advanced. It
will be advantageous to follow, for the moment, this historical sequence.

THE CIRCULATION OF THE BLOOD

The connection between food and life had been befud-
dled at the outset by the sheer efficiency of the body. Dry
fodder, when it burned on the hearth, was consumed in red
and yellow flames and surmounted by a visible plume of
smoke; but that same fodder, burning in the body, emitted

only a comfortable warmth regulated to a surprising constancy; there was visible neither flame nor smoke to attest the fundamental identity of the two fires. By what possible stretch of the imagination could blood, the warmth of the body, the motions of an animal, be linked with elemental things like earth, fire, air, and water? Animal heat and animal motion were looked upon from the first as incomprehensible and unique manifestations of life, without connection or parallel with the phenomena observed among non-living things. The terms "organic" and "inorganic" came to designate two fundamentally different kingdoms which in some mysterious way were transiently intermingled in this world. The very words *heart, blood, breath,* and *body* acquired a mystical significance.

The earliest Greek philosophers had had a thoroughly monistic view of things; but this was ruptured by the idealist Plato, and all but destroyed by Aristotle when he invented his entelechy to divide the world into two fundamentally independent parts; and between Galen and the Church this dichotomy of Nature was saddled upon men's thoughts so firmly that it still holds the reins today.

Galen had taught that food, after absorption from the alimentary canal, is converted into blood by the liver, and at the same time endued with certain very special nutritive properties which were summed up in the phrase "natural spirits." This blood still remains in a degree crude, however, and unfitted for the higher purposes of the body; so from the liver it is carried to the right ventricle of the heart, whence some of it passes through innumerable but invisible pores of the cardiac septum into the left ventricle. Here, mixing with air that has been drawn into the dilated heart from the lungs, the blood becomes warmed by heat innately furnished by the heart—heat placed there by God in the beginning of life and remaining until death—and at the same time laden with "vital spirits" and so fitted for its higher duties. In this mixing process the air served to temper the heat of the heart and to prevent it from becoming excessive.

When the heart contracts, the crude blood on the right side, endued merely with "natural spirits," surges out through the veins to nurture the body; while from the left side there surges blood endued with "vital spirits" to enable the several tissues to exercise their vital functions. Some of this left-heart blood, on reaching the brain, there generates "animal spirits," which escape from the fluid and are carried along the nerves to bring about sense and movement in the body.

As Galen taught, so men still believed in Vesalius' time, and so even Vesalius wrote, with his tongue in his cheek; not because he thought that it was "in all cases consonant with the truth, but because in such a great work he hesitated to lay down his opinions, and did not dare to swerve a nail's breadth from the doctrines of the Prince of Medicine." The implication was enough, however; for above all else Vesalius taught that inquiry was the one and proper course. His student Fallopius transmitted this precept to Fabricius, who in turn passed it on to a young Englishman who had recently taken a degree in arts at Gonville and Gaius College, Cambridge. When he received his doctorate of medicine at Padua, this young Englishman, William Harvey (1578-1657), returned to Cambridge. For a period of twenty years he practised as a physician and lectured on anatomy at St. Bartholomew's Hospital. The story is told that as a physician to King Charles I, while his sovereign's army fought at Edgehill, he sat under a hedge reading a book on the generation of men— a subject more interesting to him than their destruction. Be that as it may, Harvey is called the founder of modern physiology because he appealed, as Vesalius had done, not to authority, general principles, and analogies, but to observations and reason.

When Harvey first began to ponder the subject of the circulation, the motions and uses of the heart seemed so full of difficulties that they appeared, he said, only "to be comprehended by God." But years of thought and careful observation opened his vision to a simpler explanation. The blood does not ebb and flow in both the veins and the arteries with the dilatation and contraction of the heart, but circulates in a continuous cycle from one to the other. Beginning with the left side of the heart, it is expelled into the aorta and distributed by the arteries to all parts of the body, whence it returns by the veins to the right side of the heart; from here, through the lesser circulation of the lungs, it is driven back into the left side of the heart and thence into the aorta again.

Recall that according to the Galenic doctrine the food, after absorption from the alimentary tract, was endued by the liver with "natural spirits," and by the heart with "vital spirits"; because these "spirits" could not be seen or felt, or accurately defined, they were all the more important in the thought of the Middle Ages. However abstract and vague they might be, they were not mere words to be bandied about with scepticism, but integral and necessary parts of the living body. In contrast to all this, Harvey's description of the circulation was essentially a physical and mechanical one. He cast aside as of no importance to the immediate problem the question of whether the heart, besides propelling the blood, added something to it—heat, spirit, perfection, or what-not; this question, he said, must be inquired into later and be decided upon other grounds, and never again did he refer to it. But his demonstration was a blow to "natural" and "vital spirits," for these were based upon the supposed double supply of blood to all parts of the body and were irreconcilable with a circulatory system where the blood courses repeatedly from arteries to veins and back to arteries again. Even though the blood might undergo changes in composition at different stages of its journey, there was no room left for the hierarchy of spirits born of the old and erroneous physiology.

When one error is found in a thesis, it tends to contaminate the whole; with Harvey's discovery the entire Galenic doctrine was infected with decay. The most notable consequence of this new physiology was, however, that the virtue had been extracted from the exalted written word, and time-honoured and revered phrases were no longer a hindrance to inquiry as to what took place in an organ when it entered into activity, or what were the relations between that organ and the rest of the body. At the same time the dichotomy of matter versus spirits receded a little before the advance of science.

Harvey's work was not complete: it was not clear to him how the blood passed from the arteries to the veins, for no channels were visible. But this difficulty was not an irrational one. After all, Galen had taught that there were invisible pores in the septum of the heart through which the blood might pass, and no one was constrained to believe only what could be seen with the naked eye. But the gap was not long in filling; for a few years after Harvey's death, Marcello Malpighi (1628-1694), with the aid of a microscope, described the hitherto invisible blood-vessels of the capillary circulation. Malpighi is perhaps next to Harvey in significance to us here, because he left, or at least put, so little error into what he touched; and he touched many fields of biology. He did more in morphology, embryology, and comparative anatomy than any other man since Galen. At the close of his life, and in no small measure as the result of his labors, the human body had come to be viewed as a collection of organs, each having an individual, though perhaps little understood, function and a characteristic structure. Physiology, after Harvey and Malpighi, was pointed definitely in the direction of the analysis of the structure, function, and interrelationships of the various organs of the body.

THE BREATH OF LIFE

With this change in viewpoint the breath of life began to succumb to investigation. If Vesalius thought at all about the alchemy of his time, he probably eschewed that knowledge as being a perverted, senseless effort to discover an elixir of life or to gain incalculable riches by turning dross into

gold. But he certainly must have heard of one of its prac-
tioners, only twenty-one years older than himself, who was
stirring the world with his idle and boasting chatter. Paracelsus (Aureolus Theophrastus Bombastus von Hohenheim
to his parents) was the son of a learned physician and possessed a doctor's degree from Ferrara; yet he had a taste for
low company, and most of his education was obtained from
the astrologers, barbers, executioners, bath-keepers, gypsies,
and midwives whom he met in his unceasing travels en-
compassing the civilized world from Zurich to Samarcand.
He practised a mixture of medicine and alchemy; although
generous to the poor, sympathetic to the sick, and capable
in a measure as physician and surgeon, he was at the same
time so conceited, truculent, and forward that he made enemies on every hand.

Paracelsus knew that, as the old Arabian alchemists had
discovered, matter was of three kinds: (1) sulphur, which is
combustible and which is destroyed during combustion;
(2) mercury, which temporarily disappears or is volatilized
by heat during burning and may therefore be condensed
as a liquid; and (3) salt, which is fixed, or the ash
which remains after burning. But to Paracelsus this tri-
partite physical universe was also full of mystic things: there
were microcosms (bodies) within macrocosms (worlds), each
enlivened with gnomes, sylvans, sprites, and salamanders;
and the phenomena of life were to be attributed to a vital
force called the *archeus* presiding over a hierarchy of minor
archei. All physiological processes were alchemical but governed by the *archei*—in health, rightly, in disease, wrongly.
Death was the loss of the greater *archeus*. For every ill of the
body there was a potent plant or animal concoction that,
by opposition between antagonistic mystical forces, could
effect a cure. This alchemical principle was so potent that
Paracelsus named diseases by the drugs that cured them.
His alchemical chatter was fortunate, for it made the physicians of his day aware of the actions of plants and chemicals
upon the body; but this gain was offset by the fact that it

reinfected the newly awakened scientific thought with the virus of medieval spirits and mysticism.

Passing over a hundred years of vague, newly growing sciences, we note that by Harvey's time Kepler had described the orbits of the heavenly planets, and Galileo had invented the telescope and was bringing the phenomena of motion "down to earth by way of his inclined plane." Astrology had mothered the science of physics; alchemy, no less, had given birth to chemistry; and there were men who wanted to put these new tools to work on the problems of the human body. One of these, Jan Baptista van Helmont (1577-1644), found the lectures of his university teachers dry and unsatisfying because he had a touch of the same temper that had affected Paracelsus. So he deserted his chosen field of botany, took a course in medicine, traveled widely, married an heiress, and settled down to chemical observations and experiments with a light and charitable medical practice on the side. He was in truth a modernized Paracelsus; he had the latter's mystic impulses and propounded equally fantastic schemes wherein invisible supernatural forces ruled the world. For the greater and lesser *archei* he substituted *blas,* by which he meant a hierarchy of forces governing the heavens and the body; minor *blas* presided over muscular movements, the nerves, chemical processes, and other manifestations of living things.

Though in the matter of his *blas* Van Helmont did not advance beyond the Paracelsus he adored, he made a real step forward in another direction. To describe an invisible and seemingly imponderable substance in the air he coined the word *gas;* and in his notion of this medium and its relation to life he was not only the first of the modern chemists, but the first of the chemical physiologists. He chose the word *gas* because it sounded to him like "chaos" and because his imponderable substance had much in common with the latter. Actually, what Van Helmont meant by *gas* is now recognized to be a specific compound, namely, carbonic-acid gas or carbon dioxide; for he did not suspect that there were several different kinds of gases. He differed from Paracelsus in regard to the composition of matter; for he said that there were but two elements, air and water, and he performed many experiments to show that neither of these can ever be changed into the other. He thought that his *gas* was a form of water. Some of his experiments were so direct and simple as to be quite convincing. For example, he planted a young willow which weighed 16 pounds in 200 pounds of dry earth and watered it for a period of five years. The vessel containing the earth was carefully covered so that no dust could accumulate in it. As one might expect, at the end of five years the earth still weighed 200 pounds, but the willow had gained in weight 164 pounds. Obviously, said Van Helmont, 164 pounds of wood, bark, and roots had been elaborated out of water. His conclusion was, of course, erroneous; but one thinks he would have been delighted to learn that the willow had not been elaborated exclusively from water, but in part from his *gas*, carbon dioxide. He did, however, demonstrate the liberation of carbon dioxide when wood and gunpowder were burned, and also during the fermentation of grapes and malt. He was so impressed by the latter process that he was led to search for it at the heart of everything. Paracelsus, no doubt, had had frequent recourse to naturally fermented spirits while he was cogitating upon the supernatural kind, but Van Helmont used the bubbling, frothy process of fermentation as an actual stepping-stone to higher things. He put "ferments" to work at every opportunity, from digestion and muscular contraction to reproduction and growth—always, of course, under the direction of his *blas*.

Van Helmont made signal contributions to physiology by emphasizing the application of chemistry to this subject, by drawing attention to the process of fermentation, and by the discovery of a specific form of matter, gas. But he was, on the whole, more intrigued by his *blas* than by anything else. After all, two thousand years of philosophical and theological speculations about death had shown him its supreme im-

portance. He speculated much on the relations of the *blas*
to what he called the "sensitive and motive soul." This sensitive soul belonged to man alone, being absent in plants and
beasts; it was the master that reigned over the processes of
the body; it was the prime agent of all acts—the *blas* being its
servant, as were also the minor *blas* or *archei*, while the ferments were but the instruments of these. The motive soul
was the source of vitality and the means by which the heart
vitalized the blood, and it carried out the sensations and
movements of the body by means of the nerves. The actual
location of the motive soul was in the pylorus, the posterior
orifice of the stomach. This was evident from the fact that
great emotion is always felt at the pit of the stomach, and
that a man may have his head blown off by a cannon-ball and
yet his heart will go on beating for some time, whereas a
severe blow at the pit of the stomach can stop both his heart
and consciousness at once.[1] The soul's exact relations to the
stomach were peculiar, not as something encased in a sack
or concealed within a skin, but as the light was present in
a candle-flame. The soul was mortal, but in the living state
it was coexistent with man's immortal mind; it was the
medium through which the immortal mind directed the
major *blas,* minor *blas,* ferments, etc., to vital ends. Before
the Fall man possessed only an immortal mind which acted
directly on the *blas,* and, this being the case, he was immortal, too; but at the Fall, God introduced into man the
sensitive soul, and until death the immortal mind retired
into the sensitive soul and became, as it were, its kernel.

Thus Van Helmont tried to reconcile his chemistry with
the teachings of the Church, which had occasion to rebuke
him several times for heresy; and thus do man's speculations
about death get entangled with—and entangle—his speculations about life. Van Helmont was a great man whose opinions carried weight not only with his scientific colleagues but
with the interested laity; but in retrospect we can see that

[1] The modern interpretation of this latter fact is that the delicate nerve
ganglia of the solar plexus are injured and result in circulatory collapse.

his speculations had a more pronounced effect upon the latter than upon those who were in a position to evaluate his theories critically.

A contemporary of Van Helmont, Sanctorius Sanctorius (1561-1636), made the first rational approach to the problems of metabolism by weighing himself before and after a meal and demonstrating that he gained in weight by the amount that he had eaten. Following him, the problems of physiology were further freed from spiritistic subtleties by the teachings of Franciscus Sylvius (1614-1672), who contended that digestion, respiration, and all other processes in the body could be explained exclusively in terms of the newly born chemical science. But after Sylvius speculation went to the other extreme in the theories of Giovanni Borelli (1608-1679), that all animal movements, both external, such as movements of the legs, and internal, such as movements of the heart and blood, could be explained by the mathematics of the forces and shapes of the gross parts involved.

Had Sylvius and Borelli, and the schools of thought they started, been more easily reconciled with each other, perhaps physiology would have traveled more rapidly along a profitable analytic course; but at the opening of the eighteenth century the pendulum swung back to the mystical side again. Noteworthy in respect to error was Georg Stahl's (1660-1734) teachings of phlogiston as the essential principle of heat and soul as the essential principle of life. Stahl's theory of phlogiston was an attempt to explain the combustion in terms of a ponderable substance emitted during burning— the burned substance becoming poorer, and the air correspondingly richer, in phlogiston. Many years later, as an earlier chapter has shown, this theory was to prove a severe handicap to the progress of chemistry, as well as of physiology, before it was finally exploded. It is more difficult, on the other hand, to evaluate the consequences of Stahl's theory of the soul. This soul, unlike Van Helmont's, did not need *archei* or *blas* to manipulate the chemical processes of the body; it was not a mortal something associated in an indescribable way in the stomach with an immortal mind, but it of itself was an immortal principle, spiritual and immaterial, coming from afar and at death of the body returning into the great unknown. It occupied the body and directly motivated it, and consequently vital activities such as the movement of the limbs, of the heart and blood, bore no real resemblance to the motions in dead matter, which occurred without direct use or aim. Stahl asserted that only one link was necessary to connect the soul with matter and the body, and that link was motion. The events of the body might be rough-hewn by physical and chemical forces, but the soul, through motion, could always shape them to its ends.[2]

On the whole, however, the majority of Stahl's contemporaries were not content to accept an easy vitalistic answer to their questions concerning life, and they proceeded therefore with chemical and physiological research. In 1660 Robert Boyle, working with a pneumatic pump, demonstrated that animals shut up in a vacuum succumbed immediately, even as the candle-flame went out; but if the air were instantly renewed, both could be restored again. Thus in this one experiment it was seen that air was necessary for the maintenance of life and that respiration had something in common with the burning of the candle. Shortly afterwards Robert Hooke showed that an animal could be kept alive by artificial respiration after the chest had been opened and the heart exposed, thus exploding the theory that the mechanical movements of breathing were essential to life.

In another direction Richard Lower (1631-1691) demonstrated that the difference in color between red arterial and blue venous blood was a matter of aëration, and starting from this fact he performed experiments on the transfusion of blood from one animal to another. Although false hopes were raised by his results, his experiment affords evidence of how men's outlook had changed in a generation since Harvey's work was published. It was a common view of the

[2] There is a recurrence of a theme akin to Stahl's in modern subatomic physics.

time that air was an elementary substance, and a whole new avenue of approach was opened up when John Mayow (1645-1679) showed that it was not the whole of air that was necessary for the maintenance of life, but a special part of it. This special, vitally important part he identified with "nitre" (nitrate), an important constituent of gunpowder.[3] Nitre was, of course, a substance of great interest to both chemists and statesmen, and Mayow's emphasis on its rôle in living processes led one commentator to remark that it was capable of making as much noise in philosophy as in war. Mayow's error is easily understood: sulphur, he said, takes something essential for burning out of the air (as does the candle), and yet, when mixed with nitre (as in gunpowder), it burns in the absence of air, or even under water. Therefore, the essential "something" must be nitre.[4]

In spite of his erroneous interpretation, Mayow made more progress towards revealing the nature of the breath of life than anyone had before him. He performed his experiments in a quantitative manner and showed that the elastic force (volume) of air was decreased either by the burning of a flame or by the breathing of a mouse; and he inferred correctly that both deprived the air of particles of the same kind. He concluded that the passage of air into and out of the lungs was an entirely mechanical process, and, casting ridicule on the Galenic theory that breathing serves to cool the heart, he concluded that his "nitre" from the air mixed with sulphur in the blood and excited the latter to some needed "fermentation." Increased muscular exercise was accompanied by increased breathing to supply the necessary quantity of "nitre" to the blood, and because of the resulting increased combustion there was produced an increased amount of heat. Thus Mayow defined in current terms one side of the respiratory cycle, while many of his colleagues

[3] This was a false start, and the true identity of oxygen was not made known until Lavoisier's time.

[4] Had he said *must be present* in both nitre and air, as indeed oxygen is, he would have been more nearly right.

were still mulling around in the spiritistic theories of Stahl, Van Helmont, and Paracelsus.

But Mayow's success was a spark destined to subside into darkness; for the problem was not further advanced until nearly a hundred years later, when Joseph Black (1728-1799) rediscovered carbon dioxide and gave it a correct identification. He showed that this gas—which he called "fixed air," but which in reality was identical with Van Helmont's *gas* (minus that investigator's *blas*)—could be liberated from unburned or mild lime ($CaCO_3$) by either acids or heat; when mixed with caustic lime ($Ca(OH)_2$), this "fixed air" recombined to form mild lime again. Out of this latter fact he devised a simple test for the detection of "fixed air," consisting of the precipitation of calcium carbonate from a solution of calcium hydroxide; and he was able to prove that "fixed air" was given off by fermenting wine, by burning charcoal, and by living animals. He failed, however, to tie his "fixed air" up with that constituent of the atmosphere which supports combustion (oxygen or Mayow's "nitre"), primarily because he interpreted combustion in terms of Stahl's erroneous phlogiston theory.

Shortly afterwards, Joseph Priestley (1733-1804) hit upon a method of restoring vitiated or irrespirable air and at the same time discovered at least one of the methods that Nature uses for this same purpose. He found that air in which a candle had been allowed to burn itself out was so modified in a few days by a sprig of mint that it would no longer injure either a candle-flame or a mouse. The respirable constituent of the air which was furnished by the living plant (that is, oxygen) was also yielded in concentrated form by mercuric oxide when calcined by means of a burning-glass. Here Priestley had the experimental evidence immediately at hand for the discovery of oxygen and for revealing the connection of oxygen with Black's "fixed air"; but he failed because, like Black, he was thinking in terms of phlogiston.

Priestley's theory was such that the whole cycle of combustion, during which a fuel combines with oxygen to form carbon dioxide, could be explained without the oxygen. Air supported combustion, Priestley said, because it "took up" the phlogiston given out by a burning body, and the air itself thereby became phlogisticated and unable to support combustion any longer. New air liberated from mercuric oxide was wholly dephlogisticated. Animals whose bodies abound in phlogiston, introduced by way of combustible food, give out phlogiston so long as the atmosphere they breathe contains enough dephlogisticated air to absorb it; and when this dephlogisticated air becomes saturated with phlogiston and can receive no more, the atmosphere then ceases to be respirable. Plants, on the other hand, absorb phlogiston and thus render phlogisticated or irrespirable air dephlogisticated or respirable again. Black's "fixed air" was in this theory a decomposition product resulting from the action of phlogiston upon common air. Priestley's account of respiration in terms of phlogiston is the classic example of how a series of facts can be explained by a theory which is a complete inversion of the truth.

It remained for Antoine Lavoisier (1743-1794) at the close of the eighteenth century to turn the explanation right side up. Examining the reaction of metals in air, Lavoisier found that in this process the metal gained rather than lost in weight, a fact contrary to the phlogiston theory; and this gain in weight he attributed to the combination of the metal with a gas which he called oxygen. By his experiments it became clear that calcination was a union of the metal with this gas, accompanied by the liberation of heat rather than by the emission of a ponderable substance, phlogiston; and that respiration in the living body was a similar process, differing only in the fact that there was given out at the same time Black's "fixed air," which precipitated lime-water. The third, preponderant and relatively inert gas in the atmosphere, the one that did not enter directly into this cycle, Lavoisier called *azote*.[5] After Henry Cavendish had

[5] This was later given its English name, nitrogen, because nitre was obtained from it accompanying the discharge of an electric spark.

isolated hydrogen, Lavoisier showed that the fuel burned in the body consists of carbon and hydrogen, and that in the burning of this fuel, oxygen (Mayow's gaseous "nitre") combines with the carbon and hydrogen to form carbon dioxide (Black's "fixed air") and water, with the simultaneous liberation of heat.

Joining forces with the astronomer Laplace, Lavoisier devised methods for measuring heat-production and applied them with great skill to both the candle-flame and the living animal. The results were close enough to establish what has since been proved correct to the third significant figure: animal respiration is a combustion, slow it is true, but otherwise perfectly similar to the combustion of the same material outside the body. A fuel reacting with oxygen is degraded to carbon dioxide and water with the concurrent liberation of an exact and predictable quantity of heat. Air becomes vitiated by an animal or by a candle-flame because the oxygen is used up and carbon dioxide is given off; plants restore the air to a respirable condition because in the sunlight they absorb the carbon dioxide and incorporate it in their new tissues, while at the same time they give off oxygen to the air. The heat that is liberated when the fodder burns is stored sunlight. There is no essential difference between the burning of the fodder on the hearth or in the body, except in respect to the rate and to the ways in which the liberated heat, or energy, can be used.

Lavoisier erred in supposing that the combustion of fuel occurs only in the lungs (it was not until after his head fell under the guillotine of the Republic that oxidation was demonstrated to occur in all tissues of the body), but essentially by his work the double mystery of the breath of life and of animal heat was brought at last to a solution. "It took but a moment to cut off that head," said Lagrange, "although a hundred years, perhaps, will be required to produce another like it." One hundred and forty years have elapsed, in fact, and the sum total of progress in this subject in the intervening time must be compared questionably, in its significance for science and philosophy, with Lavoisier's contribution.

THE REPRODUCTION OF LIFE

From this summary of progress concerning the breath of life we may now turn back to the story of life's origins. Throughout the Middle Ages the subject of generation remained in a state of dismal ignorance, medieval illustrations showing the unborn infant as a miniature adult reclining

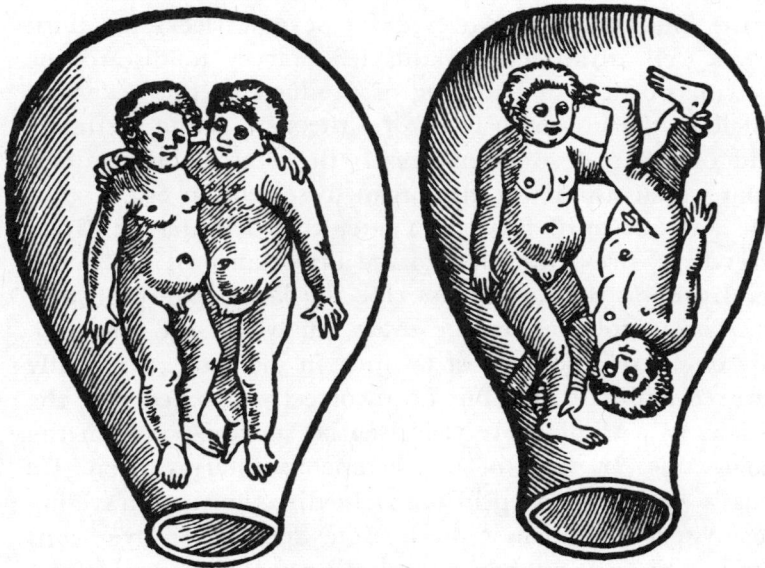

THE UNBORN INFANT AS A MINIATURE ADULT
Representations of twins in the uterus in the earliest treatise on obstetrics, Eucharius Röslin, *Der Schwangerenn Frawen* (Augsburg, 1530).

comfortably on its elbow, standing in heroic posture, or affectionately embracing a twin. Harvey was perhaps the first to resume in this field the experimental method that had to come to an end in Galen's hands. Harvey was physician to Charles I, a renowned sportsman whose royal preserves harbored many deer. Knowing that the does will accept the buck only during a four-weeks period in the early fall, Harvey saw in this fact a means of approaching, by what was essentially a statistical method, the mystery of generation. From the king he obtained twelve does during the mating-season and placed them in captivity. Some of these he sacrificed immediately and some after varying intervals of time, intent upon determining what evidence there might be found in the uterus regarding the formation of the embryo. The experiment, in its intent and plan, was the most noteworthy since Aristotle, but the time was not yet ripe for a correct interpretation of the results. No evidence of the embryo was discovered for several weeks, or until it was well advanced, and this led Harvey to discard the doctrine of an actual mixing of products of the male and female in the uterus in favor of a theory that the spermatic fluid of the male does not reach the uterus, but emits a vapor which stimulates that organ to secrete an egg.

It was not until fifty years later that Regnier de Graaf (1641-1673) drew attention to the importance of the ovary. De Graaf was the first to describe the large watery blisters, or Graafian follicles, in the ovary (in which the true egg, only one two-hundredth of an inch in diameter, is actually matured and cast off); but he incorrectly assumed that the follicles as a whole were the essential starting-point on the female side. In spite of this erroneous interpretation, De Graaf's work was a step in the right direction in unraveling the mystery of human reproduction. Speculation ever runs ahead of observation, however, and the problem at this point took a far leap from the proper course. A philosopher, Nicolas de Malebranche (1638-1715), expanded De Graaf's observations into an engaging theroy of *emboîtement* or encasement: he premised that all the ova, to create unborn generations until the end of man, had been carefully enclosed one within the other in the ovary of Eve, like a series of oriental boxes. Each female after Eve had necessarily one less egg, and the mathematicians even went so far as to calculate that after 200,000,000 generations the supply would be exhausted and the race extinct. The argument touched

the popular imagination, and once again science met the
approbation of the Church; for reproduction by encasement
traced life back to the original creation and—what was perhaps more important—offered a biological proof of the doctrine of original sin. Thus between the Church and the
philosophers the egg became all at once tremendously significant, and the male semen correspondingly insignificant.

Here the matter might have remained had not new discoveries with the microscope confounded the advocates of
the ovist theory. What the telescope did for astronomy, the
microscope did for biology; it revealed an invisible world
scarcely suspected by men. The early history of the microscope is obscure, but many contemporaries of De Graaf were
peering through it at all sorts of things—decaying matter,
pus, blood of patients dead of plague, plants, bees, snails,
frogs, and other animals. One of these peerers-through-the-
microscope, Antonius van Leeuwenhoek (1632-1723), scion
of wealthy brewers, was famous for having the largest collection of magnifying-glasses in the world, most of which
he had made himself. One day, with the aid of a powerful
glass, Van Leeuwenhoek discovered in human semen countless animalcules, extraordinarily small and wormlike creatures swimming about with vigorous motions of the tail.
That these "worms" were important for the process of
reproduction struck Van Leeuwenhoek at once, and—the
pendulum ever swinging too far—he forthwith denied the
importance and even the existence of the human egg, asserting that the sperm alone was the father of the child, the
mother being but the receptacle, the incubator, and the
nurse.

Thus another socially important doctrine, the doctrine
of the innate superiority of the male, found itself unex-
pectedly confirmed by these new discoveries. In olden days
the Egyptians and the Greeks had considered the woman
to be the nutrient soil that gave the embryo, a stranger to
her blood and personality, its growth; and in the teachings
of the Church a subtle distinction persisting from Hebraic

lore had been drawn between the sexes. The idea was even incorporated into English law in the century before Van Leeuwenhoek's discovery, when a woman of the royal family had been adjudged not akin to her own son. Now it required only a little spherical aberration in the lenses and imaginations of Van Leeuwenhoek's contemporaries to see a human visage in the head of the newly discovered spermatozoön and a closely compressed body in the tail. And, men said, within the testis of each homunculus there was another preformed homunculus, and in the testis of this, another, and so *ad infinitum*. Father Adam was the beginning of the series, and Mother Eve merely furnished a shelter by way of her primal series of eggs in which each preformed homunculus might develop in its time. To get the homunculus into the egg the latter had to be furnished with an hypothetical trap-door; this the homunculus was supposed to close behind it after it had entered, thus condemning its less fortunate brethren to death outside. How the theologists explained the fate of the innumerable homunculi who were unable to find an egg, history does not record.

The ovists immediately tried to turn the tables by asserting that these spermatozoa were useless parasitic worms such as might be found in any stagnant pool or bit of decaying matter. Resemblance of the child to the father they explained as due merely to prenatal impressions, while resemblance to the mother was due to the fact that she had nourished the embryo in her womb. The argument was further complicated by debate on whether each embryo was built *de novo* or whether it was really preformed in miniature within the egg or the spermatozoön as the case might be—a perfect being needing only to unfold and grow. The controversy found no end until the nineteenth century; it divided literature, science, philosophy, and casual intercourse into bitter schools. The compound microscope had to be greatly improved before Karl von Baer (1792-1876) could discover the minute true egg of the dog in the Graafian follicle, roughly describe its structure, and refute the theory of preformation. And men had to learn that all living tissues are made up of cells and that there is one universal principle governing the development of all animals and plants—namely, the formation of new cells by cell-division —before Albert von Kölliker (1817-1905) could establish that the spermatozoa were true body cells and that it was these which were essential for fertilization, and not the seminal fluid in which they swam.

The investigation of the reproductive process was further accelerated by the introduction of chemical methods of staining which brought into view the minute parts of the cell, and attention was soon focused on the cell nucleus. While the cell doctrine—that all cells come from preëxisting cells—had initially aided the development of a correct theory of reproduction, reproduction was now seen to be but a specialized aspect of the universal process by which all cells are formed, by which all living things grow or effect repair of injured parts. Both growth and reproduction were traced to chromosomes—minute irregular threads in the nucleus; and here too was found the key to other mysteries. Not the least of these was the problem that had so perplexed the ancients: why does the child resemble the father in some features and the mother in others? The answer to this question was first roughed out by an Augustinian priest behind the white walls of a monastery on the emerald slopes of a Moravian mountain. In a garden given over to the breeding of columbines, snapdragons, hawkweed, as well as more lowly pumpkins and peas—to all of which he referred as "my children"—Friar Gregor Johann Mendel (1822-1824), cross-fertilizing his plants by hand to prevent accidental pollination, formulated the laws of inheritance. His work remained unappreciated until thirty years ago, when investigators began making observations upon the nuclear constituents of the cell; but so advanced were the simple principles by which he described the perplexing problems of heredity that they could be superimposed upon the newly developing science of cytology, as a shadow upon the substantial pattern that casts it. It was not long before the determinants of inheritable characteristics were shown to be located in the nuclear chromosomes. The nebulous problems of similarity and difference between parents and children, which had previously been assigned to nurture, prenatal impressions, or even to more ephemeral causes, were thus placed upon a cellular basis. C. E. McClung discovered that sex itself (primeval breeder of trouble, bliss, and philosophy!) was determined by the chance presence or absence of a special chromosome; and Jacques Loeb (1859-1924) showed that the egg, at least, was self-sufficient for developing into a normal animal after artificial activation, and without the intervention of the spermatozoön (parthenogenesis). It was, however, no longer possible for startling discoveries like McClung's and Loeb's to misdirect men's ideas into fantastic errors; for it was now recognized that, in the natural process of fertilization, both the egg and the spermatozoön contribute equally to the determinative chromosomes, and therefore equally to the essential nature of the child.

In the current interpretation of biology, the inheritable, permanent features of life are determined by foci in the chromosomes called "genes." These are distributed along the chromosomes like beads along a thread; from their central position within the cell they rule the growth, formation, and metabolism of the body with a magic which would tax the credulity of any medievalist. The paternal and maternal chromosomes meet at fertilization, shuffle about, exchange genes, separate again; and the cell subsequently divides, producing successive generations of cells, successive generations of men. No vitalistic theory has as yet identified itself with this weird chromosomal dance—perhaps because the genes are with rare exceptions strangely steadfast and immutable and because these particular qualities do not invite a mystical interpretation. But there is much in the joint problems of generation and growth that is still unexplained—the reproduction of the genes themselves, the means by which they exert their determinative action upon the surrounding cell, the cause and nature of their rare but significant mutations. An open road for both sound inquiry and hazardous speculation lies ahead.

The history of man's understanding of his own reproduction has been a devious one, marked by ludicrous errors and frequently accompanied by extravagant flights of the imagination. But the wasted energy that he has put into his endeavors to understand how he begets himself is nothing compared with the wasted energy that Nature puts into the reproductive process itself. Probably not more than 1 per cent of the ova from the human ovary are ever fertilized; and at each sexual union more than 200,000,000 spermatozoa, each presumably a perfect potential being, are wasted, while only one achieves that goal which we may call personality. It is needless to multiply these figures to grasp the fact that the fruition of an adult life from the impersonal tree of the germ-plasm is a very hazardous affair.

THE MEANING OF LIFE

There is much in economics, sociology, religion, and philosophy that, resting upon ancient and worn-out foundations, now trembles in consequence of being undermined by these new facts. How will men's views be affected by the new knowledge of biology? The stars, the earth, the living body, the blood, the breath of life, the very pattern of character and personality, have been reduced to terms of atoms and molecules in a state of continuous interchange and complex interaction. What, then, is to become of that most cherished of our illusions, the sense that we are enduring personal entities? Must individuality, like all else, be reduced to atomistic fluxionary terms? Or will some sudden turn in the path of discovery reconcile the atomistic nature of life with its unitary sense of being?

The pendulum ever swings too far, and it is asserted by many that modern physiology, with its explanation of life in exclusive terms of atomism, is in error. From a personal point of view the picture is not an attractive one. The world of Galen and the medieval philosophers—a well-planned world charged with objective meaning and beauty and having a special purposive relation to every man, to his hopes and fears and destiny—has disappeared, to be replaced by a world of material particles moving in a fortuitous concourse through space and time. By accepting the relative, rather than the absolute, significance of all motion, we do not in the slightest measure restore the old-fashioned esthetic and anthropocentric elements to the universe that once were—and that still are—so precious to the human heart. Living organisms with their individualities, with everything that is so unique about them, appear to have been evolved out of a chaos of atoms no one of which is inherently distinguishable from any other. Objectively viewed, the identity, the feelings, and the purposefulness of all living things, including man, are but transient patterns in this dance of atoms and are destined to a blind and passionless decay.

This stark picture of modern atomism is not so different from that which Lucretius had of life and death, which in the paraphrase of Mallock reads:

> No single thing abides; but all things flow.
> Fragment to fragment clings—the things thus grow
> Until we know and name them. By degrees
> They melt, and are no more the things we know.

The outstanding, distinctive feature of all living organisms is the inherent impulse, tendency, organization, or what you will, to perpetuate their own structural and functional integrity. This innate character is self-evident on the lowermost plane of growth and reproduction; and it is not surprising to discover it also at the uppermost plane of sophisticated philosophical speculation. Some scientists, impelled by an egoistic desire to perpetuate themselves and guided by those vestiges of ancient and medieval cosmology which linger in modern thought, still seek to discover the metaphysical somewhere along the borderlands of knowledge. They seek to discover not only a means by which personality can escape from the impersonal substance of physical Nature, but also from death. These are the neo-supernaturalists. A favorite field for speculation is in the problem of mind and matter, to use the dualistic expression so firmly intrenched in the vulgate parlance by the theological doctrine of an immaterial soul, spirit, ego, temporarily inhabiting a material body. To many people the problems presented by consciousness and its relation to matter appear as the motions of the heart appeared to Harvey —"only to be comprehended by God." It is not enough for them to view consciousness as an evanescent flame of feeling, easily extinguished by any adverse wind—a blood-clot, a moment of asphyxia, a whiff of chloroform; instead, they would spin from it the very stuff of atoms and nebulae, so that the universe becomes a Universal Mind in which our lives are but passing thoughts. Others, more convinced of the external reality of atoms and nebulae, would see us as quite atomistic creatures aspiring and acting by virtue of a Supernal Impulse switching an electron in its quantum course. For these it is not education, memory, foresight—the integration of all our sensory experience—that guides our actions, but free will; it is the Supernatural that leads us to choose between the moral and the immoral, and that even makes for us the final choice between a blonde and a brunette.

In this usually bizarre and frequently tragic world it is sometimes harder to gain to what Thomas Huxley called "an honest agnosticism" than to adhere to a naïve faith. When we watch man speculating about himself, we all but lose hope that his critical capacity will ever get the upper hand over his naturally wishful, egotistic thoughts. But it is significant that the scientists who are in a position to speak authoritatively about this creature *Homo sapiens*—the biochemists, physiologists, psychologists, and psychiatrists—furnish few recruits to neo-supernaturalism. Most of these students of the living body conduct both their lives and their investigations in accordance with the principle that only physical causes issue into physical effects—in mice and men, if not among the electrons. They have read in the history of human error that it is this principle which has pointed the path from the first inquirer down to Harvey, Lavoisier, Mendel, Loeb; for them it still points a path leading wide and straight to the utmost limits of their vision, across the Unknown they must yet traverse.

Chapter VIII

Error in Neurophysiology

By C. Judson Herrick

NEUROPHYSIOLOGY *occupies a unique place in the circle of the sciences as the point of convergence of the physical and the psychical. It raises the comprehensive problem of man's place in Nature: whether the rich subjective psychic life of man can be brought within the scheme of interpretation applicable to the other sciences, and whether our subjective experience, as the climax of evolution, can be articulated with vital organs and vital processes on a naturalistic plane, and if so on what assumptions.*

Systematic speculations upon the relations of mind and body go back to Greek philosophy, and paralleling them are theological doctrines relating to the soul, including creation myths and the entire range of primitive philosophies. Our survey may be limited to attempted solutions in the line of ancestry to the scientific views of today. A varied panorama of error appears. On this vital issue conclusions were reached upon what knowledge was available; they could not await its completion.

Considering how crude were the methods of dissection—and these utilized only by pioneering minds—it is not surprising that early observers should have been bewildered by the curiously arranged masses within the skull and spinal canal, and should have proposed interpretations bizarre to our informed learning. It becomes intelligible that Aristotle should have regarded the brain as a cooling apparatus to offset the hot, fiery vapors of the heart, the seat of the soul, divided by Plato into rational, feeling, and other spiritual orders. The dominant philosophy transferred the physical notions of the elements to the life functions, most clearly indicated in the doctrine of the temperaments—which is essentially a neurophysiological scheme—and in selecting the fluids and the spaces they occupied as keys to the plan of the structure (see Chapter X). But in addition to these speculations there were attempts to find elementary units of action and types of neural mechanisms. The independent action of the parts of a divided earthworm came under Aristotle's observation; he extended the idea to other varieties of animal behavior. A similar mechanical scheme was the basis of the views of Descartes, who all but discovered the principle of reflex action. That notion offers a tangible point of reference to the errors of modern times and their correction.

REFLEX ACTION

THE SPECULATIONS of the pre-scientific period regarding the part played by the nervous system in the control of behavior are summarized in a later chapter. A concept that figures prominently in the argument is the reflex as a unit of behavior with a demonstrable bodily mechanism.[1] All early speculations brought into play such spiritual or occult assumptions as animal spirits; opposed thereto were mechanical explanations such as J. O. de la Mettrie's *Man a Machine* (1748). At that time the factual knowledge of bodily organization was inadequate to support so sweeping a generalization. The clear formulation of the idea of reflex action as a type of neuromuscular response with specific objective characteristics dates from Marshall Hall in 1832. It ushered in a period of rapidly increasing knowledge. Many special reflexes were accurately studied physiologically with demonstration of their nervous pathways and centers. All this promised a satisfactory mechanistic account of all human behavior.

Such mechanistic concepts then as now met great resistance from intrenched dogma. The way for their gradual acceptance was prepared in the sixteenth and seventeenth centuries by a revolutionary change in attitude among the leaders of scientific thought. The explorations of Vesalius, Harvey, and many others led the well-informed to entertain the radical notion

[1] See Chapter X; and for an excellent history of the emergence of this scientific concept see Franklin Fearing, *Reflex Action* (New York, 1930).

that some of the movements of the body might be explained without the aid of a "soul" or other spiritual agency. The mechanics of bodily action received serious attention. Descartes before the middle of the seventeenth century applied the idea of an automatic, self-acting mechanism to all animals below man. "I know, indeed," he wrote, "that brutes do many things better than we do, but I am not surprised at it; for that, also, goes to prove that they act by force of nature and by springs like a clock." Similar bodily mechanisms were recognized in mankind, but here they were regarded as inadequate; the pressure of the theological atmosphere of his time was too strong. If the brutes, Descartes said, "should think as we do, they would have an immortal soul as well as us, which is not likely." Accordingly, he invoked purely hypothetical "animal spirits," distilled from the blood and stored in the brain, "being like a wind or a very fine flame"; they are carried from the brain by the nerves to the muscles which they inflate and activate. Their function is to link the soul with the body. As part of the body they are subject to physical laws, but their movements are caused by spiritual agencies of a different order. Descartes' speculations fell just short of an intelligible scientific formulation of the problem; by virtue of his prestige his failure had momentous consequences in the direction of error.

The twentieth century introduced a new era in neurophysiology. A hundred years of intensive and well-planned physiological and anatomical investigation firmly established the reflex as an important component of behavior. Based on these concepts, German biologists elaborated an experimental mechanistic biology and a program of objective psychology. In Russia, under the guidance of Ivan Petrovich Pavlov (1849-1936), a system of reflexology was formulated, and conditioning of reflexes (which is learning) was intensively studied. In America a radical "behaviorism" carried the doctrines of reflexology to their logical *reductio ad absurdum*. The fallacy here lies not in the reflexes—they are real things —but in the apotheosis of an abstraction, a concept, into a dogma. Useful as are reflexes when recognized as components of behavior, the attempt to explain all animal and human activity and experience in terms of the rigid categories of reflexology broke down in practice. Pavlov's "analyzers" are excellent for their purpose—the formulation of responses in standardized patterns of behavior. But the vital process includes other functions even more primitive and fundamental: these are the synthetic or integrative functions, the totalizing activities which maintain the integrity of the organism as a whole and ensure progressively more efficient adjustment. These synthetic functions cannot be included in the categories of reflexology without so radical an expansion and alteration of the concept as to make it meaningless.

It is significant that Sir Charles Sherrington's notable researches on the reflex were summarized in 1906 under the title, *The Integrative Action of the Nervous System*. With the new century the interest has turned from reflex as an analytic function toward the problem of integration. Reflexology leads into a blind alley from which there is no way out except by way of the larger field of the integrating or synthetic functions.

This position is specifically stated with experimental support by Coghill,[2] who points out that the primary function of the nervous system is "the maintenance of the integrity of the individual while the behavior pattern expands." The "total patterns" of behavior are primary; indeed, they antedate the nervous system. Reflexes are local and "partial patterns"; in the animal species that Coghill has so intensively studied they are not the primitive elements of behavior. On the contrary, they emerge from the total or integrative patterns rather late in embryological development. The history of the origin and growth of the apparatus of the totalizing functions is obscure. Clearly there is a gradual elaboration of special nervous apparatus of synthetic and constructive activities. In man these organs lie chiefly in the cerebral cortex, the

[2] G. E. Coghill, *Anatomy and the Problem of Behavior* (Cambridge University Press, 1929).

part of the brain that dominates and directs all activities of the body except the most primitive physiological functious.

Conditioning of reflexes is an early step in the elaboration of the learning process. Learning is a "total pattern"; it is the whole body that learns, not the reflex arcs. The nervous apparatus primarily concerned with conditioning is separate in the brain from the reflex centers. The nervous centers and pathways of scores of reflexes have been accurately charted. Their arrangement is essentially uniform in all members of a race or species; indeed, these analytic mechanisms are remarkably similar in the brain-stems of all vertebrates. But the apparatus of conditioning and other integrative functions varies more in mass and internal texture from fish to man than any other part of their bodies. This development culminates in the human cerebral cortex, a structure lacking in the lower fishes but comprising about half the weight of a human brain.

LOCALIZATION OF FUNCTION

We turn now from the attempts to find the clue of the neural basis of behavior in the simple unit mechanisms to the opposite approach: to find them in the most developed neural structures correlated with the most elaborate phases of mental action. This attempt found expression in the parceling of the mind's activities map-wise on the human cortex and on the homologous areas of higher mammalian brains.

The accurately charted pathways and centers of adjustment of the analytic functions make up the greater part of the brain-stem, extending also into the cortex of the cerebral hemispheres, where they are known as projection tracts and centers. The cortical centers into which nervous impulses from the eyes, auditory impulses from the ears, etc., are discharged have been mapped in mosaic patterns on the surface of the cerebral hemispheres. This localization of physiological functions of analytic types is thoroughly and accurately known, and this knowledge has made possible great improvement in the diagnosis and treatment of nervous diseases.

These spectacular successes have led to the expectation that similar "centers" performing the mental functions would be revealed through further search. From the days, more than a hundred years ago, when Gall and Spurzheim founded the supposed science of phrenology until now, this quest has been prosecuted with vigor and skill; yet its results have been disappointing. The assumption that the higher psychic functions are localized in patterns similar to those of the analytic functions was an error.

The "associational" tissue involved is not aggregated in "centers," each of which performs some specific sort of perception, reasoning, or volition. This tissue is dispersed throughout the brain, but chiefly accumulated in the cerebral cortex, of which it comprises more than half. Our conscious experience is individually acquired; unlike the reflexes, it is not born with us. The details of organization of the nervous tissue employed during the acquisition of this "subjective" experience are not predetermined at birth; the arrangements finally effected depend on the experiences had. These tissues, moreover, retain some measure of their embryonic fluidity throughout life, so that the stream of conscious experience flows and ebbs in ever-changing patterns.

Rigidly fixed arrangements of cells and fibers such as we see in typical reflex arcs could no more perform these labile functions than the switchboard of an automatic telephone exchange could perform the feats of inventive skill of the engineers who designed it. The higher "subjective" functions have organs, but these organs are not localized in space or standardized in performance in patterns similar to those of the simpler analytic functions on the physiological plane.

COMPARATIVE ANATOMY AND PHYSIOLOGY

The past hundred years have seen intense and increasing activity in the study of the comparative anatomy and comparative embryology of the nervous system, in the hope of finding the principles of development and the organic agencies that have elaborated the apparatus of control of animal and human behavior. Out of this enormous labor some of the formative agencies are coming to light.

In the pre-Darwinian period the search was for some obscure pattern impressed upon the animal body in some unknown way, which established a normal form or archetype of which all other patterns are variants. The poet Goethe, who was also an anatomist, and many of his distinguished contemporaries elaborated a "pure morphology" devoted to the search for these primordial types of structure, which by some were supposed to be expressions of the Creator's designs or illustrations of a predetermined order of Nature.[3]

After Darwin the entire point of view changed. The recurrence of similar patterns of structure throughout the series of brains of all back-boned animals was explained altogether differently as an expression of the conservative influence of heredity. In embryological development the human brain passes through a succession of stages which resemble the adult brains of other vertebrates in series from fishes to apes. This serial relation was regarded by Karl Ernst von Baer (1792-1876), Ernst H. Haeckel (1834-1919), and their followers as evidence of a "recapitulation" by the individual of the evolutionary history of the race. The fundamental principles of the form of the brain, its "morphology" as taught by Thomas H. Huxley (1825-1895) and the other great comparative anatomists of his time, centered about von Baer's principle of recapitulation.

This is a conservative factor; it can account for similarities but not for differences. It is a static conception of structure, but living structures are not static. It is a partial truth which so largely dominated interest in the last third of the nineteenth century as to retard progress. Toward the close of that century the interest of comparative anatomists shifted from form to process. Experimental methods were rapidly developed with spectacular results which nullified ancient dogmas and their later modifications and brought to view

[3] See the extensive series of Bridgewater Treatises published in England under the influence of Paley's *Natural Theology*.

the vital processes shaping the growth and evolution of the nervous system.

The nervous organs are now viewed as a going concern; the interest in their structure has become dynamic. The focus of inquiry is directed to the workings of these complicated webs of nerve cells and fibers, and how changes in structure are related to shifting patterns of behavior. This interest faces forward, not backward toward archetypes as records of the Creator's design or as surviving vestiges of past evolutionary processes.

The formative agencies that determine the course of growth and evolution are now recognizable. Some of these are the physiological gradients of C. M. Child as measured by rate of activity in various parts of the living tissue, chemical messengers (hormones) spread through the body fluids, differences in electrical potential generated by local vital activity, various activators of specific kinds of function. This conception corrects older errors and suggests new problems.

FROM PSYCHOPHYSICS TO PSYCHOBIOLOGY

Other approaches have been made to the persistent problem of the relation of physical and psychic forces or events. With the methods in mind of the physical sciences proceeding by mathematical formulae, the idea was tempting that if the two members of this relation could be reduced to measurement, an equation would result which would solve the relation. The most elaborate contribution was that of psychophysics, as formulated by G. T. Fechner (1801-1887), whose personality was a strange combination of mystic and physicist. Based upon the observations of E. H. Weber that the perceptibility of a difference in two lifted weights, two brightnesses, two surfaces, two loudnesses of sound, was proportional to the amount of weight, light, sound already present, Fechner went on to formulate the law that to produce equal (arithmetic) steps of perceptible increases one must present equal ratios (geometric) of increasing stimulation, and from this he reached the abstract equation: "The intensity of the sensation is approximately proportional to the logarithm of the strength of the stimulus." It all looks exact and sounds learned, but it rests on a multiple error. These abstractions have no reality. Nature does not create senses as physical instruments, but for biological service. Obviously a pound that is one ounce short compared with a correct pound is more readily detected than the shortage of an ounce in two pounds or five; likewise an inch at the end of a man's nose makes a greater difference than in his total height. But the so-called law does not hold because Nature does not work that way; vital processes are too labile to submit to so simple and rigid mathematical formulation. Entire libraries on psychophysics (not all of it futile) bear witness to the patient energies devoted to a false premise. The error of psychophysics remains monumentally instructive.

But the search for mental measurements continues; much of it is in the direction of profitable progress, still more of it of dubious value and misleading. That results expressed in quantitative form supply an instrument of research has been abundantly demonstrated in a hundred problems of experimental psychology.[4] Similarly, in the applied field, the intelligence-quotient (I.Q.) is a useful mental measure of certain orders of capacities; but it makes no pretense to furnish a direct correlation between mental achievement and neural processes. In the studies of maturation in children in which Arnold Gesell's contributions are the most complete and significant, increasing integration and control are presented qualitatively but supplemented by measured rates and relations. Studies of learning generally proceed upon quantitative data, but they yield little additional insight into the relation between neural structure and psychic development.

The study of emotions by both neural and experimental methods proves convincingly that affective processes are specifically correlated with physiological functions. The thalamus

[4] The vast extent of these is indicated by a German compilation in eight volumes extending to over six thousand pages. E. G. Boring's *History of Experimental Psychology* (New York, 1929) is an adequate reference, including neurophysiology.

has been shown to be the cerebral station of emotional expression and its visceral components. The brilliant demonstrations of W. B. Cannon and others have set forth the close relation between the autonomic control of glandular function and such psychic conditions as fear and rage.

By all these contributions neurophysiology has been enriched, and the total view of the body-mind connection has, indeed, assumed a great transformation. Especially is the constant and diversified cerebral control of all the lower neural mechanisms made clear. Twentieth-century physiologists and psychologists have at command an extensive and detailed picture of the neuro-psychic relation as a whole and in all its component parts. All earlier views, lacking this insight, are in so far erroneous or at least defective and out of focus.

Yet that strict correlation of neural process X with psychic state Y has not yet appeared, nor any exactly quantitative index that serves as a bridge from one to the other. Their interdependence is more richly demonstrable, but no more. The reasons for this limitation are significant.

In the ordinary course of our lives, feeling, passion, sentiment, and appreciation of esthetic and other values are expressions of vital processes of great complexity, including visceral processes of diverse sorts, other sensory experiences from outside the body, and reverberations of the nervous impulses so excited throughout wide reaches of brain tissue, including cycles of action and reaction between the thalamus and cerebral cortex.[5] For these complicated and variable functions we have no single reliable numerical indicators.

In these prolonged and difficult researches in the field of psychophysics the search for a simple and exact formula to express the relation between the mental and the physical has so far been vain. The reason is that the problem has been oversimplified. Both mental and bodily processes are too complicated to yield to this sort of analysis. This does not mean

[5] George G. Campion and G. Elliot Smith, *The Neural Basis of Thought* (London and New York, 1934).

that the problem is insoluble, only that it has not yet been solved.

At the present stage of development of scientific technique it is a fallacy of method to insist that exact numerical statements of the results of observation or experiment are necessarily more accurate or more desirable than statements in terms of "intensive" magnitudes that cannot be added, subtracted, or multiplied.[6] For, as Dr. Alexis Carrel says, "In man the things which are not measurable are more important than those which are measurable.... As much importance should be given to feelings as to thermodynamics."

It thus becomes clear that the story of error cannot yet be brought to a consistent conclusion on many fundamental questions in so recent a discipline as neurophysiology. Attention is focused on radical shifts of interpretation in consequence of new insights resulting in part from new techniques, but far more from researches designed to test concepts and hypotheses emerging from increasing fullness and rightness of basic and total neuropsychic relations.

Yet the hope of establishing a registry between the neural and the psychic remains. When it was discovered that strong emotion is usually accompanied by measurable changes in the electrical potential of the skin, the hope was revived that here at last in the "psychogalvanic reaction" we had a true physiological measure of a mental process. Unfortunately the case is not so simple. These galvanic records express changes in local resistance to passage of electric current in the skin itself due to variations in moisture, blood flow, and other causes. They are expressions of various activities of the sympathetic nervous system, and their relation to the subjective experience of emotion is incidental and inconstant. This promising lead, accordingly, was a false cue leading into a blind alley.

Another and most recent technique bearing upon the problem of direct reading of psychic changes in neural terms is furnished by the oscillograph with radio-tube amplifiers

[6] See Chester L. Barnard, *Mind in Everyday Affairs* (Princeton, 1936).

which records minute electrical changes, or action currents, resulting from the excitation of nerve cells and fibers.

These instruments permit direct observation and instantaneous recording of local electrical changes of the order of a thousanth of a millivolt in time units of a ten-thousanth of a second.[7] If now a beam of light is thrown into the eye of a rabbit, the resulting nervous impulses can be followed along their devious courses from the retina through the optic tracts and centers of the brain to their discharge in the visual area of the cerebral cortex. The whole process can be recorded on a moving-picture film and the details studied at leisure and correlated with microscopic studies of the brain of the same rabbit at the end of the experiment. Thus minute anatomy and experimental physiology, which hitherto have groped their separate ways in the dark, now join hands in a coöperative effort to find answers to our puzzles.

The scope of this new electrophysiology is not limited to the level of nervous functions in the realm of the unconscious. Action currents can be recorded from the cerebral cortex and peripheral organs of conscious human subjects during the performance of various kinds of mental exercise. This field has been but little explored, but it is clear from observations already made that there is reasonable hope of some measure of success in direct observation of the correlation between mental processes as subjectively experienced and an objective record of the related activities in brain, nerve, muscle, and viscera. The great dilemma of human biology will thus, we believe, be further advanced toward solution.

MECHANISM AND SUBJECTIVITY

The problem of neurophysiology is the search for the bodily apparatus of adjustment on the physiological and psychological planes of organization. Having presented the stages of insight and proposed solutions of this problem of

---

[7] George H. Bishop, "Electrophysiology of the Brain," in *The Problem of Mental Disorder,* edited by Madison Bentley and E. V. Cowdry (New York, 1934), pp. 120-132.

problems, including the errors, the false leads, the intrusions of dogma, the blind alleys as well as the stages of increasing knowledge essential to a right approach, we find on our hands two conceptions, the one apparently incompatible with the other: that of mechanism, and what for lack of a fitter term may be called subjectivity—in one sense the conscious awareness of life-processes.

What, considered in its bearing on the body-mind relation, is the correct, what the erroneous view of a mechanism, is the first issue. The mechanisms concerned are those created, evolved by Nature; the error that rejects the mechanistic interpretation arises from an ancient and fundamental blunder in the interpretation of our own experience of Nature and especially of human nature.

Nature is not split up into separate compartments—matter and energy, inorganic and organic, body and mind. These are all interrelated; they interpenetrate; and they all inhere in a unitary natural order of things. As a normal sentient human being, I am not a hybrid offspring of natural physical and unnatural spiritual parents. A frank appraisal of our own experiences as free from dogma, prejudice, and preconceptions as possible clearly indicates that the normal human person is a biological unit. He lives one life, not two or three or nine like the traditional alley cat. The resolution of our dilemma, then, must be sought within this unitary personality.

Since the human body is obviously a mechanism, the simplest solution of our problem would seem to be to grant that all that it experiences and all that it does are products of the operation of this organic machinery. This leads to a radically mechanistic formulation of all biological and psychological phenomena.

The opposition to the acceptance of this position arises from a faulty conception of what mechanism is and how it works. Narrow-minded mechanistic biologists and psychologists are largely responsible for the disfavor that current scientific solutions of humanistic problems meets among humanists and for the accusation of defeatism which has been—quite justifiably—charged against them. Oversimplification of a problem by ignoring the troublesome factors will never take us toward its solution.

What are machines, and what do they do? They make things; they are constructive agencies. They do not make things out of nothing or perform any other miracles. A hydro-electric generator transforms the energy of falling water into electrical current. It does it automatically, and it does it actively. So every natural mechanism, like a river system, employs natural agencies to do its own work in transforming the face of a continent. It, too, is automatic and self-regulating. Because it does this in an orderly way in accordance with natural laws like our artificial machines, we call its action mechanistic. The same is true of the body of an oak-tree or a man. This means that all Nature is mechanistic. Living things are mechanistic, with self-regulating apparatus more highly elaborated than in any inorganic mechanisms, including capacity for growth, repair of injury, and reproduction. The human person is mechanistic on a still higher plane of self-control and self-direction, including both unconscious and conscious control.

Some machines run blindly. The human mechanism does not, for I have some awareness of what is going on, some thoughts about it, and some feeling of satisfaction or objection to it. This conscious experience is as truly a product of the operation of the organic machinery as is muscular movement, and it is as truly a factor in the causal complex resulting in some particular behavior.

One of the most amazing scientific aberrations of recent times is the growth of a cult of psychologists who are so impressed by the futility of traditional spiritistic explanations of mental processes and yet so completely dominated by that same traditional view of mind as a ghost that they refuse to accredit to their own mental experiences any significant rôle in shaping the course of conduct. They leave mentality out of their natural system of behavior because they find the outworn dogmas about it unintelligible and inconvenient.

These mental, including the emotional, exercises of our brains are the most significant things that we do; they control the course of human events; they have produced modern scientific control of Nature, as well as self-control and self-culture. The human brain is the organ of civilization.

These are products of mechanism as truly as the electric current is a product of a mechanical generator. The Natural Order is big enough to include human nature and all human achievement. Creative evolution is a natural process throughout. This is how Nature and human nature look to a naturalist who is a radical mechanist.

This view broadens the concept of mechanism to include all natural mechanisms resulting from evolutionary development. The chief objection to its acceptance is the hesitation to apply it to the higher products of human thought and emotion resulting from conscious reflection. This "subjective" realm of psychic operation seems too remote from any neurophysiological basis to be so included. Moreover, some aspects of subjectivity, such as the intrusion of human wishes, the development of beliefs satisfying to a human, an anthropocentric concept of the nature of the universe including man's place in it, have often led to distortion of true scientific procedure, which aims to reveal a universe as it is, regardless of human longings and aspirations. Yet these longings and aspirations are natural phenomena in their own right. The naturalist must recognize his own "personal equation," and having done so, his own interests and values may themselves be subjected to scientific study.

What Campion calls "the errorful subjectivity of human knowledge" must never be lost sight of in scientific work, and nowhere is this caution more apposite than in the field of neurology. He goes on to say that concepts that express our accumulated experience of Nature "provide by their very subjectivity the means of explaining the whole universe of human error, human illusion, and human self-deception."

Subjectivity itself is and always has been the greatest puzzle of both science and philosophy, and until recent times a scientific approach to this question has been hindered, if not thwarted, by metaphysical preconceptions.

It is only in recent decades that sufficient factual knowledge has been available to justify the hope of a scientific approach to the problem of subjectivity, and the prevalence of the older traditions is still an impediment to progress. Especially in neurology, psychology, and psychiatry mystical disembodied functions are invoked as if they were causal agencies in the control of behavior. Or else, in violent reaction against this modern demonolatry, subjectivity is ruled out of the picture entirely as an irrelevant epiphenomenon. This mysterious subjectivity is so refractory to scientific analysis that it has seemed far simpler to leave it out of account. The scientific tradition, accordingly, complacently accepted the dogma derived from an ancient metaphysics and theology that the subjective pertains to a spiritual realm that is dissociated from the world of objective experience which is the proper field of natural science.

These are questions of fact, and the truth about them can be found only by critical study of our experiences with them. This experience indicates very clearly that our bodies are the instruments of our feeling and thinking just as they are of the other things that we do. In short, the ancient riddle of the relations between mind and body is a neurological, not a metaphysical problem. Three thousand years of dialectic based on *a priori* postulates have contributed little of value toward its solution. The fundamental error here lies in failure to recognize that the distinction between the subjective and the objective in naïve experience is not primary. When the whole of experience is surveyed, they are found to be indissociable; they knit into a unitary vital process from which they cannot be torn without wrecking the fabric into which they are woven.

The nervous organs as the apparatus by which all experience is integrated and interpreted have been misconceived and will continue to present irresolvable puzzles as long as ancient mythologies or their modern offspring warp our judgment in attempting an appraisal of the meaning of what scientific research has brought to light. Preconceptions which inject mystical or unnatural agencies into the interpretation of these phenomena have been the fountainhead of errors so fundamental as to block progress in many departments of neurology and of all other sciences that in any way depend on a scientific grasp of the apparatus of control of human experience and conduct.

PART II

MAN

Introduction

THE PROCESSION OF CIVILIZATION

By James Harvey Robinson

AMONG MAN'S ERRORS the most stubborn and all-
pervading have been those in regard to himself. And
no wonder! So long as he was ignorant of his origin and
nature, he could offer only mythical explanations of his
deeper perplexities. The story of error is bound to remain
episodal and mysterious unless it is placed in the setting
of human history as a whole—a tale that had not even been
sketched a hundred years ago. Until this was done, man's
fumblings, achievements, and potentialities eluded scientific
investigation. The history of the human race is now discovered to constitute the chief explanation of errors and
of the efforts to overcome them. It is my purpose here to
recall the general views of mankind formerly accepted in our
Western world and to point out the fundamental revision
of these older notions in regard to man's "mind," "reason,"
and "progress," that seems to be demanded by the recently
acquired knowledge of his career. The present considerations will be focused upon man and the sciences dealing
with the human estate.

It seems safe to assume in the light of anthropological
data that men were wont, long before they left any record
of their thought, to ease their bewilderment in regard to
themselves and their environment by invoking myths. Insistent and troublesome questions, in short, were answered
by fascinating authoritative stories peculiar to each people
or common to many. The fact that the ancient Egyptians,
Babylonians, Greeks, and Romans reached a high degree of
civilization and sophistication did not preclude the perpetu-

ation of precious myths interwoven into their history. So it is not at all surprising that the former beliefs in regard to man's origin and nature accepted in Europe after the break-up of the Roman Empire should be based on a mythical explanation of man's beginnings. The Christian religion relied upon its Hebrew antecedents, including the tradition of the Garden of Eden. The Church taught that the first human pair had been created with immortal souls out of hand by God, able from the first to talk and tend the lovely garden in which they were placed. The whole universe was devised for their delectation—sun, moon, and "the stars also." Adam and Eve found their moral responsibilities too heavy, yielded to temptation, and were cast out into a world cursed on account of their sin. Upon their primal disobedience the Christian missionary Paul of Tarsus erected a terrifying theological structure of sin and salvation. This was further elaborated by Augustine for the Church at large and later recommended to the Protestants with ferocious conviction by Calvin. Philosophers busied themselves with man's mind or reason, the counterpart of the theological "soul." This was assumed to be an exclusively human possession which was as characteristic of man as a trunk is of an elephant. Mind was defined by John Stuart Mill as "the mysterious something which feels and thinks." Animals were guided by *instinct,* but man by something far nobler, *reason.* To the exercise of his practical reason all human achievements could be ascribed, to that of his speculative reason the detection of profound truths.

The discovery that were man's ancestry traced back far enough, it would be found to merge into that of a species of wild animal, could not but invalidate or profoundly transform the older notions of his nature and career. So recent is this discovery that the reconstruction that it implies is by no means complete—indeed may hardly be more than in its beginnings. Darwin's *Descent of Man* appeared in 1871, so that many living men of science, including contributors to this volume, have been able to follow the stages of acceptance of this most momentous revision of human history. The evidence that Darwin presented as the outcome of years of study seemed to prove that the human race is but one of the results of the whole process of organic evolution reaching back through geologic ages to the origins of life on our earth. Moreover, the human species is in body and physiological functions so like the large apes and monkeys that zoölogists have no hesitation in assigning man to the group of mammals called primates. Of these he represents one of some six hundred terminal twigs (species) still existing on the family tree. Those who accept the conclusion that man not only resembles in some respects the higher animals but is inescapably an animal whose ancestry merges into that of the vertebrates in general—is, in short, an ex-animal—can afford to rule out a good many ancient quandaries based upon misapprehensions; but they find themselves faced with a new set of problems which have to be met in terms of the new knowledge of man's past.

Of these new problems the most central is that of the origin and growth of all those human achievements which have so long disguised the fact that mankind is a species of animal which once lived as a wild, naked, speechless creature with no more traits of civilization or culture than a chimpanzee or a brown bear. Whence came all the arts, knowledge, and high ideals which seem to set him so far above the dumb beasts over which he exercises dominion? What was his noble mind doing in his primitive estate? How was "reason" deploying itself when he was glad to come upon fruit and succulent roots or cluster with his fellows around a dead deer to be devoured raw without even a jagged bit of flint to hack it to pieces? For the acceptance of our animal extraction forces us back to this squalid condition as the *terminus a quo* of human achievement.

If mankind is a species of animal derived from still lower species, our conceptions of human development, including mind and reason, must of necessity be brought into relation with what we now know of man's animal ancestry and affinities. It is profitable to make here a few suggestions in regard to the ways in which the problem is being approached.

Theologians and philosophers—professional hard thinkers —who have described most confidently the ways of God to man have rarely devoted much careful attention to what are called "the works of God." These are multiform beyond the comprehension of any human being. Only through elaborate coöperation of many observers through centuries have living creatures been fairly classified and named. There are thousands of species of animals and plants whose forms and actions can be described only by means of microscopes which exhibit them five hundred, a thousand, or two thousand times as broad as they would appear to man's unaided eye. Then in public aquaria bigger creatures may be observed, bewildering in variety, who live in water and successfully propagate their kind from generation to generation. The soil, the woods, and prairies are inhabited by hundreds of thousands of kinds of living things, each kind surviving in terms of its particular equipment for making a living and starting a new generation. Paleontologists find that great numbers of species and genera have lost out: they once flourished but failed to meet new conditions. The graceful ginko tree, for instance, reminds us of many allied species which no longer exist.

Of all the manifold and multiform creatures on the earth, *one only,* a species of primate, was so constituted that it was able to introduce something new in the ancient game of getting a living and starting a fresh generation. It was so made that under favorable conditions members of a group might adopt a trick which one of their number had happened upon. For example, an enterprising fumbler might find himself improving a sharp fragment of flint by chipping its edges with a convenient roundish stone used as a hammer. It might happen that some of the group were not so inert and unobservant as to fail to see through the process. The example once set might be followed and become so common that the children would acquire the art and transmit it to their children. This would be an element in a new type of heritage peculiar to the species of primates to which we belong. Like all other living creatures, we have our animal heritage, our particular bodily form and functioning, but in addition we have what Graham Wallas calls a "social heritage," now so vast and varied that no single person can comprehend or master it. This social heritage, which is sometimes called *civilization,* sometimes *culture,* is superadded to man's animal equipment. We are not born with it but into it. It is our man-made artificial environment.

There is no satisfactory evidence that a single element of civilization can be transmitted hereditarily—can get into the blood, as the expression goes. If this is so, every human child is born absolutely uncivilized but usually qualified to adjust itself to any culture in which it finds itself, whether that of a prosperous New York family or of a small tribe of Papuans such as Margaret Mead has so carefully described. The animal most nearly resembling man in his make-up and possibilities seems to be the chimpanzee. With careful training he takes kindly to bicycles and roller-skates, can light a cigarette and smoke it; but the three R's seem to be entirely and forever beyond his reach.

We have seen that before the discovery of man's animal ancestry, it was assumed that his superiority over all other creatures was due to his unique possession from the first of a *mind* and his use of it in discovering truth, through *reason.* Our thoughtful predecessors were still under the influence of the ancient idea of spirits of one kind and another—agents to which a great variety of acts were ascribed, on the ground that if something was done, someone must do it. Among the successors of primitive spirits was mind. This is clear in French, for *l'esprit humain* means "mind." One may say paradoxically that the latest development of mind is the denial of its own existence. This is a phase of recent criticism which suggests the condemnation of a number of decayed old words as public nuisances. Not only *mind*, but *consciousness, will, emotion, memory, sensation*, are now viewed as activities or processes rather than active agents. "All such nouns are properly verbs or adverbs," Professor Woodworth warns us. He rightly says that "we forget that our nouns are merely substitutes for verbs, and go hunting for the *things* denoted by the nouns; but there are no such things; there are only the activities that we started with." This seems to the writer a most important modern discovery. One who accepts it is exonerated from tracing the history of a mythical agent formerly known as the human mind; but quite as difficult a task remains in explaining the origin and development of those activities seemingly peculiar to mankind which led to the former assumption that he had a mind.

The human species very closely resembles in bodily form and physiological functioning its nearest relatives, but it has, as has been pointed out, accomplished things altogether unprecedented in the whole animal world. The peculiarities of mankind are not, taxonomically, very striking; but while they appear to have been "marginal," they were just sufficient to make possible, when given time enough—several hundred thousands of years—and leeway for innumerable experiments, failures, and lapses, the altogether astounding results we see on every hand. Perhaps the most obvious advantage man had over anthropoid apes was his ability to walk securely on his hind legs; this left his finely formed sensitive hands free to feel, fuss, and fumble. Aristotle, I believe, ranked touch as our most important resource; certainly by touching and feeling many important things can be learned not revealed by sight or hearing. Man's central nervous system with its intricate cerebral cortex is always assigned a decisive rôle in human achievement. Its workings are, however, still obscure and subject to widely divergent conjectures. It is certainly dangerously uncritical merely to substitute *brain* for *mind*, as is sometimes carelessly done. It took the whole human organism to initiate a social heritage and to accumulate civilization. It is safe to guess that had horses brains like human beings, they would never turn into Houyhnhnms, for their hoofs are rude explorers when compared with hands.

Before the discovery of man's animal descent, his mind was set off sharply from his body and conceived as a sort of immaterial spirit which controlled and guided the material body. Philosophers did their best to attenuate the mystery of this fundamental dualism. Now the human body seems to have a history reaching back to the beginning of life on our earth, while the human mind, as we know it, seems a very recent emergence of previously unsuspected potentialities of our ancient and altogether marvelous organism. We are faced by a new dualism, that of bodily as over against mental (or mind) processes—physiological activities as contrasted with psychological. God forbid that a mere historical student should venture on the treacherous bogs across which comparative psychologists and experimental psychologists and analytical psychologists, and even bold biologists and physiologists, are striving to construct a reliable trail. I mean no more than that this is proving a very hard job; but there is now far more hope of reaching some generally acceptable "analysis of mind," as Bertrand Russell calls the enterprise, than on the old hypothesis that mind is an immaterial spirit sojourning in a mortal body.

Some years ago, as a student of history, I wrote about *Mind in the Making* in a mood of what I hoped was judicious generalness. Since the book appeared, a number of attempts to review the ups and downs of civilization have been coming out; and gradually we may get clearer notions of what those processes embraced in the expression "human thinking" have actually done. The historian's yardstick for measuring mind is past human achievement as he finds it recorded. A comprehensive sketch of human accomplishment fairly up to date is, let us say, the last edition of the *Encyclopædia Britannica*, which in twenty-four volumes, with delightful pictures, gives a brief account of man's present varied arts, knowledge, institutions, and past experiences. At the end of the articles there is often a bibliography to warn the reader that there was space only to tell a very little about the subject in hand; often there are no more than hints concerning vast fields of scientific investigation—astronomy, chemistry, physics, geology, anatomy, history, and so on. Compare this or any other excellent current encyclopædia with, let us say, that compiled by Pliny the Elder as the result of a sedulous examination of the Greek and Roman sources of information available some nineteen hundred years ago—just a little way back in the history of the human species. How could so-called mind in his day carry on thinking in the terms of present knowledge and available instruments of investigation, to say nothing of the jocose discredit into which so many of the worthy Pliny's solemn reports have fallen?

So it seems apparent that those specifically human activities which we call mental and rational, previously regarded as manifestations of mind or reason, have varied mightily in nature, scope, and intensity during man's existence. In the beginning there was no more than what is now studied under the caption of "animal intelligence" and "wisdom of the body," very marvelous indeed and still ill-understood, always essential to our survival and probably deeply affecting our mental operations.

It should be the business of the ever-increasing experiments in the history of civilization to trace the process of civilization in its augmentations and diminutions, of which there are many striking examples in the recorded past of our race. But our knowledge of these fluctuations is often scanty, and any analysis of the elements involved is necessarily conjectural—as, for instance, the reasons why the Greeks developed the civilization they did at the time they did, and the circumstances that brought this development to a seeming standstill. The older explanation based on the assumption that the Greeks possessed innately exceptionally fine minds and were an instance of a sort of mutation, as

Galton seems to have believed, makes little appeal to anthropologists and historians today. Franz Boas and Paul Radin find little evidence that there is any marked inherent inferiority of mental ability among so-called primitive peoples as contrasted with those living in a more elaborate civilization, and they believe that wondering and philosophic spirits —always very rare—may be discovered among the savage and the barbarous. If in human groups there are always potential discoverers and inventors with a chance of creative achievements, the character and extent of their operations depend on the accumulations of the past upon which they can build. The lasting results of their original contributions depend upon the readiness of their fellows to appreciate and perpetuate their findings, from generation to generation.

Not until the last three or four hundred years has the accumulation of civilization been sufficiently rapid and spectacular in our Western world to beget the idea of unlimited advances in knowledge and its application to man's convenience. Previously, the appeal to the ancient and established was well-nigh omnipotent and freely exploited by classes with vested interests. While the conception of "progress" appears in the works of Roger Bacon, its modern herald was Francis Bacon (1561-1626), who eloquently sets forth the measures for correcting old errors and augmenting information and insight. It is perhaps safe to say that the existence of slavery among the Greeks served to discredit mechanical inventions and exalt purely intellectual feats. On the other hand, the modern developments of machinery, mechanical devices, and scientific instruments, particularly microscopes, telescopes, and spectroscopes, have carried human knowledge and speculation into wholly new realms and revolutionized all the old conceptions of living creatures, and ultimately of matter itself.

Any number of examples could be given of discoveries that have in modern times deeply influenced mental operations and ambitions. Contrast, for instance, Franklin's thinking about electricity, when in 1752 he sent up his little kite to see if lightning was an electric spark like that which might be had from the recently invented Leyden jar, with the kind of thinking Pupin and Edison were able to carry on, in which they could utilize a vast amount of accumulated information and all sorts of ingenious devices undreamed of in Franklin's day. Consider the point of departure for the experiments of Harvey as over against those of Walter B. Cannon. A hundred years ago the cell had not been discovered to be the unit of organic life; now thinking about cells in various new ways is the ambition of hundreds of biological laboratories and the theme of numberless treatises.

There seems at present to be no term set for the accumulation, clarification, and application of knowledge to the relief of man's estate, or to new ways of thinking about mankind and his surroundings. One somewhat familiar with man's mental history is impressed with the fact that he has not until recently been interested in true and false statements about objective facts. He was preoccupied, as he still is, with safe and dangerous, familiar and strange, nice and nasty, good and bad, beliefs. We know now that notions acquired in childhood, as a result of the talk that youngsters hear going on about them, are apt to enjoy what has been called an *a priori* certitude, which insures their persistence through life. It is not hard to find illustrations of this among professional critical thinkers, whose mature speculations are obviously influenced by childish preconceptions.

Among the prime sources of error are *words,* which are now being examined with unprecedented suspicion. Man has for many thousands of years believed in spirits; his language is permeated with animistic expressions which cannot but be a potent factor in influencing his thought. As has been said, he has not been much concerned with scientific truth and error, and so his ways of expressing himself are ill-adapted to careful description and discrimination. Indeed, most people most of the time are sadly bored by attempts at accuracy. In their ingenious work, *The*

Meaning of Meaning, Ogden and Richards have made a most illuminating collection of bad ways of saying things of which supposedly careful thinkers have been guilty. Our speech is full of anachronisms recalling ancient misapprehensions of all kinds. So one of the tasks that is imposed upon us by the increase of knowledge and the consequent revision of old ways of thinking is further study and investigation not only of these faults of language, but of how they have influenced and continue to influence even scientific thinking. We must devise better and better ways of talking about things as they now appear to us; for thinking and language are inextricably interwoven and interdependent, and both must be constantly considered in any attempt to illustrate the story of error.

Section IV

THE HUMAN REALM

INTRODUCTORY NOTE ON THE SCIENCE OF MAN

By Joseph Jastrow

THE ADAGE that "the proper study of mankind is man," rather barren in its ordinary connotation, is readily infused with meaning under the illumination of mankind's past—unknown to Pope's generation. This interpretation gives the humanities the central place in the temple of learning, with anthropology the key science to the group of disciplines and interests recognized in the present volume by its division into the two parts, "World" and "Man." By reason of its unique place in the league of sciences, the approach to anthropology calls for special consideration.

Such a hybrid term as "humanology" would be useful to designate the larger bearings of the knowledge of mankind. This broader concept should not be lost sight of in the special researches of the anthropologist, to which, in keeping with the present project, the chapter to follow is devoted.

Occasional reflections through the classic ages, eventually incorporated into anthropology, concerned the varieties of race, the diversities of custom, and the rise of peoples from primitive to civilized status. Such reflections, though springing from philosophy or shaped by religious doctrine, were those of a dominant people exalting their own race, their own traditions, their own ideals—as, for the Greeks, the elect Hellenes, all others being barbarians. To itself, if racially self-conscious, each group in history has similarly been the chosen people, the others just the nation—"gentiles." In neo-modern days, when thinkers had become anthropology-conscious, this inevitable subjective position, with its assumption of superiority or rightness, was aptly called the "cultural compulsive."[1] The phrase expresses the difficulty in studying the ways of men with the objectivity that is readily attained with respect to the world, to the ways of Nature. The allegiance to our own standards forms an ever-present and many-sided source of error when the student and the specimen are equally products of mankind.

It is indeed remarkable that the most engaging of all stories—that of man's past, his place in Nature, his various designs for living—should have been but sparsely considered; it may be accounted for by the constant and active concern of human energies with dominion. The growth of science in the seventeenth and eighteenth centuries, which were also politically alert, formed a foundation for the movement that was to burst into fuller flower later on. The sense of the significance of man's place in Nature had to await the stimulation and direction of the principles of evolution; the aftermath of Darwin created anthropology. Its development was necessarily dependent upon advances in the collateral sciences; to assign man's zoölogical place was of first consequence.

To set in proper perspective the story of man there had to be established his great antiquity and the long epochs of primitive condition. Reaching far back of historic record stretches a vast prehistory which correlates obscurely with geologic time. When remains of extinct animals were found in undisturbed sites associated with stone implements shaped by men, the evidence became conclusive. To recognize crudely flaked and chipped stones celts and hammers—for long defined as "thunderstones"—as products of human workmanship was itself a "discovery." The perspective of human occupation was momentously extended.

But skepticism, as always, died hard. Cuvier (1769-1832), the father of the science of comparative anatomy, had denied the presence of man in association with the mammoth, and for most of the nineteenth century there were plenty of

[1] V. F. Calverton, 1930.

unbelievers in the plain evidence of skeletal and artifact remains. In 1879 were discovered the most famous examples of prehistoric art—the bison, mammoth, and other extinct forms so vividly portrayed in polychrome on the ceiling of the cave of Altamira in the Spanish Pyrenees, now a show-place where the fortunate visitor gazes with a sense of kin-ship at the work of the earliest of artists. Still, the authentic character of these remarkably well-preserved pictures was questioned; they were suspected of being fraudulently pro-duced, until similar drawings on other caves (La Mouthe, 1895, and other sites in the Dordogne), undisputably dated, established them as the work of a gifted prehistoric race, to which the name Cro-Magnon was given.

Until our own and the preceding generation the atten-tion of all but a few scholars was confined to the nearer historical epochs and ignored the far larger spans of gen-erations that lived in primitive and prehistoric times. Now anthropological man has been added to historical man. With the rapid accumulation of relics of early man in all por-tions of the globe, which today fill scores of important museums, not scholars alone but the educated public gen-erally command a picture of human antiquity, human di-versity, human inventiveness, and human psychology of hitherto unsuspected dimensions. The picture is not only rich in detail, overwhelming in its mass of evidence of prehistoric man, but through the labors of a growing num-ber of specialists—today enrolled as anthropologists—has be-come consistent in perspective and vitally significant for the understanding of the processes of human culture. All the views of the story of man previous to this enlightenment may be regarded as variously and markedly erroneous. An-thropology stands as the correction of a great long-enduring error.

To place man in a correct biological perspective formed one great task of anthropology. The other and far more com-prehensive undertaking was the establishment of the per-spective of human culture. The past was but incompletely interred with the bones and artifacts of early man; it lived on in "fossil" stages of development among primitive peoples. The conception that the study of peoples living in primitive conditions would supply the data of our cultural ancestry was a further momentous discovery. These living survivors of our remote precursors became authentic documents of anthropology.

The Tasmanians were studied and described by nineteenth-century anthropologists as people still living in the Paleolithic culture-epoch, the Old Stone Age. The term "culture" came to mean the total status of a group in terms of their design for living. The data of such observations, like the collections of objects and photographs illustrating tribal culture that accompanied them, came to be systematized under distinctive headings. First and foremost was the social organization—the marriage relation, the family, lines of descent, caste, property rights, and the relations of the sexes. Language, religion, art, medicine, magic, inventions, industries, trades, professions, warfare, ceremonies, all yielded their quota to the total concept of primitive life and primitive mentality, of which the objects displayed in museum cases were but the visible, tangible embodiments.

There thus entered the intellectual horizon the fertile idea of evolution—of developmental stages through which the human mind and its intellectual, moral, and social embodiments in mores and institutions proceeded to their present estate. Darwinian evolution created a ferment in all phases of biological thought; it keyed up interest and, like every new instrument of interpretation, opened new horizons. Fertile as was the principle of evolution within its proper domain, in some of its applications, it introduced new error.

In the fervor of acceptance of the evolutionary creed there was a confusion between biological evolution, operating under Nature's forces, and successive stages of sociological development, determined by the complexities of human psy-

chology combined with circumstance. Cultural stages show
change and progress. It was, however, a very dubious inference that these could be set in an orderly series paralleling
steps in biological evolution. A brisk and lengthy controversy
arose in regard to marriage: how monogamy developed,
whether it possessed analogies among the life-habits of anthropoid apes, whether the course of social growth was from
promiscuity to polygamy to monogamy, how incest taboos
arose, how exogamy and endogamy; whether there were
matriarchal as well as patriarchal orders of society, how
castes and tribal expansion arose from family life, and so
on indefinitely. The diversities of social practice could
hardly be fitted into a single evolutionary pattern. Thus,
again, errors had slowly to be corrected, and anthropology
was called upon to supply the antecedents and early parallels
of what was to become the science of sociology.

The concept of mores, or cultural expressions, has become the center of active study. The traditional view of
the social contrasts gave us such loose terms as "savage"
and "barbarian" and, in discussions of their religious rites,
as "heathen" and "idol"—again reflections of a cultural compulsive applied to customs wholly out of range of the ideas
and practices of peoples of the modern, particularly the
Western, sophisticated world. The problem of cultural anthropology was to remove these crude, vague, and falsely
focused concepts and build up in their place an intelligent
interpretation of the ways of primitive men and mind. It is
of interest to note that in contrast to the prevailing superiority complex, which looked down upon "savage" ways as
unemerged degradation (with cannibalism as the extreme
of offensiveness), there was an occasional conception of the
"noble savage," popularized by Rousseau (1712-1778) as the
embodiment of virtues lost in the fierce struggle of civilization—a doctrine reflected from the then current belief in a
lost Paradise.

Pervading the mass and maze of custom under the sway

of a primitive mentality and a primitive social structure are certain habits and attitudes that constitute a primitive psychology. There has thus arisen a distinctive sector of cultural psychology which shapes the entire modern view of human nature and mind in the making. Within this frame certain concepts emerge of wide bearing, such as totemism and taboo, animism and "psychic participation," magical influence, and sympathetic magic. All of these primitive habits and attitudes survive and among more advanced peoples are ranked as superstitions.

One school of psychologists, the Freudians, find in these ancient cultural products the patterns of psychic conflicts in our sophisticated personalities; in their view, myth, symbolism, primitive psychic residues generally, form an archaic level of operation still explicitly influential in our psychic life.

The scope in all its glory—to our thinking, of glorious error—of the primitive mind was conveyed in the classic work of Sir James Frazer, *The Golden Bough,* which began with two volumes and required ten more in successive editions to bring it to completion. Its theme is "the gradual evolution of human thought from savagery to civilization." Its focus within that comprehensive orbit is the mythical-magical-religious pattern that pervades the products of belief —rite and custom—in endless diversity of expression but a common esprit of motive and solution of the meaning of life, continuous in the stream of historical time and geographical space. The priest as personal embodiment of the mystic force, totem for protection, taboo in avoidance of evil, purification for offense, vicarious sacrifice in the scapegoat, omens, portents, propitiation, readings of fates, charms and amulets, evil eye and witchcraft, fears intense and ecstasies no less—all this mentality which abounds "at a low level of social and intellectual culture" is far more characteristic of the human psyche in its total career than the limited areas of emancipation through which we trace the intellectual history of our own segment of mankind. It delineates a primary and primeval anthropo-psychology, from which we emerged.[2]

It will thus be evident that the "anthropological" view of human origins, human mores, human mentality, human social forces, has tremendously enriched the entire group of humanistic sciences, contributing a new insight, a new dimension to our understanding, as in the nearer perspective anthropology becomes history.

The momentous significance of racial alliances and antipathies in international affairs, bringing with it problems of conflict and assimilation as well as ideal fostered by historical tradition, gives anthropology an added importance. Its counsels would mitigate prejudice and bring tolerance into deliberations. That man is his own worst enemy, that feuds and wars are maintained on differentiations on which anthropology has a decisive voice, still further enhances the value of a study whose ramifications go deep into the biological past and extend far toward the adjustments of the present and the shaping of the future. The racial problem alone, in its bearings on history and politics, would make anthropology the intensely significant study of man.

In the technical sense, modern anthropology arose only when all these streams of knowledge converged upon man as the basic study of mankind. Like all sciences, it had to develop its techniques, and did so under the warning influense of ancient errors and the specific impact of contemporary cultural compulsives. Like all modern sciences, again, it developed special and varied problems, to some of which the ensuing chapter is devoted.

[2] In this contribution to "humanology" the significant work of Frazer does not stand alone. Taking its start in the "animism" of Sir Edward Tylor, it finds a penetrating restatement in the writings of Levy-Brühl, in the analysis of Boas, in the contributions of Westermarck, Radin, Malinowski, Briffault, and many another follower of the psychological clue in the early story of man.

Chapter IX

ERROR IN ANTHROPOLOGY

By Ralph Linton

ANTHROPOLOGY *shares the common aspiration of all the sciences to consider its subject-matter objectively, but it can hardly attain that ideal. With the best intentions it remains impossible to view man and his works with the same detachment with which one observes the geologic system or the reactions of amœbae. Differences between breeds of men are felt to be more important and more significant than differences between breeds of dogs. To the student of human behavior, the habits and values to which he has been reared seem more natural or better than those of alien groups. He may learn to avoid the more naïve expressions of this attitude, but he does not escape its influence completely.*

With the decline of supernaturalism and divine sanctions, the ruling classes turned to anthropology to justify belief in the superiority of their own physical type and institutions, to justify the exploitation of other groups and the rightness of the status quo—*or, conversely, the necessity of changing the* status quo *in accordance with dominant social or political theories. The development of anthropology is closely related to the course of modern European events. Social and political factors only remotely connected with the advance of science have determined its trends as much as have genuine advances in scientific knowledge. European ethnocentrism runs through its history in a continued thread. Had anthropology developed in another culture or under different historic conditions, it might well have encouraged other social and political implications.*

In terms of its major problems the development of anthropology follows that of its component subdivisions: the two great comprehensive departments, so distinct in procedure and pursuit as to constitute coördinate sciences, are physical anthropology

*and cultural anthropology. Physical anthropology deals with
man as an animal; its advance follows upon the findings of
biology, zoölogy, anatomy, and genetics. Cultural anthropology
deals with the works of man and his behavior in groups, and
it is thus closely correlated with sociology and psychology in
content, methods, and concepts. Except in so far as there may
be established a significant correlation between forms of culture
and physical types, the two departments focus upon distinct
problems.*

THE STORY of man's ways of living, to meet his needs
and adjust to his habitat, is a continuous tale, from
the obscure past of the caveman to our own complex designs
for the same ends. The total patterns thus developed constitute the field of cultural anthropology. *Archæology* is
the term applied to the study of the civilizations of the
past, particularly of those within historic periods which
have left abundant relics and records and stand in the line
of ancestry of civilizations still extant. The study of culture-
patterns among living peoples, more particularly those of
lower levels of civilization down to primitive status, not of
European or comparable traditions, are assigned to *ethnology*—in its descriptive phase called *ethnography*. Where no
written records are available, as for the long primitive preliterate epochs, the anthropologist's methods of securing
data and interpreting them are necessarily conditioned by
such limitations. His purpose, however, always goes beyond
the mere record of customs, myths, traditions, inventions,
social organization, and mode of life generally, to an interpretation of the complex as an integrated whole. In this
pursuit cultural anthropology overlaps and calls upon the
aid of sociology and psychology.

CLASSICAL CONTRIBUTIONS

The several branches of anthropology emerged within the
modern period of scientific investigation. Precursors for each
branch may be traced in the classical period, but these be-

ginnings were slight and are little related to present positions. The early and persistent problem was the origin of man and the diversities of race. Even in classical times the origin of man had become a matter of religious dogma or traditional myth, quite regardless of observable fact. Thus, the brilliant Lucretius (first century B.C.), whose speculations on the beginnings of culture came rather close to the truth, wrote fantastically of man's origin: "Then must you know did the earth first give forth races of mortal men. For much heat and moisture would then abound in the fields; then, therefore, whenever a suitable spot offered, wombs would grow attached to the earth by roots; and when the warmth of the infants, flying the wet and craving the air, had opened these in the fullness of time, Nature would turn to that spot the pores of the earth and constrain it to yield from its open veins a liquid most like milk. . . . To the children the earth would furnish food, the heat raiment, the grass a bed rich in abundance of soft down." [1]

To the study and classification of races the Greeks and Romans contributed little. They were not keenly conscious of the existence of race, and there was little to make them so. The ancient world was a very small world, mainly populated by racially mixed groups of the same Caucasic stock. The clearly observable differences in physical type were slight and not readily associated with differences in tribal or national affiliations or with social status. Thus there were many tall blond Gauls, but there were also some short dark ones. The Roman, whose neighbors were predominantly short and dark, would find among them a sprinkling of tall blonds similar to the dominant type in Gaul. A Roman or a Gaul was characterized by his speech and dress more than by his physical type. The wars were mostly local, and conqueror and conquered, master and slave, looked much alike.

The only people of non-Caucasic stock known to the Greeks and Romans were the Negroes of the Upper Nile.

[1] Lucretius, *De Rerum Natura*, Book V.

They were too distant to figure largely in the slave-market, and the Greek attitude toward these Ethiopians was much the same as that toward other foreign groups. Still, their marked difference invited an explanation, which is embodied in the legend of Phaëton. The black skin and frizzy hair were referred to the action of the sun. The philosophers may have doubted the legend, but they accepted the explanation which—a shot at a venture—came close to the mark, as modern knowledge finds a relation between pigmentation and the actinic rays of the sun.

With few exceptions, the sense for the past which fostered the study of archæology was unknown to the ancients. An early Mesopotamian queen, it is true, preserved in a room of her palace inscriptions and bits of statuary unearthed in laying the foundations of the building—which royal hobby may have originated the earliest archæological museum. But the common attitude toward the past is indicated by the fact that fifty centuries ago tomb-robbing was a common and lucrative vocation in Egypt. Each successive occupation raked the soil for the stone implements and objects of metal left by the more remote inhabitants. The Egyptians mined their ancient tombs in search of precious metals and salable art-objects, while the men of the Iron Age assigned magical virtues to the stone implements of their ancestors. They believed them to be thunderbolts, a misconception which seems to have spread over the whole of the Old World. It still survives among European peasants, in China, and even in West Africa, where stone celts are placed on altars of the Thunder God.

The Greeks and Romans, so far as they had developed an interest in history, relied on written records and oral traditions for their reconstructions of the past; it was not in the temper of their study to verify such historical materials by actual excavation. So it is rather startling, considering their remoteness from realistic data, to find in Lucretius a reference to the use of stone, bronze, and iron implements in their correct chronological order: "Arms of

old were hands, nails and teeth, and stones and boughs
broken off from the forest.... Afterward the force of iron
and copper were discovered; and the use of copper was
known before that of iron, as its nature is easier to work
and it is found in greater quantity. With copper they would
labor the soil of the earth, with copper stir up the billows
of war and deal about wide-gapping wounds.... Then by
slow steps the sword of iron gained ground and the make
of the copper sickle became a byword; and with iron they
began to plow through the earth's soil, and the struggles of
wavering war were rendered equal." [2] Since Lucretius could
have had no direct evidence of the accuracy of this sequence,
it seems probable that he was drawing upon tradition still
active enough to be valid. Both the Etruscans and early
Romans were Bronze Age peoples; their successors, partly
familiar with iron, would know that iron was a later introduction.

When expeditions of venture or trade, and in due course
colonization, were undertaken, contributions to ethnology
accumulated. Men were naturally impressed by the strange
behavior of peoples of other tribes. Among the very ancient
writers, Herodotus (fifth century B.C.) was constantly occupied with such descriptions; for the later period Tacitus'
account (first century A.D.) of the ancient Germans is a good
piece of ethnography; and many a traveler or historian from
that day to the advent of the modern era scattered bits of
such information through his works.

The early writers were inclined to be overcredulous and
to emphasize those features in which foreign groups differed
from themselves; yet they approach modern standards of
description more closely than do writers of the next thousand years—the dark ages of learning generally. Medieval
Europeans were reared in an atmosphere of miracles; the
literate public had a keen taste for marvels. The favored
travelers' tales of this period include composite beasts, men
with faces in their chests, and similar wonders. Marco Polo,

[2] *Ibid.*

crudely credulous in some respects, was for the most part a shining exception; his fanciful "marvels" were accepted without question, while his sober and surprisingly accurate account of the China of the late thirteenth century earned him a reputation as a colossal liar.

Some classical writers went beyond description to interpretation, and by way of the Renaissance their views were directly transmitted to modern thought. Plato (fourth century B.C.), in his speculations upon the nature and origin of society, was the first propounder of ethnological as well as sociological theories. His analyses were based on Greek social systems, not upon comparative data. He may have regarded the study of barbarian institutions as beneath his dignity. Yet Plato and his near forerunner Protagoras must be ranked as pioneer contributors to ethnology and history.

Many a Greek city traced its peculiarities of belief or custom, of which it was inordinately proud, to some allegedly historic incident—usually the mandate of some early hero. Frequently a god was invoked to lend prestige to a traditional ceremony. Such legendary explanations were acceptable even to fairly critical historians. Plutarch, for example, seems to have believed that the complex social and political organization of Sparta originated in the mind of Lycurgus and was imposed by him upon the Spartan State. It was many centuries before this type of error succumbed to the concept of societal evolution.

SCRIPTURAL DOGMA

The meager beginnings of anthropology in classical lore were completely swept away during the Middle Ages. The Church anticipated modern anthropology in centering its interest in man, but in man as recorded in the Scriptures. The Scriptural account included specific pronouncements as to man's origin and place in the universe; in the story of Noah and his sons it offered an explanation of the varieties of races. It establishes the length of time that man had been on earth and the reason for his being here. Building upon the Scriptures, the Church provided a complete set of values and a canon for living; in accordance with these it divided all cultures into Christian—that is, approved and proper—and heathen—that is, disapproved and improper. All heathen cultures were at best examples of human error, while at worst they were devices of Satan, devised to keep damned souls securely in his net. In either case it was the duty of Christians to destroy them, not to study them. In fact, to study any aspect of man or his works was to reopen some question which had already been closed and locked with the keys of Saint Peter. The Church was firmly convinced that it had received the truth through revelation; upon this truth it had built up a consistent system, and it regarded questioning of its position in any respect as heretical.

As the Middle Ages merged into the Renaissance, men turned their attention from the contemplation of dialectical theology to a study of the tangible world without. As the objective attitude grew in importance, it included a clearer perspective of man himself. Slowly the idea grew that man was simply another natural phenomenon to be studied by the methods of science. With this idea established, anthropology in the modern sense was in the making.

THE ORIGIN OF MAN

Physical anthropology followed upon the development of zoölogy and comparative anatomy. As these sciences gained ground, the Biblical or other accredited versions of human and racial origins appeared less and less satisfactory, though it remained for the discovery of the principles of evolution to place the entire body of data upon a scientific basis. Abundant observations proved the close similarity of man to the other primates. As early as the Renaissance, progressive thinkers realized that man and primates were closely related. In 1489 Leonardo da Vinci made a notation regarding "the description of man, which includes such other creatures as are of almost the same species, as apes, monkeys, and the like." This view was greatly strengthened when, in the late sixteenth and early seventeenth centuries, Europeans generally became familiar with the anthropoid apes. In 1699 Edward Tyson, an English anatomist, published an anatomical comparison of a monkey, a "pigmy" (actually a chimpanzee), and a man. In this study the close structural similarity of these three species was fearlessly pointed out, although Tyson drew no conclusions as to their possible relations. Until the publication of Darwin's *Origin of Species* in 1859, there were no adequate principles of interpretation, certainly none on a naturalistic basis. In his *Descent of Man* (1871) the doctrine of the development of our own species from some subhuman primate was clearly stated. Darwin's conclusions were based entirely on comparative anatomy and embryology; no human fossils of the oldest periods were then known, with the exception of the Neanderthal specimen of early man, which Darwin, however, did not use as evidence in his argument.

The recognition of human evolution inaugurated a frantic search for man's ancestors, misleadingly popularized as the quest for "the missing link." Though not in the same sense, this search still continues, and with no final conclusion. The primate species immediately ancestral to our own still remains to be discovered; even the locale of human origin is still conjectural. Although each of the early human or subhuman fossils that have come to light has found enthusiasts to support its claims to the ancestral distinction, none of these attributions has been generally accepted. Similarly, the point at which the hominoid stem diverged from the other lines in primate evolution is still a matter of dispute. Anthropologists are confident that man was evolved from some subhuman primate and that he was not evolved from any living species; but further steps toward solving the problem of his origin must wait upon new findings in the field of paleontology.

VARIETIES OF RACE

The lack of positive evidence of man's ancestry, combined with the uncompromising attitude of the Church on the dogma of special creation, delayed the development of the study of racial origins until after the study of physical differences—that is, of race—had got well under way. The concept of race had taken form at least fifty years before the publication of *The Origin of Species*—indeed, by that time was already the subject of a large body of literature. Eighteenth-century zoölogists had included man and his types in their systematic surveys of the animal series. In 1735 Linnæus assigned man to a place in his zoölogical classifications—calling him *Homo sapiens*—and tentatively divided the human species into several varieties. Likewise Buffon in his *Histoire Naturelle* (1749), two volumes of which considered man, recognized varieties of race. The first investigator to devise an exact scientific approach to the determination of racial differences, however, was Johann Friedrich Blumenbach (1752-1840), often called the father of physical anthropology. Blumenbach first employed cranial measurements as an important clue in racial classification. Between 1790 and 1828 he published measurements and comparative descriptions of sixty skulls from widely different parts of the world. On the basis of his observations he divided mankind into five distinct varieties or families: the Caucasian or white, the Mongolian or yellow, the Negro or black, the Malayan or brown, and the American or red. He called the white race Caucasian after an especially perfect type-form in his collection, the skull of a woman from the Caucasus.

To supplement observations of superficial differences in appearance—mainly coloring and facial features—such resort to measurements was necessary, particularly to skeletal measurements that could be applied to remains of bygone races. *Anthropometry* was the name given to this branch of research. It brought into vogue many types of indices—such as the cephalic index, the ratio of breadth to length of head, the basis for distinguishing between dolicocephalic, or long-headed, and brachycephalic, or short-headed, individuals. These cranial characteristics were correlated with stature, pigmentation of iris, and hair, and upon such composite determination the European races were divided into Nordic, Alpine, and Mediterranean. These groupings crisscross national affiliations, so largely the issue of political movements. They still play their part in the dissensions as well as the disparities of race.

There was little in these initial stages of the study of human varieties that came into serious clash or that carried implications incompatible with the traditional Scriptural account. It is true that Blumenbach recognized five primary races and the Bible only three—named for the three sons of Noah; but this discrepancy was not seriously disturbing. Not so, however, the concepts of race that followed Darwin and became involved in the furious controversy over the principles of evolution which still sporadically engages the Biblicists.

RACE AND EVOLUTION

The problem of racial origins remained in oblivion until about the beginning of the eighteenth century. Then historical circumstances emphasized the physical differences of racial strains. With the rapid rise of Europe to world domination these differentiations of human groups became socially important. Travelers landing in America or West Africa or the islands of the Pacific met individuals of a physical type markedly different from their own. As such white men came increasingly as conquerors and rulers, jealous of their prerogatives, and as the native could imitate white dress and manners but could not change his skin, physical type became an easy and certain way of determining social status. Any white man was a member of the ruling group, any native a member of the subject group. The increasing importance of Negro slavery during this period also influenced the European's consciousness of race. Any black man seen in Europe was a chattel, while any white man was free.

The study of race thus began in a period when the doctrines of separate creation and fixity of species were unquestioned, when there was no inkling of the biological meaning of race. Racial classifications were all based on the assumption of a small series of originally distinct and clearly marked human varieties. Groups that were intermediate between these hypothetical pure and original types were explained as the result of crossing, a process which could be observed when disparate groups were brought into contact. These serious errors had to give way to a more correct view of human origins and racial divisions.

By the time the doctrines of evolution were enunciated, the existence of original, widely different human physical types had become a tenet of anthropology. It still dominates thought in this field, though the position cannot readily be brought into line with the principle of evolution. To posit a small group of racial types as the parent stems of the many human varieties now extant involves one or other of two assumptions: either that these original races were evolved from distinct subhuman primates, or that the evolution of these original races was completed at a remote period; that, furthermore, the members of such original groups, as long as they remained pure-bred, retained their distinctive physical characteristics unchanged through the vicissitudes of migration and the resulting environmental changes. Of these alternatives, the first is certainly in best agreement with evolutionary processes, yet it has found few supporters. Aside from the close similarity of all racial groups in all but a few and rather minor characteristics, the assumption meets the almost insurmountable difficulty that all living races belong to the same species by the simple test of hybridization. Different races, no matter how diverse in physical type, produce hybrids which are not only as fertile as, but in many cases more fertile than, the pure parent strains.

The second alternative, that of early completed human evolution, is now tacitly accepted by many anthropologists, although the evidence for it scanty. The fact that a physical type, such as the population of Egypt, has survived in a particular region for several thousand years without noticeable change is not a proof that such a stabilized type has lost the ability to change when placed under different environmental conditions. When a species or variety has reached a satisfactory adaptation to its environment, the necessity for further structural modification ceases. Boas has shown [3] that the children of certain immigrant groups in the United States differ from their parents in head-form, and that the extent of these differences, though very slight, correlates with the years of residence in America of the parents. If this result can be accepted as an indication that human evolution is still in progress and that new racial types can still be developed in response to changing environment, it is in conflict with the "primary race" theory.

The believer in primary races must face the fact that modern man is markedly diverse in physical type. Many groups, apparently pure-bred or nearly so, cannot be fitted into the primary-race classification because they present a mixture of physical characteristics which ally them to two or even three racial stocks in almost equal measure. Formerly such intermediate groups were assigned an independent status as primary races; but by this process the number of primary races was increased to a point where the concept became meaningless. Unwilling to abandon the old pre-evolutionary doctrine, modern students of race have shown a tendency to retrench and to reduce all mankind to three primary stocks, Caucasoid, Mongoloid, and Negroid, passing lightly over the peoples who cannot be fitted into this scheme. The evolutionary biologist may prefer to abandon the idea of primary races altogether, or at most to retain it

[3] Franz Boas, *Changes in Bodily Form of Descendants of Immigrants* (Washington, 1911).

as a basis for purely objective classification with no implications of a common ancestry for all the members of each stock.

PRIMITIVE TYPES AND PRIMITIVE LIFE

Early man must have had tools and fire from the moment that he became distinctly human. With these aids he could penetrate and exploit a wide range of environments. At the beginning of the Pleistocene the whole of the Old World formed an almost continuous land mass over which recently evolved man could have wandered freely. Unless we accept the evolution of primary races from different subhuman species, we must suppose that the first men carried in their genes potentialities for the development of all the physical varieties now extant. The first *Homo,* an anthropological Adam, was an undifferentiated ancestral form. Presumably early man, like the hunting peoples of today, lived in small inbred groups, the size of which was limited by the food-supply, dividing into further groups as the food-supply became too sparse. Such inbred groups offered an almost ideal condition for the fixation of physical varieties advantageous in a particular environment, while there was a constant urge to the formation of new groups as long as promising territory was available. It seems safe to conclude that early man spread rapidly and widely, that from these inbred bands arose many human varieties. As natural environments are not sharply divided and as closely similar environments may occur at widely separated points, the adaptive variations which were a response to the environments would show gradual changes throughout continuous areas, and also a certain degree of parallelism in similar though widely separated environments. Modern conditions could thus be accounted for by the assumption, not of a small group of primary races, but of many originally isolated groups whose physical types would diverge or converge in response to dissimilar or similar environmental conditions. While superior strains might be evolved at certain points and expand their range, as soon as this expansion brought them into markedly different environments, the process of physical adaptation would begin again, resulting in the production of still other and new types.

Such an explanation of the present diversity of racial types and their frequently irregular distribution is unacceptable to most physical anthropologists. The doctrine of primary races has become such an integral part of European ideology that it cannot be abandoned without a severe wrench. It has been seized upon and amplified to provide Europeans with intellectual and emotional justifications for the domination and exploitation of other groups and for maintaining a social distance between themselves and the members of other races. Beginning with preëvolutionary reliance upon the Biblical text in which the sons of Ham were condemned to menial labor, it has accumulated pseudo-scientific rationalizations favoring dominance and superiority of the white race. Having set up his ideal racial stocks, the white man has attempted to range these stocks in an evolutionary series with himself at the top. To this end he has had to exaggerate the importance of those features in which he was least anthropoid and to pass over lightly such details as hairiness and massive brow-ridges in which he more closely resembles the anthropoids. When this line of rationalization collapsed in the face of the increasing evidence for disharmony in human evolution and for the like evolutionary status of all living groups, he turned to claims of superior intelligence for his own stock. Since he usually judged the intelligence of other groups by tests and appraisals devised by white men for white men upon a knowledge of modern European ways, the results were more gratifying than scientific—again an error of intrusion unfair to other ethnic groups.

If the European mind were prepared to abandon the assumption of primary races, substituting for it that of an indefinite number of graded lines of resemblance carried in the heredity, these laboriously accumulated proofs of white superiority would have to go by the board. Nor would an appeal to history be of avail: for history proves conclusively that Europe was a marginal area to which civilization came late; that, even in 1600, its peoples were on the defensive against the steadily advancing Asiatics. If the concept of primary races were abandoned, there would remain only the assumption that certain small groups of related individuals—family lines, in fact—were less anthropoid or physically stronger or more mentally alert than others. This original basis of differentiation, to which must be added the large equalizing through mixture and migration, would provide no justification for the exploitation of one people, one nation, by another.

In this connection a word may be said about the recently resurrected Nordic myth. The Nordic race is as vaguely delimited as any of the other races that have been set up as primary groups. Ideally the Nordic is tall, blond, and powerful, with a long head, high nose, and narrow face, also an abundant beard. He appears typically in German sacred art, but in his ideal type he is not very common even in Germany. All over Europe may be found blonds with broad heads as well as with long heads, blonds with broad faces as well as narrow faces, short blond strains and tall blond strains. As to the Nordic's inherent abilities, raiders from the region of the North Sea and the Baltic—the presumed center for Nordic dispersal—have contributed much to the confusion of European history and have provided feudal aristocracies for several of the countries of the south; yet the Nordic homeland has contributed little to the growth of European civilization. In fact, it has rather uniformly shown the culture-lag characteristic of marginal areas.

CLASSICAL AND PREHISTORIC ARCHÆOLOGY

Of the two main branches of cultural anthropology, we have already noted that the beginnings of archæology came with the Renaissance, which ushered in a vogue for the classical. Simple tomb-robbing gave place to a systematic search for ancient works of art, while the profound respect of Renaissance engineers for classical builders and the growing interest in ancient inscriptions led to an increasingly careful study of everything uncovered. Classical archæology was well under way in Italy by the beginning of the eighteenth century. Egyptian archæology began with Napoleon's expedition to Egypt. Mesopotamian studies were inaugurated in the middle of the nineteenth century. In all these regions the investigators were dealing with highly civilized cultures whose history was already partially known. They centered their investigations at first on the finding of inscriptions and objects for museum display. They failed to appreciate the importance of recording the location and the level of excavation at which specimens were found, and they did not know how to handle the evidence thus offered; hence they often destroyed more information than they gained. Moreover, they gave undue weight to the statements of classical historians and tried to bring their own findings into agreement with them. It was only when the classical archæologists began to borrow the techniques developed by investigators working with the remains of unknown preliterate peoples and to reconstruct history on the basis of first-hand evidence that their work took on a really scientific aspect.

The archæology of preliterate peoples began its career as a sort of poor relation to classical archæology. The exploration of early human occupation sites yielded no precious metals and no works of art acceptable to the pseudo-classical tastes of the seventeenth and eighteenth centuries. It was left to local antiquarians—school-teachers and country gentlemen of inquiring mind. Also, it came to lie somewhat under the ban of the Church, since its findings threw doubt on the Scriptural chronology. The first modern European to recognize the true nature of stone implements was Mercati, toward the close of the sixteenth century. His writings, however, were not published until more than a century after his death, by which time a comparison of European stone implements with those found actually in use in other parts of the world had finally established their status as human artifacts instead of thunderbolts.

During the whole of the eighteenth century the development of prehistoric archæology was rapid. In 1715 Conyer, a London antiquarian, found worked flints in association with the skeleton of an elephant in undisturbed gravel deposits. In 1797 John Frere, another Englishman, made a similar find in Suffolk. Frere believed that his finds belonged to "a very distant period, much more remote in time than the modern world"; but few of his contemporaries accepted this conclusion.

THE AGE OF MAN

It precipitated, in fact, the problem of the age of man and his possible association with extinct animals in Europe. Those prepared to admit that the finds were authentic accounted for the presence of the elephants by explaining them as war elephants, imported by the Romans for use against the Britons, and supported their view with a reference in a classical work on military strategy. A considerably larger and more scientifically minded group, among whom Cuvier was outstanding, denied the authenticity of all such finds. As late as 1823 Cuvier declared, "All evidence leads us to believe that the human species did not exist at all in the countries where the fossil bones were found at the period of the upheavals which buried them." It was not until 1846, when Boucher de Perthes published his *Antiquités celtiques et antédiluviennes,* that the great age of both man and his implements came to be generally recognized.

The later workers in prehistoric sites, belonging to the Stone, Bronze, and Iron Ages, encountered dwindling opposition. They went forward steadily, amassing data and developing techniques for their interpretation. As early as 1750 J. G. Eccard had suggested a sequence of prehistoric periods, based upon his study of sites in Germany. In 1758 Goguet established the European culture-sequence of stone, copper, bronze and iron ages; this was soon verified and developed by the Danish archæologists C. J. Thomsen (died 1865) and J. J. A. Worsaac (1821-1885).

The paucity of material and the complete lack of inscriptions forced prehistoric archæologists to pay a great deal of attention to methods for studying their finds, including particularly means of estimating the age of human relics. When possible, the early prehistorians used geologic methods, studying the river-terraces and glacial deposits from which stone implements were obtained; but the spans of geologic epochs were too vast for measuring the brief story of man. Archæologists turned next to intensive studies of the successive deposits or levels of occupation (stratigraphy) and of evidences of evolution in implement types. The position of objects *in situ* in accumulated deposits afforded clear proof of the sequence of cultures; the depths of the accumulations provided a rough idea of the time intervals involved. Once special forms and decorations of typical objects (typology) such as axes, scrapers, spears, etc., were associated with well-established periods, the forms themselves, wherever found, became evidence of relative and approximate age and place in a culture-sequence. They were, however, of little use in the establishment of exact chronologies. They made it possible to say with certainty that one type of artifact was older than another, but they gave no clue to the length of time that separated them. By combining the geologic, stratigraphic, and typological techniques, archæologists are now able to assign with fair accuracy the sequence of prehistoric cultures in any locality; but they do not yet venture to date even the end of the Old Stone Age in Europe more closely than within a range of several thousand years.

A distinctive and, within limits, revolutionary technique for determining the chronology of remains of preliterate cultures was contributed by A. E. Douglas as recently as 1929 in *The Secret of the Southwest Solved by Talkative Tree-Rings*. As has happened before, he applied the findings of another science to the advancement of anthropology. His method is derived from the study of the annual growth-rings of timbers from ancient ruins in New Mexico and Colorado. Their variations were correlated through other data with variations in rainfall, and the count of the annual rings indicated the years in which the trees had been cut and the ruins built. While the method is applicable only to regions arid enough to preserve wood, the peoples in such regions often traded far and wide; hence their chronology can be made to serve for the regions in which objects entering into their commerce can be found. It seems probable from this research that the entire development of Southwestern culture in the United States took place within about 2,000 years, instead of the 4,000 or 5,000 years previously assigned. The utilization of tree-rings as an anthropological clock may well take its place in the annals of novel techniques in science.

MAN IN AMERICA

This account of the rise of archæological research and the errors regarding man's past that it served to correct may stand as representative of the course of progress in this branch of anthropology. Its major reference has been to European occupation. American archæology offers additional points of interest.

Anthropologists are agreed that man arose somewhere in the Old World. How and when he arrived on the American continent is altogether uncertain. Some make his arrival late —perhaps only 10,000 years or so ago; others grant him a period of 40,000 to 50,000 years and find confirmation in the recent discovery in Minnesota of a skeleton beneath undisturbed deposits of glacial age.

In the historical period numerous problems are raised by the remarkable ruins of Mexico and Central America; also in the interpretation of the Indian mounds—some being occupation sites, others burial-grounds, and frequent examples exhibiting symbolic, animal-like forms. The contact of European settlers with Indian tribes through three centuries gave to American archæology a distinctive and rich opportunity. Nevertheless, there were in the early periods fantastic speculations to account for architectural remains on American soil, including migrations from Egypt and a mysterious lost Atlantis. These theories soon died, as far as scientists were concerned, but some fifteen years ago their ghosts rose to walk again under the stimulus of G. Elliot Smith and W. J. Perry. According to their "diffusionist" school, the origins of all the world's civilizations can be traced to bands of hardy adventurers who set out from Egypt in ancient days and wandered over the face of the earth in search of gold and pearls. Wherever they touched, they left behind them the seeds of civilization, including such diverse elements as building with large stones, worshiping the sun, and tattooing. If this heliolithic theory is correct, the Egyptian adventurers must have been able to wander in time as readily as in space. Central American chronology is known with considerable accuracy, and by the time the Mayas built their first stone temples, Egypt had become a Roman province.

While early Spanish chronicles proved beyond doubt that Indians had built the great monuments of Mexico and Central America, no such records existed for the building of the earthworks that white pioneers found scattered over most of the Mississippi watershed. Until about 1880 it was generally believed that these remains were the work of a mysterious race, the Mound-Builders, who were supposed to have passed from the scene before the Indian arrived. This idea was completely exploded by the work of Cyrus Thomas and his associates in the Smithsonian Institution, but it is still firmly entrenched as a popular belief. It has had an important effect on the development of Mississippi Valley archæology. Since it was taken for granted that the mounds were not built by Indians, the study of them was carried on for many years without reference to Indians, their known habits or artifacts. In this way a demarkation developed between the fields of archæology and ethnology in America, to the detriment of both.

CULTURAL EVOLUTION

However improved the methods of archæology may become, they can only imperfectly reveal the life of long ago. It is the paleontology of culture, uncovering the skeletons of dead civilizations but giving us only hints of what these civilizations were like when they lived and moved. To understand the vital processes of culture we must observe these processes in action, in living societies and the individuals who compose them. Such is the task of ethnology.

From early times curiosity inspired ethnographic description. Travelers had always brought back interesting tales of the strange peoples they encountered; but toward the close of the seventeenth century these tales began to acquire a practical value for missionaries, merchants, and the rest of the vanguard of European expansion. Their demands for accurate information led to more systematic and more enlightened studies, out of which gradually developed the techniques of modern ethnography in observation, selection, and presentation of data. These techniques, however, are still so imperfect that it is doubtful whether any observer has ever been able to see a culture as a whole, much less record it in its entirety. Even after the language barrier has been overcome and the observer has been admitted to participation in the life of the native group, he remains at best alien and subject to the limitations of his civilized background and education, often further restricted by a selective interest in some few aspects of primitive life.

Its imperfections notwithstanding, descriptive ethnology was already well developed before analytical ethnology came into being to assume the task of interpretation. Not only were the scientific methods devised for the study of the material world extremely difficult to apply to social phenomena, but for generations scientists were content to leave the explanation of such phenomena to the philosopher and the theologian. Hegel (1770-1831) initiated the emancipation of the social sciences by his fruitful postulate of stages of history, but ethnology did not emerge as a distinct science until the latter half of the nineteenth century—so late, indeed, that one of its greatest pioneers, Sir Edward Tylor, died so recently as 1917. The first workers in the field were confronted with the Herculean task of sifting the vast accumulation of unorganized data about primitive peoples. They had first of all to develop some system by which the facts of culture and also cultures as wholes could be classified and arranged. They had also to develop techniques for the analysis of culture and theories to account for both observed similarities and differences. None of these problems has been completely solved as yet, and none of the attempted approaches to them has so far been proved entirely worthless or been completely abandoned.

The earliest approach to ethnological problems was that of the evolutionary school. As aids toward science's prime requisite of a system of classification, they had both the theory of the stages of history, by this time well developed, and the newly enunciated theory of biological evolution. For classificatory purposes these two theories did not clash, while the evolutionary theory offered a naturalistic explanation of progress to replace the supernaturalistic one of the earlier historians. It is hard to say which of these doctrines was drawn upon more heavily in the formation of the evolutionary school. Sir Edward Tylor (1832-1917) and Herbert Spencer (1820-1903) admitted the truth of biological evolution, but the American Lewis Henry Morgan (1818-1881), whose approach to the study of culture and society was almost identical, was a conservative Biblicist strongly opposed to the theory of biological evolution and carefully avoided the use of the term in his writings.

The evolutionary school built its theoretical structure upon the postulate that all human groups have much the same intellectual and emotional equipment and will react similarly to similar situations. Thus, if there are two tribes to both of whom water transport would be advantageous and both of whom have suitable wood and tools for working it, both of them will spontaneously invent canoes. When such tribes have advanced somewhat farther in their cultural equipment, they will both inevitably take the next step, say the invention of oars or sails, and so onward and upward. Working from this basic postulate, the evolutionists developed a series of evolutionary sequences for various institutions. They then assumed that certain combinations of institutions represented stages in the development of culture, thus arriving at a basis for the classification of cultures as wholes. They outlined three main stages, savagery, barbarism, and civilization, later subdividing and amplifying these, and assigned all the cultures of the world to one or another of these stages on the basis of content. Uncivilized peoples were regarded as living fossils, preserving into the present institutions which the ancestors of civilized peoples had once had but had outgrown.

In their establishment of stages in the evolution both of institutions and of cultures as wholes, the evolutionary school resorted to a curious and thoroughly unscientific device. Since the actual history of most cultures was unknown, they assumed that the institutions characteristic of Victorian England (the first evolutionists, with the exception of Morgan, were English) represented the last evolutionary stage in every case, and that the corresponding institutions of other societies could be arranged in a descending scale, their position in this scale being determined by the degree in which they differed from the Victorian institutions. Thus, promiscuity, being most unlike the English idea of marriage, must be the first or lowest form of mating. Group marriage, which was only slightly less objectionable, must come next as a method of limiting promiscuity and starting man on the right path which, after many vicissitudes, would lead him to the happy, moral monogamous life. Coupled with this arrangement of stages went a belief in the unilinear evolution of all institutions and cultures, that is, that all cultures had passed or were passing through exactly the same stages in their upward climb.

The theory of an inevitable, unilinear evolution of culture has been abandoned by practically all ethnologists. It cannot stand in the face of known facts. Stages of culture take on a varying form in their many adaptations. There is, of course, no lack of data of certain kinds to support the underlying idea that common practical needs, common emotional needs, common mental habits, will lead to similar issues in inventions, customs, beliefs, and that these will more or less form a series of lower, cruder, earlier, and of higher, elaborated, and later expressions. Besides our example of the canoe, the history of invention is full of parallels of culture-stages. The story of fire-making, of drilling and polishing, of weaving, of pottery, of forged and alloyed metals, would all fall within a similar pattern. Wherever one touches the stream of culture, one finds evidences of "evolution."

To apply a similar concept of sequences to social products and intellectual interests was a tempting and, within limits, a valid procedure; but it readily led to the common error of forcing data into a theoretical frame. It overlooked the great diversity of solutions possible within a common psychology and a comparable environment. The possibility of copying and of mutual influence of cultural products likewise enters. Every people has mating customs and develops a cult which is also an idea of the marriage relation, including the family as an institution. But the variety of its expressions is endless, and the illusory simplification of the evolutionists under the Victorian cultural compulsive—though in one form it has the authoritative support of Edward Westermarck—seems incompatible with the disparities of customs within the groups in comparable stages of culture, and especially with the fact that monogamous unions may occur at almost any stage. The evolutionary interpretation led to varied errors, including, in this instance, a false analogy between anthropoid and human relations—the one biologically, the other culturally, established. Moreover, it ignored many forms of sporadically distributed ethnic or tribal customs that do not conform to any evolutionary pattern, such as the custom of *couvade,* in which the husband takes to bed when the wife is confined.

The evolutionary theory posited fishing, hunting, pastoral, and village stages of culture, and likewise a matriarchal as preceding the patriarchal form of tribal government. Yet among American Indians the groups of lowest culture are almost uniformly patriarchal; indeed, we can trace certain actual shifts from patriarchal to matriarchal institutions. Moreover, the American Indian had no pastoral stage and, except in Peru, no domestic animals to herd. Nor does hunting or food-gathering always precede agriculture. Even such a general concept of the evolutionary principle as that which pictures the ascent of man from savagery to semicivilization to civilized culture stages does not accord with the varied patterns of material, social, and intellectual development that societies present. The rigid application of evolutionary principles is inimical to the profitable study of the primitive mind, of primitive society, of primitive life generally, and its varieties of procedure. There is no single line of evolution, no generally applicable sequence of stages, no inevitable series of successive developments, but a variety of distinct solutions open to unbiased analytical investigation.

There can be little doubt that the evolutionary school would now be extinct, or at least modified beyond recognition, if there had not been certain political developments which had nothing to do with the advance of scientific knowledge. Karl Marx (1818-1883) and Friedrich Engels (1820-1895) took the doctrine of stages in the development of civilization as a corner-stone in the building of communistic philosophy. Although Marx had fully outlined his theories of economic determination before the inception of the evolutionary school of ethnology, the theories of this school fitted his postulates so perfectly that they were at once adopted by his followers and have become an integral part of communist doctrine. When the Marxians came into power in Russia, evolutionary ethnology received a new lease of life. All the ethnological work now being done there, aside from purely descriptive studies, has to conform to its theories. It is curious to note how these theories, which were welcomed by the nineteenth-century Englishman as an excuse for *laissez-faire,* have become in a different setting the intellectual and even emotional justification for the most vigorous attempt to remake a culture and a society that the world has ever seen. The communist not only believes that the evolution of culture is unilinear and inevitable; he also believes that he knows what form that evolution will take. In creating the communistic state he is simply speeding up the evolutionary process, working along the lines of immutable law.

As the inadequacies of the evolutionary school became increasingly evident, various new approaches were developed. One small group of workers, the environmentalists, tried to account for the diversity of cultures and the lack of uniformity in their rate of evolution on the basis of differing natural environments. In part and in extreme cases environment is a controlling factor; for the Old World civilizations, as Ellsworth Huntington has shown, flourished in temperate zones, while Arctic severity reduces life to a minimal social basis and the tropics fail to stimulate energies. In general, however, the environmentalists were few because it was obvious that their interpretation simplified the complex situation that surrounds the development of any culture by ignoring a large part of the factors involved. The school collapsed in the face of the obvious fact that people living in identical natural environments, say Indians and white pioneers in the American Middle West, may have profoundly different cultures and adapt themselves to environment in quite different ways.

CULTURE DIFFUSION

In Germany, where the scholastic tradition was still imbued with the idea of historic stages, there developed under the leadership of R. F. Gräbner what is known as the *Kulturkreis* school. From the early eighteen hundreds the Germans

had believed in the mass migration of peoples, especially
their own ancestors, and the new school was a cross between
this concept and that of the stages of history. It substituted
for a series of cultural stages growing out of one another a
series of complete cultures each of which had been carried
over all or at least a large part of the world by actual migrations. Current diversities in culture were explained partly
by limited migrations, partly by the blending of cultures
which had been superimposed one upon the other and the
consequent loss of certain elements once present. A profound
knowledge of the variety and distribution of culture traits
was drawn upon to establish the content of each of these
hypothetical original cultures, but since none of them was
supposed to exist today in its original form, the content assigned to each of them rested, in the last analysis, upon the
subjective judgment of the investigator. The accuracy of the
results obtainable by this method was thus roughly comparable to the achievement of an engineer who guessed at
his measurements and then computed his results to the
fourth decimal place.

The *Kulturkreis* school still has a few active adherents, but
it has never taken root outside Germany and Austria, where
it appealed to the developing cult of Teutonic nationalism.
While its assumptions can neither be proved nor disproved,
the weight of evidence is all against them. Aside from the
difficulties involved in long-distance mass migrations and the
improbability that any uncivilized people could carry their
culture intact through generations of wandering which
brought them into a variety of natural environments, it is
unnecessary to advance such a hypothesis to account for similarities in culture. The independent invention of the same
simple appliances, say dugout canoes and paddles, may very
well have occurred in several different parts of the world.
Also, we have abundant evidence that single culture traits,
or small groups of related traits, may diffuse over wide areas
without any migration at all. Thus, tobacco spread over most
of the world within two hundred years of the time it was

first brought back to Europe, yet it carried with it only those elements of Indian culture, the pipe and cigarette, which were functionally associated with it.

The *Kulturkreis* school has not been the only one to attempt to explain the wide distribution of certain elements of culture on the migration hypothesis. We have already referred to the heliolithic theories of Smith and Perry, who hold that the main foundations of civilization were laid everywhere in the world by a single cultural group who came from Egypt. These people spread far and wide, establishing themselves as rulers of native tribes and introducing the aborigines to the advantages of civilization, including war and the laborious erection of megalithic monuments. The inclusion of war in the heliolithic complex has done more to win support for this theory than all the scientific evidence that its founders have been able to advance. For if war is not an original human institution but something foisted on the rest of mankind by wicked heliolithic immigrants, it should be fairly easy to do away with, and pacifists have accordingly flocked into this school. Actually many of the traits assigned to the heliolithic complex do show definitely linked distributions in the Old World and may very well have been diffused from the same center, but, again, it is unnecessary to call upon actual migration to account for them.

Much that we have reviewed must be assigned to the category of error, but the workers in discredited schools have none the less added greatly to our knowledge of the great diversities of ethnic products and helped to define the problems relating to the agencies concerned in their development, spread, and decline. To envisage a people and its culture as a whole and achieve an insight into its meaning we must know its origins (that is, its historic background), the functions and interrelations of its parts, and its basic psychological patterns which shape its organization and determine the direction of its growth. Work on the determination of origins has gone on longest, and we already know a good deal about where new elements of culture come from and how cultures grow. The overwhelming importance of borrowing in determining the content of any culture has been proved, but we still know very little as to why certain elements are borrowed and others rejected or as to how such borrowings are incorporated into the new culture. Theories of culture diffusion occupy much of the arena of discussion. Cultural products in some instances, like the distribution of fauna and flora, show the effects of contiguous geographical areas; we know, however, that contrasted systems in almost any respect may exist within neighboring territories, and that the decisive factor is the acceptability of ideas within the frame of what may be called cultural ideology. The temper and content of primitive society brought into operation a different range of effective "ideas" from those prevailing in high-grade social structures. Yet the analogy holds between the processes of assimilation in such mixed populations as those of the United States and the coherence of primitive tribal groups. Ways of thought and belief, reinforced by custom, form the media of cultural unity as well as of diffusion. The emphasis is placed upon socio-psychological patterns, the study of which, now barely begun, is the major task of present-day cultural anthropology. Their interpretation for past and present brings into play the entire resources of the social sciences, with psychology and sociology contributing fundamental concepts for the understanding of men's varied solutions of the nature of the world they lived in, the forces that prevailed in it, their social mores, and the philosophy requisite for expression of their needs.

Only in recent days have there been established a critical attitude and a right interest that could be focused upon these complex problems. The attainment of this position has been a slow growth, obstructed at every point by established errors of tradition and awaiting the data of the allied sciences and their interpretation. Now, despite its many imperfections, anthropology takes its place among the important resources of knowledge available for the understanding of man. Every important problem of human relations has benefited by the illumination of the inquiries and conclusions that constitute the most recent of the sciences dealing with the most ancient and comprehensive, most intimately engaging of all topics, the scientific study of humanity.

CHAPTER X

ERROR IN PSYCHOLOGY

By George Malcolm Stratton

MICHAEL PUPIN, *the late eminent physicist and inventor, one morning was coming down the steps of the Cosmos Club in Washington. In his friendly manner he stopped as I met him, and soon he was speaking with charm about the study of the mind and about the uncommon difficulties in understanding this great side of us.*

The lifeless world, he said, is hard enough for us to grasp—electricity, chemical reactions, the nature of space. Much harder is it to master the full truth about living bodies, of plants and animals. And in this arduous region where life is, the hardest of all to understand is the nervous system, the brain.

Now psychologists, he continued, must go on to a more difficult level even than this. They must include in their studies a still more intricate and elusive set of facts, the facts of conscious life. No wonder, he concluded, that psychologists find the work hard. Their task is the most difficult of all the tasks that confront the scientist.

Pupin, in my judgment, was clearly right. Mere difficulty, however, never dismays all men—though it dismays some. Indeed, there are psychologists who see so clearly the difficulty of a scientific study of our conscious life that they give up the task as hopeless. This side of us, they declare, either does not exist or it lies quite outside the pale of science. For such students the proper study of mankind is man's glands, man's muscles, man's nerves and brain—these only. If there be aught beyond, it is swept aside as of no interest to scientific psychology.

Such men appear to me to be defeatists. They give up the fight when it has hardly yet begun. They give over to hopeless ignorance what to many men is the most interesting side of our life.

Their surrender, though, is not general today, nor has it ever been general. While scientific psychology has never turned its back fully on the muscles and nerves, and indeed never can afford to do so, it has as little turned its back on still other facts observed in living individuals. It cannot for long ignore such observed facts as that men perceive, have ideas, joys, sorrows, purposes. Such events are, without question, linked up with glands, muscles, nerves and brain. But never yet have they been observed to be identical with what is observed in these organs of ours, by dissection, chemical analysis, stained slides, and miscroscope. Among the many keys needed to unlock the secret doors of knowledge, scientific men have never widely been ready to leave unused our mental acts. Such acts are too fascinating in themselves and are too effective keys to other knowledge. The doctrine that our mental life cannot be brought within the field of science would appear to be an important mistake of some psychologists.

In all that follows, then, psychology will be understood in its large historic sense. Its field will include all the behavior of living creatures—behavior human as well as below the human level, behavior observable both within us and without. The task of the moment is to observe some of the wanderings of men's minds as they have sought the truth in this wide and diversified field.

The field of psychology can easily be located by such well-recognized facts as these: we see, we hear, dream, learn, understand; we become excited, fearful, or angry; we are pleased, we desire; we act at times almost like machines, without intention; at times we act with clear purpose. These activities are observed in men and to some degree in animals. And long they have puzzled the scientist. What are they? What causes them? How are they connected with one another? How are they connected with the body--with the eye, the ear, the blood, the heart, the brain?

It is impossible to tell of all the questions to which the answers have gone astray. One must choose. And the choice is hard where so much invites us. But the questions that follow are chosen because they are significant and have held the interest of many investigators and—have led them astray. And the errors made will be fair samples of the numberless errors which, like bleached bones, mark the trails that the explorers here have followed.

Where does the mind have its place in the body?

How do we see; what is this power we call sight or vision?

What causes our dreams, and what do our dreams mean?

Why are some creatures more intelligent than others? And where do our ideas come from?

And, finally, what of our mental life as a whole? Is it, or is it not, made up of sensations, or perhaps of muscular reflexes? Is it, indeed, a whole at all, or is it only a casual assembly of events?

In the answers to these problems we perhaps may see the partial failure, the partial success, thus far, of one of the greatest of scientific endeavors, the attempt to understand ourselves.

THE BODILY SEAT OF OUR MENTAL POWERS

THE BRAIN was not, from the first, considered to be assuredly the chief bodily organ of the mind. The blood, so Empedocles (fifth century B.C.) thought, has first importance; it is because of the happy mixture of the blood in their hands or tongue that some men are artisans and others orators; and so for all the other forms of ability.[1] Diogenes of Apollonia (fourth century B.C.), with a different view, gave prime importance to the air within the body.[2]

And yet the brain itself was early believed to hold this office. Alcmæon, Anaxagoras, and Democritus, among the Greeks before Aristotle's time, thought the brain to be effectual for our mental life. When Plato (fourth century B.C.) came to pronounce on this question, he distinguished, and would have the brain to be the seat of the divine reason in us; while the thorax is the place of that part of us which is endowed with courage and passion and the love of strife; and in the abdomen, between the navel and the diaphragm,

[1] Theophrastus, *De Sensibus*, 11; see pp. 75 *ff.* of the present writer's *Theophrastus and the Greek Physiological Psychology before Aristotle* (London and New York, 1917), cited hereafter as *Greek Physiological Psychology*.

[2] Theophrastus, *De Sensibus*, 39; *Greek Physiological Psychology*, pp. 100 *ff.*

are our desires for food, drink, and other bodily needs.[3] Aristotle, great scientist though he was, took a step backward in this regard, holding the brain to be, rather, an organ of cooling.

Later Greeks—and above all, Galen,. that eminent figure in early medicine—were not dominated here by Aristotle, but returned to the early if vague recognition of the part played by the brain. And from Galen's time on, the gross error that our psychic life has its seat wholly in those parts of the body outside the brain gives way to lesser errors. Some wrong part of the *brain* is now chosen as the true place of mental activity, rather than some part of the body outside the brain.

The story of all the false scents followed within the brain itself might be made as breathless as a chase, but it would lead us too far. Galen's own view, right in attaching prime importance to the brain, was wrong in giving emphasis to the ventricles. And Galen's view deeply influenced Arabic science, which in turn deeply influenced science in medieval Europe. Albertus Magnus (1190-1280) assigned to the various cavities or ventricles of the brain quite an array of mental powers, including sense, movement, memory, imagination, and appraisal.[4]

Nor did the errors end with medieval Europe. In the sixteenth century, Leonardo da Vinci—a great modern: painter, anatomist, student of aviation, and much besides—examined the brain's ventricles by injecting hot wax into them and, when it hardened, dissecting this out. And we still have his sketches of these little waxen casts, with notes—in his odd reversed or "mirror" script—telling quite mistakenly the mental use of the parts of these several hidden cavities in the brain.[5]

The error took a novel form for another great modern,

[3] *Timæus*, 69, 70, 73.
[4] *Alberti Magni Opera*, ed. Jammy, Vol. III, *De Anima*, and other writings in other volumes; see the present writer's "Brain Localization by Albertus Magnus and Some Earlier Writers," *American Journal of Psychology*, Vol. 43 (1931), pp. 128 ff.
[5] "Leonardo da Vinci," *Quaderni d'anatomia*, Vol. 5 (1916).

René Descartes (1596-1650), who chose as the place where the mind acts on the body a small portion near the brain's center, the pineal gland. Descartes' error is so ingenious and was so well worked out by him that one almost wishes it were true. It is supported by everything but experiment, which unfortunately sweeps it quite away. But earlier, and in spite even of Descartes' prestige, the belief continued that the brain's ventricles were its parts most weighty for the mind.

But all earlier errors were quite overshadowed in the eighteenth and early nineteenth centuries by a new and striking aberration connected with the names of Franz Joseph Gall (1758-1828) and Johann Kaspar Spurzheim (1776-1832). Hardly anything to us of today seems scientifically less inviting that that tissue of mistakes known as phrenology. Yet Gall was an anatomist, a man of some standing, and made a certain show of evidence by observation, although his evidence was never so controlled as to prevent most glaring blunders of induction. Gall is an example of scientific talent almost brilliantly astray.

For the highly original view of these two men was that the bony elevations and depressions seen and felt on the outside of the head indicate a man's brain and thus his mental powers and weaknesses. In due time this idea was developed into an elaborate method of personal diagnosis and advice, a method which enjoyed some popularity for vocational guidance, and which even yet has not quite lost favor in unprivileged circles. Its foolish charts of the head are still displayed to the credulous in cultural backwaters of our land.[6]

Phrenology has the distinction of being wrong in every particular, and yet right in an important general respect. All its details as to the cerebral location, for example, of inventiveness, amativeness, or philoprogenitiveness—indeed its very dividing of the mind itself into these powers as though they were distinct and separable—all these details have for scientific psychology gone by the board. But the

[6] See in this connection "The Skull Science of Dr. Gall" in Joseph Jastrow, *Wish and Wisdom* (New York and London, 1935), Chap. xxii.

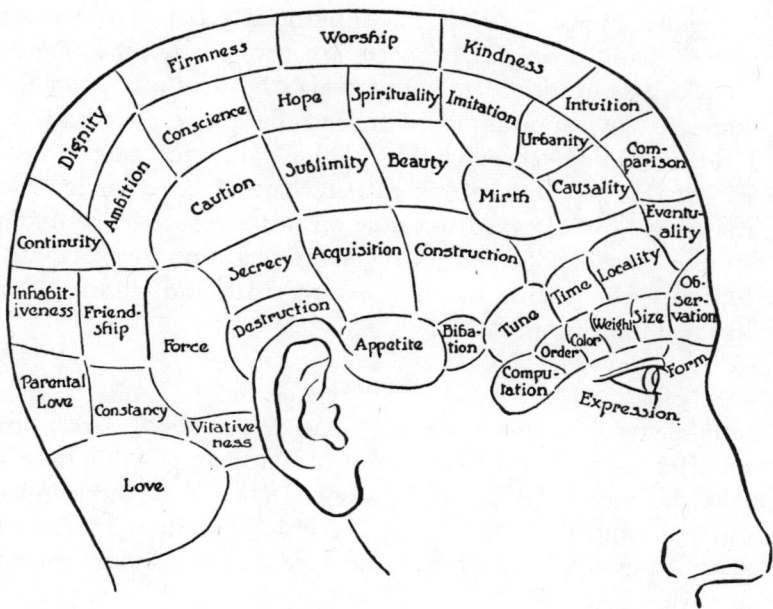

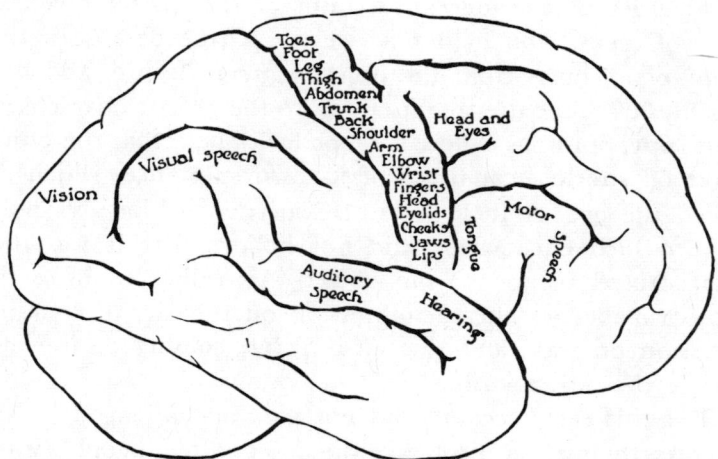

PHRENOLOGICAL AND MODERN VIEWS OF BRAIN FUNCTIONS

Above, a typical phrenological map; below, the human cerebral hemisphere with the functional areas of the cortex indicated. From J. A. Harris and others, *The Measurement of Man* (University of Minnesota Press, 1930).

general assumption, that the superficial parts of the brain rather than its deep-lying parts are psychically most significant, this assumption by phrenologists has had a happier fortune. It has importantly influenced the entire course of the later physiological psychology. The value for our mental life of the cerebral cortex, rather than of the ventricles or the pineal gland, was from now on never lost to view. Later studies have confirmed this general account and are still busied with clearing up its inherent truth and ridding it of its encumbering mistakes.

SIGHT

It seems so simple a matter for us to see trees, stars, automobiles, men, that it would appear almost impossible to mistake the way in which this is done. Do we not merely open our eyes, and the object there is seen? Yet a most interesting array of errors have been made regarding so simple an act as this.

First let us consider an account of the matter given by early Greeks, who held that we see a tree because a tiny copy of it comes from the object, enters the eye, and thus enables us to see the tree itself. As to the precise character of this copy, opinions differed. Empedocles held that the object gave off subtle emanations of its own substance, which entered the eye. A little later, Democritus, while also holding to such emanations, did not believe that it was they that caused us to see the object. According to him, the object makes a delicate impression on the air, like an impression on wax, and this air-imprint, coming to the eye, makes the object visible.[7]

But this early account did not pass unchallenged by the Greeks themselves, probably the most gifted people, intellectually, our world has ever had. Plato noted defects in the explanation, and cured some of them, by pointing out that nothing material enters the body by way of the eye,

[7] Theophrastus, *De Sensibus*, 7, 49; *Greek Physiological Psychology*, pp. 70 *ff.*, 108 *ff.*

but only motions; by pointing out also the important part played by *light*, and by indicating that the eye itself is active when we see.[8] And, still later, Theophrastus (late fourth century B.C.), who succeeded Aristotle in Athens as a professor and college president there, called attention to Democritus' failure to explain, by a tiny air-imprint in the eye, our power to see the object in its full size and at its true distance. Moreover, such air-imprints, Theophrastus held, are a sheer absurdity; for air is not like wax; air has no consistency to hold an impression. If it really could hold such impressions, said he, we ought to see objects by night as well as day; for the cool air of night also would take impressions and convey them to the eye. And when persons face each other, the air-imprints passing from the first person to the second would clash with those passing from the second to the first, and the clear vision they actually have of each other would be impossible.[9]

These are but samples of the sharp attacks made by later Greeks upon the errors of their earlier kinsmen in attempting to explain vision. Nevertheless the early errors lived on, and the idea that a little copy of the object passes from the object to the eye continued erroneously to be regarded as the last word of Greek science on the problem of sight.

Descartes in the seventeenth century, consequently, faced the ancient error as though it were still living and refuted it afresh. Imagine ourselves tapping our way at night with a staff, said he; we thus would be able to recognize objects in our path—now a tree or rock, now some grass, mud, or water, all perceived by means of the staff in the hand. It is clear, however, that the staff does not transmit to us any material likeness of the object. No substance enters the staff and, passing through its length, enters the hand. Each object merely causes, as we tap it, certain different motions and resistances in the staff which affect the hand—motions

[8] *Timæus*, 45, 64.
[9] Theophrastus, *De Sensibus*, 51; *Greek Physiological Psychology*, pp. 110 ff.

and resistances which we have learned to recognize, so that they signify to us a tree, a rock, or whatever else.[10] These signs need not at all *resemble* the object they signify, even as the word *tree* need not resemble a real tree in shape or substance.

Our seeing, for Descartes, is a mental act of a like sort. Motions,. and motions only, enter the eye—motions of light which of themselves may be quite unlike the object we perceive by their means, but which are signs to us whose significance we have learned to know. Thus Descartes laid bare not only the error in the belief that objects send copies of themselves to the eye, but also the error in any explanation of sight that leaves the mind out of account and fails to make clear that we must learn the *meaning* of the various impressions made upon our eyes by the light coming to the eye from objects. The process is like learning a language.

But a particular puzzle which led psychologists astray lies in the fact that while we see things right side up, the picture of the object on the retina of the eye, made there by the eye's refractive power, is upside down. When the problem created by this strange state of things was faced, it was boldly declared that the inverted image in the eye offers no difficulty; indeed, the inversion is *necessary* to our seeing things in their proper position; unless the picture within the eye were inverted, we should see the world wrong side up.

This view of the case, which impressed psychologists for a time, was refuted quite simply by an experimenter who for a number of days wore glasses that turned the retinal picture around until it stood right side up, like the objects it pictured. The trend of his experiment showed that in due time and with due training one could see the world upright by an upright image in the eye, quite as well as by an inverted image.[11] The inverted image is physically more convenient. Optically it is simpler to have it so. And we have learned

10 *Œuvres*, ed. Cousin, Vol. V, *La dioptrique; Discours premier.*
11 *Psychological Review*, Vol. 3 (1896), p. 611; Vol. 4 (1897), pp. 341, 463.

to get our mental results from it with perfect ease. But we could quite as easily have got these results from an upright image in the eye. There is no psychological need that the retinal image be upside down. It could have been in either position, so far as the mind is concerned.

Even a brief account of vision must not omit the doubt and mistake as to what would be the experience of a man blind from birth if he were at last given his sight. Would he be able at once to recognize by vision alone the familiar things around him? There were those who said he would, and those who said he would not. The whole matter remained in the limbo of mere debate until there could be an actual experiment with such a person. The earliest and most famous evidence comes from William Cheselden, who, as reported by him in 1728, surgically removed a cataract and made the needed observation. The patient, a boy, when now he was shown his pet cat, was asked what it was. He could not tell. Not until he was allowed to touch the cat did the sight of it come to have its proper meaning.[12] And later, when Raehlmann's patient was shown a large bottle, he said it might be a horse! He was twitted on not knowing the difference between a bottle and a horse, and he sheepishly answered that it was not so easy after all.[13]

The psychology of vision has thus run a troubled course. It has through difficulty escaped some of the errors in any purely physical and any purely physiological explanation of sight, and has come to recognize how complex and important is the mental action here. There is in vision a complicated process of learning—rapid learning and at an early age though it may be.

DREAMS

Dreams have always stirred men's curiosity, and the interpretation of dreams often has outrun by far the plain facts observed.

[12] *Philosophical Transactions*, 1728.
[13] *Zeitschrift für Psychologie und Physiologie der Sinnesorgane*, Vol. 2, p. 53. For other cases of this sort see *Mind*, Vol. 8, N. S., No. 32.

Folklore is here the beginning of psychology, and it is full of the thought that dreams are a nightly miracle, that the sleeper is not having mere vagrant imaginings, but somehow is aware of a reality beyond himself. Many a primitive people has believed that in sleep the soul visits a distant land, perhaps the abiding-place of the departed ancestors, and there holds intercourse with them. The dream, for such a belief, is the experience of an actual journey, from which the soul returns upon awakening.

Or the soul, itself not making the far journey, receives a message from the distance. In Homer, where folklore doubtless is preserved, Zeus may send to the sleeper a form which simulates some person known to the one asleep, and this fictitious person communicates what the god wishes said; and for the ancient Jews a dream might have a hidden and prophetic meaning, its purport being concealed as in a cipher which an expert in such things could understand. The dreams that Joseph interprets—of the sheaves, or of the kine —illustrate this belief.

But in India folklore passed over into a studied account of dreams, for in India there was an early psychology which is still too little known in the West. The views offered by the two thinkers Sankara and Ramanuga, who in many respects stand in contrast to each other, show error mingled with sound work. For Sankara, dreams are quite unreal, are mere illusions, as he makes clear in some detail; yet they may point to reality, and they support the idea that the self is separate from the body.[14] And for Ramanuga, dreams are unreal and yet are *not* illusions. They are marvels produced by the Supreme Person only, and not by any wish of the dreamer himself (thus opposing Freud some centuries before Freud could reply), and are intended to reward or punish the soul of the dreamer.[15]

But the Greeks who came long after Homer's time were ready to look on dreams with something of a critical mind.

[14] *Sacred Books of the East*, ed. Max Müller, Vol. xxxviii, pp. 134 *ff.*, 272.
[15] *Ibid.*, Vol. xlviii, pp. 75, 602 *ff.*

Plato suggested various natural causes for them, although some dreams, he seems to have thought, might be a means to truth through divination. But divination he regarded in general as a gift to the foolishness of men and not to their wisdom. Dreams come when we are not in our wits; and for any suitable interpretation they need a clear use of reason.[16] The virtue of Aristotle here lay chiefly in his shrewd observation and in his caution. For him dreams are nothing but illusions, quite of a kind with the waking appearance of the sun, which looks small even when we know it is large. To his mind it is absurd to ascribe a divine origin to dreams; and he offers acute naturalistic suggestions to explain the occasional agreement between dreams and future events.[17]

From these early attempts to brush off the superstition so ready to attach itself to dreams, let us come at once to our own day, and particularly to the work of Sigmund Freud. For whatever the errors in his view, he has given to the study of dreams a new life, a new direction. Freud and his followers have imparted to dreams a fresh scientific interest and have restored something of their old practical importance. The dream is again a sign; but now it is a sign of the past rather than of the future, being an index to forgotten experience and to hidden and thwarted desires. The interpretation of dreams thus becomes a respectable scientific aim.

Of Freud's questionable statements regarding dreams, one may be mentioned. He has an assurance, not wholly unlike that expressed in some popular dream-book, that a particular kind of object in a dream has usually, although not invariably, a definite and fairly constant symbolic meaning. The objects to which Freud attaches these definite meanings include stairs, sticks, tree-trunks, stoves, and many more.[18] But to many students who are ready to be led by the facts

16 *Timæus*, 71 *f*.

17 *Parva Naturalia*, pp. 462 *ff*. W. A. Hammond, *Aristotle's Psychology* (New York, 1902), pp. 247 *ff*.

18 Sigmund Freud, *The Interpretation of Dreams*, tr. Brill (New York, 1913), pp. 246 *ff*.

and the ordinary rules of induction, it seems highly doubtful that these objects have such definiteness and constancy
of symbolic meaning in different dreams and in different
dreamers.

Closely bound up with this is the specific error that all
these objects have a sexual meaning and motive. No one
need doubt that sex is of profound influence on human
imagination, both by night and by day; but it is only one
among several profoundly important influences. In his
theory of dreams Freud has kept his attention away from
the full circle of facts.

The error in any interpretation of dreams by sex alone,
as well as other errors and omissions in explaining them,
can perhaps best be illustrated by considering a particular
dream. One reported by the great Swedish explorer Sven
Hedin may serve. He and his men at the time were in
Tibet, where he had wished to push on into regions that the
Tibetans themselves were unwilling he should enter. A relatively large force of them, five hundred strong, armed and
not too friendly, were now "escorting" him, causing him
deep anxiety lest he and his band should be set upon and
perhaps destroyed. One night during this time of strain,
"about an hour after midnight," he tells us, "I was awakened by an unpleasant dream"—a dream "that the Tibetans
had flung a dead body into my tent. In a moment I jumped
up and struck a light, and found the tent empty; then, as
soon as my thoughts cleared, I saw how matters stood." [19]

Now Freud might easily offer an explanation of this dream
in terms of sex. Sven Hedin's own explanation shows, I
believe, a more excellent way—though we may feel inclined
to go farther in the same direction. He had been lying, he
says, on his right arm, which had been stretched out along
the ground, and the hand had grown numb and without
feeling; it was icy cold. By chance he happened to touch
this numb arm with his left hand, and, as he was only partly
awake, his fancy gave his impressions the form just described.

[19] Sven Hedin, *Central Asia and Tibet* (London, 1903), Vol. ii, p. 476.

Quite in accord with the findings of scientific psychology today, Sven Hedin here lays stress on his sensations present as he half-slept and on the lack of sensations in his numb hand. And he indicates that these were mentally elaborated into the illusion that a corpse had been thrown into his tent.

We may now go on one step farther and inquire why his sensations and lack of sensations were worked up into the precise form he gave them. There were other ways open to his dream-imagination. Why did he not dream, for example, that one of his own men had thrown a piece of ice into the tent? Or why not the drab commonplace of a dream that he had thrust his hand into a basin of cold water near his bed? The reason against such tame imaginings lay in the special force behind the dream, which in this case was the besetting fear of what might befall his party and himself. Anxiety for the safety of his men and himself here gave to this dream its special dramatic form. There is no need to attribute sex to it.

Or again, when the dream fulfills a wish rather than a fear, the wish may not be of sex at all. In a certain orphanage, for example, the children often dream of having a home. The desire here fulfilled is not sexual; it is a desire, rather, for the place and surroundings of children more fortunate than themselves—for house, home, brothers, sisters, companionship, and affection: for these literally, and not for some dark thing that must not be allowed to come to consciousness, that may not even be self-confessed and of which the actual dream-objects are but symbols.

It is an error shown to be an error by careful study, then, that dreams have only one motive, that of sex. They have this and many other motives, even as our imaginings by day have many motives. Freud oversimplified the causes of our dreams and thereby gravely misinterpreted many of them.

OUR INTELLIGENCE, OUR ABILITY TO KNOW

In the very early days of Occidental psychology, the question why some creatures are more intelligent than others

336 THE STORY OF HUMAN ERROR

was answered by naming a particular substance whose varying amount or mixture in the body caused the difference.
The air in the body, as we have already seen, was given the
place of first importance by Diogenes of Apollonia. And it
is the character of this air in the body, according to him,
that explains why men are more intelligent than animals;
for animals breathe air that comes from the earth, he said,
while men breathe purer, drier air. An excess of water thus
was regarded as an obstacle to intelligence, which explained,
for Diogenes, the low ebb of understanding when in one's
cups. And young children are impetuous and flighty, he believed, because the air is expelled from their small bodies,
and they are overmoist.[20]

Such things were said even by acute Greeks. But this was
in early times. Later acute Greeks saw the crudity of such
accounts. To explain the difference between men and animals by the air they breathe is childish, said Theophrastus.
If the explanation were true, said he, our intelligence
should change when we go to another climate, and moun-
taineers should be brighter than lowlanders, and birds
should surpass us all.[21] This unfeeling treatment of an honest
blunder reminds one of the fate that befell an equally crude
blunder of a later day. Phosphorus, it had been said in the
nineteenth century, is a most important substance for the
brain; and this explains why fishermen are more intelligent
than farmers. For they eat much fish, which contain much
phosphorus. William James dismisses this inexpert theory of
intelligence with the laconic remark, "All the facts may be
doubted." [22]

And now as to the *size* of the brain, which has been
thought to be a most important sign of intelligence. That
the brain's size is the prime factor in determining a man's
intelligence is clearly a mistake. It is true there is a slight
connection, when large groups within the same race are con-

20 Theophrastus, *De Sensibus*, 45; *Greek Physiological Psychology*, pp. 104 *ff*.
21 Theophrastus, *De Sensibus*, 48; *Greek Physiological Psychology*, pp. 108 *f*.
22 *Principles of Psychology* (New York, 1891), Vol. I, p. 101.

sidered, between the average intelligence of the group and the average brain-weight of the group—a fact once used against votes for women. But such a fact tells nothing of the individual. Men of high intelligence have had small brains —Erasmus, for example—and men of low intelligence large brains. And in some of the lower animals—in the elephant and the whale—larger brains are found than in men. Nor is the error avoided by saying that the important feature is not the brain's absolute size, but its size relative to the body —an argument *favoring* women: for in proportion to her body, which is smaller than man's, woman has a slightly larger brain than man's. Unfortunately, certain rodents and monkeys have brains larger in proportion to their small bodies than even women have.

The *shape* of the brain in normal persons has similarly offered a false clue. The long head, one heard some years ago, marks a more intelligent person than does the broad head. But no support has been found for this by careful survey. Indeed, any attempt to derive intelligence from the mere size or shape of the brain, and without emphasis on the finer structure and total organization of the brain, should, according to our present knowledge, be counted a mistake. The whole matter is still obscure. Native intelligence, it now seems probable, is importantly linked with an obscure difference in the native constitution of the brain; and acquired intelligence, or understanding, is connected with difference in training, and especially with the training due to the social environment. Beyond this narrow circle of light the darkness deepens and all manner of false assertions are easily made. So much for our first question about intelligence, why some creatures are more intelligent than others.

To a second question, what is the source of all such knowledge as we have, two main answers have been given. All our ideas, it has been said, arise from sense-impressions on our eyes, ears, and other senses. Such was the belief of Thomas Hobbes (1588-1679). But John Locke, also in the seventeenth century, noted the defect in this account, so dear to many even today, by pointing to the many important ideas that come by observing our own inner activity, especially of thinking and of willing. This Locke called reflection. All our ideas come from *experience,* he held, but experience is not wholly of the *senses;* we have an inner experience, a power of inner observation, as well.[23]

That the senses are our only source of knowledge has been denied also on other grounds. Plato had insisted that certain of our ideas cannot have arisen from our senses only. We have an idea of a perfect circle, for example, which our eyes have never seen, for every circle actually seen is always in some respect imperfect. And Descartes in modern times was confident that our thought of God must have been implanted in the mind by God himself, as a mark of the workman impressed upon his work.[24]

But as it is an error to suppose that our senses give us all the knowledge we have, so it is an error to suppose that any knowledge is given us quite without our senses or inner experiences. The clearest light we have today is that none of us at birth is supplied with ideas ready-made, or with ideas undeveloped but ready to develop of themselves, regardless of the actual experience we may come to have. Impressions without or within are indispensable.

But these impressions are not enough. Unless we had the power to give to our impressions a form or organization not inherent in them, we should be without the knowledge we have.

Two things are needed, then: impressions and the inner power to give them form or patterning. The error in holding to either of these factors alone was clearly announced by Immanuel Kant (1724-1804), who was both psychologist and philosopher. And the power in us that he indicated, the power to give form or pattern to our impressions, has been an object of special psychological study in recent years—

[23] *Essay concerning the Human Understanding,* Book II, Chap. v; *Works of John Locke,* ed. St. John (1889), Vol. I, 234.

[24] Meditation III; *The Method, the Meditations, etc.,* ed. Veitch, 9th ed., p. 131.

indeed, has been the corner-stone of a "school" of psychology known as *Gestalt* psychology. By support of novel experiments the psychologists of this school have accented the peculiar formative activity to be discerned in perception and in other functions of the mind.

In agreement with Locke and all his followers, we now can detect the error in any account of knowing that omits sensory and inner experience. With Leibnitz, Kant, and the *Gestalt* psychologists we can detect the error likewise in any account that fails to include among our many mental activities these organizing powers. The truth here does justice to two half-truths and allows us to see that knowing is a complex whole in which there are contributions both from our senses and from an inner activity not to be identified with our senses, not to be identified even with the somatic senses hidden in our muscles, joints, and other inner parts of the body. Organized thinking, such as the scientist carries on, depends on certain things which are *not* born in us—on external stimulation and on the sensations connected with this stimulation. But it depends also on certain powers which *are* innate—on the innate sensitivity to light, sound, and so on, and on the innate activity by which our sensations are given order and meaning. Whenever psychology has neglected either of these two groups of factors—the factors that *are* innate and those which are *not*—it has been in error.

THE MIND AS A WHOLE

We have just had before us our powers of sense and our powers to organize into experience and science the sensations arising from our senses. And this may bring us to the mind as a whole, the present idea of which has come up through much tribulation.

To begin with, our mental life is of far wider compass than was once believed. For it had been stoutly held that this mental activity of ours was quite open to inner observation and included only such acts as we could directly notice thus. This idea is now almost completely displaced by the conviction that even the most expert observer of himself can give no adequate report of what goes on within him. Of much that occurs in his mental life he is immediately quite unaware, and he can only recover it, if at all, by circumstantial evidence. Let me illustrate this. The phrase "Guinea pigs or babies?" came to my mind one evening, I knew not how. But the next day I noticed a poster I had passed the day before, aimed against "antivivisectionists," a poster bearing the phrase just quoted, and which had mystified me. Now I must have seen the words the day before but quite without noticing them, my attention being then on other things. And this unnoticed experience, on the very margin of consciousness, caused the phrase to appear, hours later, in the field of my attention. Or again, the motive for some practical decision may take its form from a forgotten kindness or slight, now quite out of reach of observation. Only by indirection, by use of hypnotism or word-associations, or by means of instruments that record the pulse and blood-pressure and breathing, or by the delicate evidence of galvanometer or chronoscope, do we lay bare the evidence for a host of deep connections which quite elude the direct notice of the person in whom they exist.

But to accept subconscious factors is a dangerous business. It induces some persons to accept a whole brood of current ideas about the "subconscious self" or "the unconscious." Popular assertions in this region now go the rounds, assertions which are highly imaginative, highly colored, and without a shred of scientific evidence for their truth. In protest against these wild imaginings there are psychologists who would almost die rather than yield an inch for subconscious mental happenings. But the ancient belief that mental activity is only such activity as we can directly notice becomes more and more clearly an error. The evidence against it increases daily, showing that psychologically we are far more complex than once was thought, that each of us includes an unnumbered host of events and forces which are quite beyond our direct observation.

A person, moreover, as we have now come to know, is not his intelligence only, nor his glands and muscles only. He surely includes these, but these are not all, nor are they all that is important. He has also his activities of liking and disliking; he has desires, sentiments, ready to be expressed in anger, fear, sorrow, or affection; he has, perhaps above all, his organizing aims, his purposes. Acts of intelligence are woven into these, but these other operations which are primarily not intellectual are, in turn, woven into our acts of intellect. The person has, in this direction too, a more varied constitution than many have believed. Not only does much go on in him subconsciously, but much also goes on of which he may be fully aware, but which has a structure and quality quite different from acts of intelligence pure and simple.

But in turning away from an account of the individual in which he appeared to be chiefly a creature of intelligence, another factor has come into prominence. For many writers in recent years sex has seemed almost the sole force in human life. The idea that sex dominates our dreams appeared far from true, we have seen, in the light of fuller evidence. The still larger error is that the sexual passion rules not only our dreaming, but our thinking, our feeling, our emotions, and our purposes.

Shell-shock in the World War and other forms of evidence, however, have made us recognize how wide of the mark such a theory falls. Persons display, first and last, a wide array of motives, including a desire for muscular activity and rest, a hunger for food and drink, a curiosity for much that is quite apart from sex, a love of hunting, adventure, and companionship. A man can be trained to die for his country, or to risk his life to put out a neighbor's fire, or to be the last man to leave his sinking ship; and in a score of other ways he may show how weak, relatively, even so strong a motive as self-preservation may be. The person as a whole usually is far more complex than is dreamt of in this handy insistence on sex.

Finally, one is tempted to inquire what is the relation of the mind as a whole to its particular functions. We face here a living question to which psychologists give contradictory conclusions, both of which cannot be wholly right.

An older psychology, now known as "faculty psychology," regarded the mind as a fairly loose aggregate of distinct faculties, such as memory, imagination, and reason. Phrenology, as we have seen, assigned these and a score more of faculties each to its distinct area of the brain. And the downfall of phrenology, which was a special outgrowth of faculty psychology, helped to discredit the kind of psychology that had given phrenology birth. At the present day scientific psychologists to a man have renounced faculty psychology and all its works. The mind is not an aggregate, a mosaic, of large powers, like memory, imagination, and reason.

But is it an aggregate of small powers? Is it a composite of powers more restricted, more nearly simple, than those just named? There are psychologists today who believe our mental life is a composite, even as there were those in Greek psychology who held this view—a view held at a far later day by the "associationists" in England. In considering the source of our ideas some pages earlier, there was noticed an inner power which gives form to our sensations, organizing them, setting them in order. For this reason today psychologists are less inclined than once they were to explain our full mental life by means of sensations and their associations.

An analogous view, however, today finds favor in some quarters. Instead of stress on sensations, stress now is laid on muscular reflexes. Reflexes are today regarded by some as the sole material out of which all our behavior is composed. Each of us, according to this account of the person, is a composite of reflexes which are united by a process called "conditioning." Instead of being a multitude of associated sensations, as once was held, we are thought to be a multitude of conditioned reflexes.

There is, however, an account of the mind's constitution which is fundamentally opposed to this. While this opposed account accepts wholeheartedly the fact of sensations and their associations, and accepts wholeheartedly reflexes and their conditioning, it believes there are evident facts beyond these and not to be explained by these alone. Instead, the mind is seen to be not a composite at all, but a whole which displays an almost endless variety of operations. It is a whole which may, when diseased, become disorganized, at times profoundly disorganized into what are called dissociated personalities. But even these deep disruptions never result in completely separate persons. The individual, mind and body, with all his tensions and dissociations, still appears a living whole. Normally our various functions vitally affect one another; they interlace, they are interdependent.

Psychology announces no decision as yet between these opposing disputants. Which of them stands for the truth and which for error is undetermined. They are in the lists, they tilt today and tomorrow. But the outcome is fairly clear. Already there is experimental evidence to indicate that the conditioning of reflexes has a quite restricted field and by no means explains everything. Nor does it require a seer, it would seem, to foretell the downfall of any account of our mental life as a whole that fails to do justice not only to our relatively elementary functions, such as sensations, reflexes, and their associations, but also to those peculiar functions which appear in reasoning, in holding to a conscious purpose, and in applying standards of value to all such acts of ours. As a builder takes brick and stone and gives them an architectural form, so there are forces in the individual which take his sensations and reflexes and organize them into a living whole of personal experience and personal purpose.

In this devious way psychology moves toward its far-off goal of scientifically understanding our mental life, its structure, its functions, its springs and consequences. Their own errors should be studied by psychologists. But these wanderings have not been like the wanderings of one who has no goal nor any sense of the general direction in which he must press on. The mistakes in psychology—like those in other sciences, I presume—have their rhyme and reason. They are such as are possible only where there is intelligence, where there is guiding and intelligent purpose. They spring from thoughtful curiosity insistently questioning. The errors arise in approximations to the truth, the errors' size being gradually lessened.

The errors' size is lessened, but not their *number.* Paradoxically, psychology—like other sciences, I imagine—makes a larger number of mistakes today than ever before. It asks more questions. For every question in psychology asked by the Greeks we ask a thousand. And the number grows. Every discovery breeds ten new problems. And every new answer is likely to be in some measure a mistake.

In general, then, more blunders of smaller size. And with them more truths and larger ones. The yearly balance-sheet shows a profit. But there is no greater error than that psychology now moves in clear truth, where earlier it moved in blind error. It always has moved in both, and always will. But our opportunities of discovery are better than ever before. More psychologists, better trained, are at work. They use better technique. The experimental methods and the statistical methods greatly add to the power of the science. And each of these methods is daily refined and reshaped for the special problems of the field. New devices reduce the size of the errors that arise from personal prejudice and precipitance. And while there will perhaps always be a lunatic fringe to psychology, and a rebellious fringe and a reckless fringe, yet the central sane province is enlarged, extending our knowledge of the most important object open to science—the mind, and particularly the mind of man.

Section V

THE APPLIED REALM

Chapter XI

ERROR IN SOCIOLOGY

By Harry Elmer Barnes

IT IS PROBABLE *that error has been more rife and prevalent in sociology and social thought than in any other of the sciences. This has been due in large part to the complex nature of the subject and to its immediate association with nearly every type of human prejudice and passion. Sociology involves such matters as the origin and nature of man and human groupings; the rise, character, and excellencies of the various forms of government; the evolution and nature of moral systems; the possibility of improving human society; the effect of biological, psychological, geographical, and economic factors upon human society; and the rôle of the supernatural in social control.*

Much of the specific knowledge that is necessary to carry on a scientific study of human society has become available only in the last half-century, and it is still inadequate in many respects. Yet sociological thought goes back to an extreme antiquity. Most of the generalizations concerning man and society prior to the twentieth century were based on conjectural premises, though here and there the shrewdness and intuition of one social philosopher or another approached the truth. Moreover, the errors in all the special sciences that were drawn upon by sociology were taken over and reworked into the development of sociological thought. For example, the myths respecting human origins, religious rationalizations, biological and psychological errors with respect to the nature of man, political errors and prejudices, historical mistakes, and the errors and exaggerations of students of astrology and geography, all found their place in the building up of the social thought of the past. Preceding chapters in this volume provide a veritable arsenal of facts which explain the prevalence of error during the growth of the science of sociology. In our treatment of sociological errors we shall combine the chronological and topical methods. We shall consider the major types of errors in a roughly chronological order as they appeared during the history of sociological thought. But we shall bring the treatment of each topic down to date.

MISTAKEN VIEWS OF SOCIAL ORIGINS

WE MAY CONSIDER first the errors connected with sociological doctrines bearing on the origin and nature of man and human society. Early social thought accepted the prevailing myths with regard to the origin of man and his creation by a divine act in the relatively recent past. Social philosophy among Jewish and Christian thinkers took over the Biblical tale recorded in *Genesis,* which placed the creation at about four thousand years before Christ. Oriental thinkers and pagan philosophers very generally accepted a comparable view of the recent origin of mankind as a result of divine creation. Indeed, we now know that the Jews and Christians derived their views of the origin and destiny of man from the Babylonians and the Persians. So far as we know, only a very few classical thinkers, such as Plato and Lucretius, possessed even an inkling concerning the long human past and the gradual development of man and human institutions. It was not until the nineteenth century that the sciences of physical anthropology and prehistoric archæology gave us an accurate and substantial picture of human origins and development.

The supernatural hypothesis concerning the derivation of man himself gave color and character to most of the early thinking in regard to the origin of human institutions. The latter, like man's physical being, were supposed to have been the product of God's creative ingenuity. Religious institutions, moral systems, and governmental forms were believed to have been devised by the gods. Primitive man attributed the whole complex of social institutions to supernatural interest and intervention. This hypothesis was elaborated in historic times; but the general nature of the interpretation remained the same, whether among primitives or among the social philosophers who rationalized and defended the Roman Catholic Church and the medieval Holy Roman Empire. Indeed, this view still finds support, not only in popular belief, but in the works of erudite apologists of the supernatural view of things.

Very early, however, certain social philosophers began to conceive of social institutions as a product of conscious social invention. One of the earliest of such doctrines was that associated with what we call the "social contract." This appeared first among the Epicurean philosophers of ancient Greece. They held that man had established government because he perceived the direct utility of orderly political relations. In the later Middle Ages and early modern times it came to be believed that not only government but also social institutions as a whole had been created by the conscious will of man. Perhaps the first to suggest this idea was the humanist Pope, Pius II, Æneas Sylvius Piccolomini (1405-1464), who provided, however, ample scope for divine direction.

According to this view, which was elaborated in the seventeenth and eighteenth centuries, mankind had originally lived in a "state of nature" without any authoritative form of social control. This worked well enough until the rise of property. Then, however, came the struggle to amass greater material possessions; the anarchy and strife that resulted made the state of nature unendurable. Hobbes held that it meant the war of all against all; his oft-cited statement was that the life of man in the state of nature became "poor, nasty, brutish, and short." Such a state of affairs led more thoughtful persons to consider the possibility of escaping from this intolerable situation by establishing civil society, which would bring about law and order. This deliberate determination to create orderly social relations is known in the history of social philosophy as the "social contract." But it was further necessary to set up a specific form of government in order to make good this desire to attain peace and order. Hence a second contract followed between the governing element and the governed masses, which is known as the "governmental contract." Thus it was held that society and government were both the outgrowth of man's deliberate decision to escape from anarchy and to establish social order.

The classical statements of this social-contract doctrine are associated mainly with the names of Richard Hooker (1552-1600), Thomas Hobbes (1588-1679), Samuel Pufendorf (1632-1694), John Locke (1632-1704), and Jean Jacques Rousseau (1712-1778). Even in our own day the French sociologist Alfred Fouillée has insisted that all social relationships and institutions rest upon a contractual basis, either real or implied. As is well known, this social-contract idea was used as the justification for the right of revolution preached by the liberal philosophers of the seventeenth and eighteenth centuries.

The social-contract theory was first comprehensively and vigorously assailed by the brilliant English philosopher and publicist David Hume (1711-1776). Hume criticized it on both psychological and historical grounds. Psychologically the doctrine was untenable because it implied knowledge prior to experience. It assumed that men who had never known the advantages of government could be conscious of these advantages and seek to secure them. From the historical point of view Hume insisted that the social-contract theory did not square with any observed facts. He challenged its supporters to show one single instance in which society and government had been created by a deliberate act of the social will as expressed in a contractual relationship. Hume declared that society arose from the social instinct of man and that governments originated most usually in force and war. They might be supported in time as the result of a perception of the utility of government, but in the beginning they were a product of superior force. Subsequent writers have expanded Hume's criticisms but have added nothing substantial to them.

Hume's attack on the social contract was sufficient to ex-

pose its glaring theoretical errors as well as its lack of historical foundations. But the theory not only dominated
social thinking for several centuries; it also had a large influence in the realm of practical affairs. The English,
American, and French Revolutions derived their theoretical
foundations and justification from the social-contract doctrine. The American Declaration of Independence embodied
a condensed version of the notion.

Another popular pre-scientific view of the origins of social
and political institutions was that which attributed the rise
of orderly society to war and the application of physical
force. This interpretation of social and political origins was
hinted at by the Roman Epicurean philosopher Lucretius
(first century B.C.) ; it was definitely stated by Thomas
Hobbes, who held that the state might possibly arise from
the application of physical force as well as in a voluntary
contract. Hume's position that the state originated in war
and force was also that of the Scotch philosopher Adam
Ferguson (1723-1816), especially in his notable *Essay on the
History of Civil Society* (1765) , which elaborated the thesis.

A new and powerful impulse to this interpretation of social
and political beginnings followed upon Darwin's formulation of his theory of organic evolution through the operation
of the principle of the struggle for existence. A school of
writers developed the doctrine known as "social Darwinism."
They held that war plays the same rôle in social evolution
that the struggle for existence occupies in the evolution of
individuals and species. Hence there was a tendency to praise
war as a major factor in social progress. An early and brilliant
upholder of this position was the English publicist Walter
Bagehot (1826-1877), whose *Physics and Politics* (1873) was
a frank effort to apply the principles of natural selection to
social and political phenomena. More elaborate were the
writings of the Austrian jurist and sociologist Ludwig Gumplowicz (1838-1909), who is regarded as the classical exponent of social Darwinism. He laid special stress upon the
part played by war in the establishment and mutations of

states. Gumplowicz was a serious and important writer, but social Darwinism soon degenerated into a vulgar defense of the war cult in the writings of General Friedrich von Bernhardi in Germany, of J. A. Cramb in England, and of Hudson Maxim in the United States.

While there can be little doubt that war played a predominant rôle in the creation of early states, its place in social evolution was greatly exaggerated by the social Darwinists. They were especially in error in contending that biological processes and mechanisms may be transferred directly to the social realm without due and proper qualifications. Indeed, even biologists have come to lay less stress than Darwin was wont to do upon the part played by the struggle for existence in the development and alteration of species. Moreover, the social Darwinists failed to understand that, however important war may have been in the early days, it has now become an anachronism and a menace. Finally, they were wrong in their contention that man is, by his physical nature, exclusively a fighting animal. The weaknesses of social Darwinism were ruthlessly exposed by pacifistic sociologists, especially Prince Kropotkin, Jacques Novicow, and Georg Friedrich Nicolai. These writers have stressed the importance of mutual aid, sympathy, and social instinct, and peaceful commercial relations in the emergence of social and political institutions.

At the opposite extreme from social Darwinism lay the views of those writers who held that society arose primarily out of the social instinct, which rested in turn upon an innate sense of sympathy and altruism. This view was first expounded by Hume and by his contemporary Francis Hutcheson (1694-1746). It was most elaborately developed in early days by the great economist and philosopher Adam Smith (1723-1790) in his *Theory of Moral Sentiments*. Smith's views had a profound influence upon the eminent American sociologist Franklin Henry Giddings (1855-1931). The latter derived the fundamental psychological fact in his system of sociology, namely, the "consciousness of kind," from Smith's doctrine of reflective sympathy. The most thoroughgoing exposition of the alleged sympathetic basis of social relationships appeared in the voluminous work of the Australian philosopher Alexander Sutherland, *The Origin and Growth of the Moral Instinct* (1898). Closely related to this interpretation was the emphasis upon mutual aid, coöperation, and altruism by the pacifistic writers above-mentioned, Kropotkin, Novicow, and Nicolai.

This group of writers who extolled sympathy and coöperation erred, like the social Darwinists, as a result of their one-sided emphasis upon a potential trait in human nature and social development. They went to the other extreme in denying the latent pugnacity of man. Social scientists today reject either extreme interpretation. They recognize that man is not innately warlike or pacific. His behavior is the result of response to stimulation, and he is capable of extensive adaptation to particular forms of conditioning. He can be conditioned by training to become a bellicose super-patriot or a fanatical pacifist whose cheek-turning reaches the point of exhibitionism.

The social Darwinists were not the only group of sociological writers to attempt to adapt the doctrine of evolution to the explanation of social development. An even more famous school, known as the classical evolutionists, sought to apply the evolutionary formula to the explanation of all phases of social development. They assumed that social development proceeds from the simple to the complex in essentially the same form the world over. They were the first group of writers on social origins to be deeply affected by the new archæology and anthropology which had revealed the great antiquity of man and human institutions. They believed that social evolution passed through essentially the same stages everywhere on the planet.

Instead of carefully studying the data in advance and trying to find out what had actually taken place, the classical evolutionists took for granted the validity of the evolutionary hypothesis when applied to human society and worked out a preconceived scheme of social development. They then sought for information, from whatever source, that would confirm and illustrate their scheme of social evolution. They held, in general, that men were originally grouped in small and ill-disciplined societies—hordes. Then came kinship or gentile society, in which relationships were first traced through the mother. Finally the part of the father in procreation was discovered, and this, together with other factors, led to the development of patriarchal society, in which relationships were traced through the father. In due time kinship society was replaced by the origins of civil society and the state. The outstanding writers of this school were Lewis Henry Morgan (1818-1881), Herbert Spencer (1820-1903), Andrew Lang (1844-1912), Sir James G. Frazer, E. S. Hartland, J. F. McLennan, Charles Letourneau, and Julius Lippert. The classic work was Morgan's *Ancient Society* (1878).

The devastating criticism of this early type of historical sociology has been mainly the work of Franz Boas and his disciples, who established critical anthropology in the United States. In the first place, they showed that the evolutionary sociologists were unscientific in their method of procedure: they had reversed the sequence of scientific method and were essentially deductive in their approach. In the second place, Boas and his disciples put to test the specific theories of evolution presented by the evolutionists and showed that they did not square with the actual facts drawn from a careful study of primitive society. They proved, for example, that there is little evidence that mother-descent ever preceded the tracing of descent through the fathers and demonstrated that it is very rare indeed that a system of tracing descent through mothers has ever been transformed into one of tracing ancestry through fathers.[1] One of the most important contributions of this critical school to our understanding of

[1] The best summary of this critical attitude is contained in Robert H. Lowie's *Primitive Society* (New York, 1920), which is comparable to the position attained in the older thinking by books like Lewis Henry Morgan's *Ancient Society*.

social origins lay in its emphasis upon the complexity of primitive social groupings and relationships. They showed that there is no such order and uniformity in primitive social organization and social developments as the older evolutionary school had asserted.

THEORIES OF THE STATE OF NATURE

Closely related to the errors connected with sociological views of the origin of society and the state, especially to those of the contract theory, was the notion of the "state of nature." This implied the belief that there had been a long period of human existence anterior to the origin of orderly social relations. Man had then lived without government, law, or fixed moral codes. This conception first arose in systematic fashion among Greek and Roman social philosophers, who generally identified the state of nature with an assumed early "golden age" of mankind. It was first elaborately set forth by the Roman philosopher Seneca (first century A.D.). He believed that life in the state of nature was almost idyllic until the appearance of private property. When Christian writers came in contact with these pagan theories, they tended to identify the pagan golden age with the Biblical concept of the blissful state of man before the temptation of Adam and the Fall. The pagans had held that the golden age of nature was destroyed by the origination of property. The Christian writers held that the destruction of pristine felicity was the result of expulsion from Paradise. With these qualifications, medieval Christian social philosophers tended to carry over the pagan theory with respect to the state of nature. The doctrine received its most thorough development, before Rousseau, in the writings of a Spanish Jesuit, Juan de Mariana (1536-1624).

We have already called attention to the basic dependence of the social-contract theory upon the idea of the state of nature. The writers of the contract school differed somewhat in their views of the character of human life in the state of nature. Some, like Hobbes, believed that it became utterly intolerable as a result of the struggles for dominion and possession. Others, such as Pufendorf and Locke, held that existence in the state of nature was inconvenient rather than unendurable. It was this inconvenience rather than absolute necessity that led to the social contract and the termination of the state of nature.

Down to the period of explorations and discoveries following the voyages of Columbus the so-called "natural man" had been mainly a product of the philosopher's imagination. Few had seen any savages, and even fewer realized that savage life was the prelude to developed historical society. Then extensive geographic and commercial expeditions brought Europeans into first-hand and wide-spread contact with primitive peoples all over the world. The social-contract doctrine was especially popular at this time, and writers eagerly accepted the stories relative to the newly discovered primitive peoples as a veritable confirmation of the actual existence of a state of nature. The philosophers believed that, at last, they had surprised the natural man in his true habitat.

This, in turn, led to the near-worship of primitivity and the supposed virtues of the noble savage which are popularly associated with the writings of Jean Jacques Rousseau and with the sentimental affectation of the French salons of the eighteenth century. Such veneration of savage life is customarily referred to in historical and sociological literature as "primitivism." There has been considerable error, however, in interpreting Rousseau's actual views about the virtues of savage life. It is true that he considered the life of the savage preferable to the existence of civilized men of his own day. But, as Schinz and Lovejoy have made clear, Rousseau did not maintain that the life of the savage was the most perfect that had ever been enjoyed by man. He believed that the ideal period of human existence was that of patriarchal society, such as that of the Biblical patriarchs Abraham, Isaac, and Jacob.

Directly associated with the state-of-nature doctrine was the related theory of natural law. The notion of a sociological law of Nature is an old one; it goes back to Socrates and the Stoics. But in the early modern period the conception was clarified and related more closely to specific political and legal applications. The law of Nature was regarded as the body of rules and principles that had governed men in pre-political days. Natural law was the norm by which to test the soundness of civil laws that were drawn up by the government after the state had been established. The state should not terminate the law of Nature, but rather should provide for the enforcement of its benign principles. The state should not restrict our natural freedom. It only frees us from the terrors and anarchy of unorganized pre-political society. Hobbes, Pufendorf, Spinoza, and others contributed to the development of the doctrine of natural law, but it was John Locke who gave it the particular interpretation that has made it of such great significance in legal history and business operations. He found that the major tenets of the law of Nature were the sanctity of personal liberty and of private property. The state was doing its supreme duty when it assured their protection and perpetuity. This notion was seized upon by the rising capitalistic class, embodied in the constitutions that it wrote, and introduced into the jurisprudence that it fostered. Here we find the legalistic basis of the contemporary reverence for property and the impregnable defenses that have been erected about it.

This whole body of doctrine, which prevailed in differing degrees from the time of the Greeks to the opening of the nineteenth century of our era, was demolished by the rise of the theory of evolution and the emergence of the modern sciences of anthropology and prehistoric archæology. In the place of the state of nature there was established the long epoch of preliterary or prehistoric culture, compared with which the period of civilization or the so-called historic age has been of very short duration. This prehistoric period was not one of uniform savagery or barbarism. It encompassed the passage from the very crude and near-bestial condition of the earliest men to the highly developed culture and institutions of the complex forms of primitive society. Far from being without government, law, or social control, primitive society came to be extremely well disciplined and devised elaborate forms of property rights. Moreover, the transition from primitive society to civil society was not a sudden event, conjured up in the consciousness of man, but a gradual development produced by the operation of many factors. In short, in the place of theories concerning the state of nature and the social contract, we now have the well-organized body of anthropological science, which tells us in detail of the life of man in the long stretch of time preceding the establishment of civil society and the mastery of the art of writing.

IDEAL SOCIETIES

Another prominent fiction of the sociological imagination was the devising of ideal societies or utopias. These were, for the most part, escapes from and compensations for the harsh and unsatisfactory social conditions that existed at the time of the several writers. Their utopias implied that man, by taking thought, can create an ideal society free from the imperfections of the existing social order. The utopians more or less assumed that the dominant rôle in social affairs is played by the social environment. They believed that if an ideal set of social relations were created, human nature would permit man to operate in a pacific and altruistic fashion.

The first of the important utopias was embodied in Plato's *Republic* (fourth century B.C.). Like the writers on the state of nature, Plato attributed most social evils to private property. He was also a believer in the natural aristocracy of the intellect and held that philosophers should be the kings. His utopia was constructed on these two foundations. He proposed communism and a system of eugenics for the small governing aristocracy, in order to promote harmony and the development of a superior race of leaders. For the masses he would permit private property and the monogamous marriage in order to continue disunity and mediocrity on the part of the "mob" ruled by the philosopher kings.

Plato himself realized that his *Republic* was a sociological error so far as actual practice was concerned. So he composed another book, entitled *The Laws,* which outlined what Plato conceived to be the best possible society and state that had any practical prospect of being brought into existence.

Plato's was the only important writing of the sort down to early modern times. Then Sir Thomas More (1478-1535) composed his famous *Utopia.* More wrote during the Tudor period, when English society, particularly the lower classes, was feeling the impact of the breakdown of the manorial system and the rise of large-scale sheep-farming. The suppression of the English monasteries by Henry VIII also increased the misery of the people. In the opening section of his famous book More presented a vivid picture of the disorder and suffering of the time. In order to suggest a remedy for these sorry conditions he portrayed an ideal society on the fanciful island of Amaurote. Here wealth was to be divided equally, so as to put an end to that avarice and covetousness which More, like Plato, regarded as the root of all human evils. The whole society on the island was to be a well-organized community based upon coöperative principles. Altruism would prevail, and each would have in his mind the interests of others. Everybody was to engage in agriculture and, in addition, to learn some trade. The government was to be a combination of aristocracy and the force of public opinion. More did not believe that many laws would be required, since equality and the coöperative principle would automatically bring to an end most of those evil desires and acts which are the result of corrupting institutions and require legislative restraint.

Francis Bacon (1561-1626) advocated the betterment of human society through the application of natural science. His ideal society, described in *The New Atlantis,* was located on an island off the coast of South America. Its central feature was the House of Solomon, a great laboratory of coöperative scientists. It also sent out travelers to visit the rest of the world and gather advanced scientific knowledge. All this was

360 THE STORY OF HUMAN ERROR

to be applied to increasing the happiness and welfare of the population. All superstition was to be rooted out, and social improvements were to be assured through the knowledge acquired by the scientists.

Much more radical were the proposals embodied in *The City of the Sun,* written by an Italian friar, Tommaso Campanella (1568-1639). He maintained that society is based upon the principles of power, love, and intelligence, and he contended that there could be no desirable social system unless these received due recognition in the organs of social control and political administration. Campanella argued for the complete abolition of all slavery, for the dignity and importance of labor, and for the elimination of the leisure class. Everybody was to work, but a short day would suffice to produce all the required necessities of life. He favored communism in property. He believed that the home and the family were the chief cause of the urge to acquire property; hence, property could not be done away with so long as the individual home was maintained. Community of wives and children he held essential to the elimination of the acquisitive tendency.

The utopia of James Harrington (1611-1677), an English publicist, had a much more aristocratic cast. In his *Oceana* he held that society must be organized on psychological principles, so as to ensure the leadership of the intellectual élite. Further, political organization must be so arranged as to secure the predominant influence of the landholding classes, which he believed to be roughly identical with the intellectual aristocracy. He sponsored the equal division of landed property and a wide use of the elective principle in government.

The *Télémaque* of François Fénelon (1651-1715) was a long and fanciful pedagogical novel which utilized the simple Homeric society in order to inculcate ideals suitable for the education of a prince. It endeavored to teach sound principles of government which were at wide variance with the tyranny and exploitation that prevailed under the French

Bourbons. Among the novel principles that Fénelon suggested was the education of women.

A new crop of utopias was stimulated by the beginnings of the Industrial Revolution. The man who started the movement was an enthusiastic disciple of the new industrialism, Count Claude Henri de Saint-Simon (1760-1825). He envisaged a new social and economic order in which material advance would be brought about by scientists and inventors. The practical direction of society was to be taken over by the captains of industry and finance, and the whole system was to be given moral fervor by a revamped Christianity which would revive the alleged radical social gospel of Jesus. Saint-Simon's notions were carried further by such disciples as Enfantin, Bazard, Leroux, Comte, and others.

Another famous utopian was a fellow-Frenchman, François Marie Charles Fourier (1772-1837). He proposed the reorganization of society on the basis of coöperative communities or "phalanxes" of about eighteen hundred individuals each. He provided for work at congenial occupations and for what he believed to be the proper distribution of the social income between capital, labor, and management. A number of communities based on his ideas were formed in Europe and the United States.

A leading utopian inspired by the new industrial conditions was the Englishman Robert Owen (1771-1858). Owen's personal credential as a utopian was his successful operation of his own cotton-mill at New Lanark, where he built up an ideal industrial community. Owen proposed the establishment of semi-communistic coöperative industrial communities which would combine productive efficiency with social justice. Like Fourier, he had a number of disciples who made practical attempts to establish such communities.

Utopianism persisted into the twentieth century, later echoes being William Morris' *News from Nowhere*, W. D. Howells' *A Traveller from Altruria*, Edward Bellamy's *Looking Backward*, and the utopian novels of H. G. Wells, which would reconstruct human society on the foundations of science, invention, internationalism, the world state, and social justice. Bellamy's work was especially important for its day. Nearly a half-century ago (1888) it clearly forecast the problems of our own advanced industrial civilization on the eve of the Third Industrial Revolution. It was the definite forerunner of present-day Technocracy. The latter advocates the turning over of the control of our material civilization and productive processes to industrial engineers who would plan a system of production for human service rather than private profit.

Utopians were inspired by the highest sentiments, and many of their conceptions were sound. Yet much of utopian thought was wishful thinking, the rationalization of the personal biases of the writers, or compensatory escape from oppressive existing conditions. This type of thought also tended to minimize or ignore the social stability and inertia which make it extremely difficult to achieve the direct and sudden transitions that utopian ideals require. But utopianism had one great virtue, namely, that it was the forerunner of social planning, which we have now come to recognize as necessary if civilization is to be sustained.

THE THEORY OF PROGRESS

A leading conception of historical sociology which had its origins in the period when the doctrines of the social contract and the state of nature were in full blast was the theory of social progress. The ancient Jews, holding to the doctrine of the Fall of Man, believed perfection to be found in the past rather than to be sought in the future. Other ancient peoples shared to some degree a comparable notion, namely, that of a decline from a golden age. Even more popular with the Greeks and Romans was the conception of the cyclical nature of human development. Culture would rise to a certain point and then decline to a level comparable to that which had existed at the beginning; then the process would start all over again, and the cycle would be repeated. The

Christians took over the Jewish notion of the Fall of Man and combined it with the pagan view of the decline from a golden age. Man could never expect any utopia here on earth. The state of blessedness was to be attained only in the world to come. The final judgment and the end of things earthly was, according to the Christian view as stated in the *Book of Revelation,* to be preceded by unusually horrible and devastating earthly portents.

Gradually, however, there arose the conviction that better things might be in store for humanity here on this earth. Back in the thirteenth century Roger Bacon had a vision of what applied science might do for man. Montaigne had a glimmering of a new idea when he suggested that philosophy should be concerned with human happiness here on earth rather than with salvation in the life to come. Francis Bacon and Descartes united in decrying the authority of the past. Bacon contended that the moderns were superior to the ancients and suggested that utopia might be attained through applying science to human problems.

The doctrine of progress as it is conventionally understood, however, began with Jonathan Swift (1667-1745) and Bernard de Fontenelle (1657-1757). Swift's *Battle of the Books* (1698) was a telling satire on the defenders of the authority of the ancients and the worship of the classics and classical scholars. In his *Dialogues of the Dead* (1683) Fontenelle hardly went beyond the contention that the ancients were no better than the moderns, but five years later, in his *Digression on the Ancients and the Moderns,* he took a more advanced position. He held that the ancients and the moderns were essentially alike in a biological sense, there being no progress in this respect. In the fine arts, which are chiefly a spontaneous expression of the human spirit, there seems to be no law of progress: ancient peoples had great achievements to their credit, but the best modern works in art, poetry, and oratory equal the most perfect ancient examples. On the other hand, in science and industry we find an altogether different story. In these fields development is cumulative.

There has been vast progress since antiquity, and even greater things may be looked for in the future. Moreover, Fontenelle proceeded to assert that unreasoning admiration for the ancients is a major obstacle to progress. It is doubtful if anyone, even in our own day, has more intelligently stated the general principles involved in the problem of what we call progress than did Fontenelle.

Charles Perrault (1628-1703) was a contemporary of Fontenelle and expressed very much the same view in his *Parallel of the Ancients and Moderns* (1688-1696). But he was so much impressed by what he regarded as the perfection of the culture of his own generation that he was not much concerned with future progress—if, indeed, he would have conceded that anything could be better than his own age. A more positive attitude towards future progress was taken by the Abbé de Saint-Pierre in his *Plan for Perpetual Peace* (1712). He contended that progress was real and that the achievements of his own age were more notable than those of the era of Plato and Aristotle. He was particularly interested in social progress and believed in the desirability of an Academy of Political Science to guide social advance. He placed great faith in the power of a wise government and was a forerunner of Helvétius and the Utilitarians. Claude Adrien Helvétius (1715-1771) was the foremost of the French social optimists of this period. He believed thoroughly in the possibility of human perfection and thought it could be achieved effectively through universal enlightenment and rational education. He believed in the equality of man and held that existing inequalities could also be eliminated through education.

In the first half of the eighteenth century the Italian Giovanni Battista Vico (1668-1744), a philosopher of history, developed his conception of progress. He held that human progress does not take place directly or in a straight line. Rather, it takes the form of a spiral. There may seem to be cycles of development, but they never go back to the original starting-point. Each turn is higher than the preceding. A little later in France a more realistic historical theory of progress was proposed by Anne Robert Jacques Turgot, Baron de l'Aulne (1727-1781), himself an eminent contributor to the philosophy of history. He laid great stress upon the continuity of history and the cumulative nature of progress. He contended that the more complex the civilization, the more rapid human progress. Thus, advance was very slow in primitive times, but had been greatly accelerated in the modern epoch. Even more optimistic was the distinguished writer of the French Revolutionary period, the Marquis de Condorcet (1743-1794). He not only affirmed his belief in the reality of progress, but presumed to divide the history of civilization into ten periods, each representing a definite stage in the development of mankind and human civilization. Nine of these periods had already been passed through, and the French Revolution and modern science were leading us to the brink of the tenth, which would produce an era of happiness and well-being the like of which had never been known.

There were other men who contributed variously to the notion of progress. The German philosopher J. G. von Herder (1744-1803) attempted to discover laws of progress based on the joint operation of Nature and God. Immanuel Kant (1724-1804) sought to prove the reality of moral progress. The English publicist William Godwin (1756-1836) believed that perfection might be attained through the abolition of the state and property and the inculcation of reason through private instruction. The Count de Saint-Simon followed the line of the Abbé de Saint-Pierre in holding that a definite social science must be provided to guide human progress. These notions culminated in the historical philosophy and sociology of Auguste Comte (1798-1857). He elaborated a comprehensive system of "laws" concerning intellectual progress and formulated an expansive philosophy of history, embodying the division of the past into a large number of periods and subperiods, each characterized by some phase of cultural advance.

While the theory of progress has received enthusiastic support from many since the time of Comte, pessimistic or chastened attitudes have also appeared. Some, like the German philosophers Friedrich Nietzsche and Oswald Spengler, have reverted to something similar to the doctrine of cycles characteristic of classical thinking. More common, however, has been the tendency to substitute the notion of change for that of progress. The latter implies that things are certainly getting better. Of this we are not now so certain, but we are aware of change in many phases of life and thought. Most important has been the recognition that change takes place rapidly in the realm of science and material culture and very slowly in institutions and morals. This discrepancy in the rate of progress as between material culture and social institutions —now called "cultural lag"—seems to have placed modern civilization in particular jeopardy.

Enthusiastic belief in the theory of progress thus appears to have been an incidental product of the optimism of the period of "enlightenment," middle-class revolutions, early scientific progress, and the rise of capitalism. We can now take no such dogmatic position on the subject as did the writers of the eighteenth century. The future is uncertain. It may bring the realization of dreams far beyond those of the most fantastic utopians. On the other hand, it may bring collapse and reversion to barbarism. Only the triumph of reason and knowledge over tradition and passion can prevent the latter outcome.

GEOGRAPHICAL DETERMINISM

One of the most persistent, popular, and interesting errors in sociological thought was that involved in the doctrine of geographical determinism, that is, the theory that geographic factors are of primary importance in the shaping of social customs and institutions. This belief has persisted throughout the entire history of social theory and has powerful champions even in our own day. The first writer to suggest the notion was the ancient Greek physician Hippocrates (C.

460-370 B.C.) , who set it forth in his work on *Airs, Waters, and Places*. While immediately interested in ascertaining the effect of geographic factors on disease, he suggested a general theory of geographical determinism which revolved around the idea of the superiority of the races of the temperate zone. He believed that the peoples of warm climates were clever, but morally weak and wicked; those of cold climates were strong and virtuous, but stupid. The inhabitants of the temperate climates, especially the Greeks, combined the good qualities of both the northerners and the southerners without taking over their respective weaknesses. The inhabitants of the temperate zones were, in short, strong, brave, wise, and good. Hippocrates intermingled astrology with geography in arriving at his conclusions, a tendency which was shared by all other important writers on geographic influences until the eighteenth century of our era. Thus, to overemphasis upon geographical factors there were added the preposterous errors involved in astrological lore.

This view of the superiority of the races of the middle climates dominated the writing on geographic influences until modern times. Aristotle accepted the notions of Hippocrates and elaborated upon them in order to vindicate his theory of Greek superiority. The same conception of the superior character of the inhabitants of temperate zones was adapted by Cicero and Vitruvius to demonstrate the superiority of the Romans and to justify the triumph of Roman arms and imperial administration. The popularity of Aristotle in the Middle Ages led to the revival of his theory of the ascendancy of the people of temperate climates. St. Thomas Aquinas was one who set forth this notion; it found its echoes among the Muslims in the writings of Ibn Khaldun.

It is obvious to the impartial observer that this whole idea rested upon unscientific foundations and was no more than the rationalization of the biases or patriotic fervor of the writers promulgating it. As a fact, in the beginnings of civilization the first great advances took place in semitropical areas where the bounties of Nature were most profuse. As civilization has progressed, it has geographically followed a north-ward course.

The first extended and detailed discussion of the doctrine of geographical determinism appeared in the writings of the famous French publicist Jean Bodin (1530-1596). His work represented the final summation and elaboration of astrological and geographical lore in defense of the notion that peoples of intermediate climates are those who build the highest civilizations. He added few new ideas, but his treatment was far more systematic and extensive than that of any previous writer on the subject. Yet there was scarcely a paragraph in this portion of his work that rested upon a sound factual basis. It was a congeries of errors rarely matched in the whole history of sociological thought.

The beginnings of a scientific attitude toward geographic influences manifested themselves in crude fashion in the eighteenth century, following on the heels of the remarkable scientific advances of the sixteenth and seventeenth centuries. The first of such works was that of the eminent English physician Richard Mead (1673-1754), author of a *Treatise Concerning the Influence of the Sun and Moon upon Human Bodies*. While his conclusions were crude and dogmatic, Mead reflected the advances in the study of gases and atmospheric pressure made by Boyle, Newton, and others. A further elaboration of this point of view appeared in the book by another English physician, John Arbuthnot, *An Essay Concerning the Effect of Air on Human Bodies*, published in 1733. This applied the new scientific concepts to a study of the effect of climate upon human traits. It inspired a far more famous work, *The Spirit of Laws*, by the French Philosopher Baron Montesquieu (1689-1775).

Montesquieu abandoned the old dogma of the comprehensive superiority of the peoples of the intermediate climates. In its place he attempted to develop a general philosophy of geographical determinism, indicating the effects of geographic factors upon all types of institutions. He assumed that climatic influences produce distinctive traits in the inhabitants of each climate. Those institutions are best which conform most closely to the characteristics of the inhabitants of a particular climate. For example, of political institutions, a despotism is best adapted to the needs of the inhabitants of warm climates; a constitutional monarchy to those who dwell in temperate zones; and a republic to the inhabitants of the cold areas. There can be no absolute standard of superiority in human institutions, their excellence depending upon their adaptation to the qualities developed by climatic effects in the inhabitants of different zones. In spite of its semblance of the use of a scientific technique, the work of Montesquieu was mainly dogmatic and deductive. It took for granted, without proof, the dominating influence of climatic factors and upon this erected a philosophy of history, law, and government in accordance with his preconceived dogmas. But there was one sound contention in the position of Montesquieu, namely, that institutions are good or bad according to their adaptation to the character of a particular population.

The remarkable progress of exploration and geographic science in the nineteenth and twentieth centuries led to a vast improvement in the character of writings on geographical determinism. The modern theory found its most eminent representatives in such comprehensive writers as Karl Ritter (1779-1859), Henry Thomas Buckle (1821-1862), Friedrich Ratzel (1844-1904), and in our own day Ellsworth Huntington, Ellen C. Semple, and others. These writers have broadly considered the dominating influence of all types of geographic factors—climate, topography, rainfall, isolation, contacts, routes of travel, and atmospheric conditions. Moreover, they have understood that geographic influences exert widely varied effects upon humanity in different eras of historical development. Great bodies of water, which were once a barrier to travel and progress, have since 1500 become major instruments for the transmission of civilization. There has been a vast improvement both in the range of the geographical knowledge possessed and in the caution with which it has been applied to sociological interpretation.

No student of sociological theories can safely ignore the very important contributions of this school of thought. Yet, as the cultural historians and anthropologists have made clear, these writers were guilty of a major error in contending for geographical determinism. Geographic influences exert a very important conditioning effect upon human culture, but they do not determine the course of its development. This fact has been well stated by Robert H. Lowie: "Environment cannot explain culture because the identical environment is consistent with distinct cultures; because cultural traits persist from inertia in an unfavorable environment; because they do not develop where they would be of distinct advantage to a people; and because they may even disappear where one would least expect it on geographical principles." In short, the geographical determinists have erred in over-emphasizing an important consideration in sociological thought. It is not that geographic influences fail to play a very important rôle in human affairs, but rather that they exert a conditioning instead of a determining influence.

REASON VERSUS EMOTION: REVOLUTION OR QUIETISM

The epoch-making debate between social philosophers of the rationalistic and romanticist schools from 1750 to 1850 gave rise to numerous errors in sociological thought. The controversy centered around the question whether social institutions are amenable to rapid and revolutionary changes through the application of human reason, or whether they are the outgrowth of the unconscious forces embodied in national genius. To a considerable degree the issue was whether revolution is a legitimate and desirable type of social change, or whether we must allow matters to make headway slowly through a process of natural and spontaneous development. This debate occupied much of the social thinking during the eighteenth and early nineteenth centuries.

The leaders of the rationalistic and revolutionary school were the Abbé Sieyès (1748-1836), the Marquis de Condorcet, William Godwin, and Thomas Paine (1737-1809). They emphasized their belief in the possibility of abolishing old abuses and introducing a new social order by applying human reason to social problems. Paine produced the most eloquent defense of this point of view in his famous work *The Rights of Man* (1791-92), written as a reply to Edmund Burke's ferocious attack upon the French Revolution. These writers minimized social inertia and cultural lag and were optimistic regarding the amenability of social life and institutions to rational analysis and artificial change.

Upholding the other extreme view, to the effect that we cannot artificially alter institutions through the application of reason but must wait for the gradual development that grows out of national genius, were Edmund Burke (1729-1797), J. G. von Herder, J. G. Fichte (1762-1814), and F. K. von Savigny (1779-1861). These men looked upon social institutions as the outgrowth of somewhat mysterious psychic forces, operating mainly on a subconscious level and expressing themselves collectively through what is known as national genius. They fiercely attacked the prevailing rationalistic and revolutionary theories. They advocated political quietism and social patience. The best-known example of this type of sociological thinking was Burke's *Reflections on the French Revolution* (1790). So extreme were some of these writers that von Savigny even warned against trying to codify law. Law he held to be an ever-growing and expanding product of national culture; to codify it would kill it—would be as reasonable as to try to embalm a growing plant. The reactionary and conservative view of social development was set forth in even more extreme form by Louis Gabriel de Bonald (1754-1840), Joseph de Maistre (1754-1821), and Ludwig von Haller (1768-1854).

It is obvious that both schools of writers erred in their extreme emphasis upon one interpretation to the exclusion of the other. Social inertia, cultural lag, vested interests, customs, traditions, and the like rest primarily upon an irrational basis and offer tremendous resistance to the inroads of reason and change. A rational and sensible scheme for an ideal society can certainly be formulated, but it is another matter to assert that it can be speedily and easily set up in the face of the opposition of all the forces that resist social change. On the other hand, those who relentlessly oppose reason and revolution are in danger of doing no more than rationalizing social stagnation, social abuses, injustice, and the dominion of the existing powers. Social change may have to be slow and take account of the historical past and the existing social environment. But this is something quite different from holding that it is illegitimate or wicked to plan intelligently for the future and to seek to bring about better conditions as rapidly as possible.

Rather closely related to the extreme rationalists was the philosophy of Jeremy Bentham (1748-1832), perhaps the most fertile and inventive mind among social philosophers. It was Bentham's desire to become "the Newton of the moral world." He proposed the famous "felicific calculus," which embodied his efforts to reduce social philosophy to a scientific discipline. This calculus of human conduct was based upon the assumption that man is guided and controlled in his behavior by "two sovereign masters, pain and pleasure." Bentham contended that man is absolutely rational in his procedure and conduct in society. He seeks the utmost possible pleasure and the minimum possible degree of pain. Human conduct is the outcome of this desire to secure pleasure and avoid pain. If, sometimes, our conduct does not seem to conform to such a test, it is because we do not completely or properly understand the situation—in short, a proof of inadequate information. Hence, what is needed to ensure the perfect operation of the pleasure-pain principle is education, which will clear up our understanding of all human situations.

Bentham was both a prolific writer and an ardent social reformer and was himself active in promoting this point of view. But even more significant was the effect of Benthamism upon the political and economic psychology of the nineteenth century. The works on political psychology down through the time of James Bryce were based upon Bentham's premises. It was not until the first decade of the twentieth century that Graham Wallas exposed its superficiality in his notable work *Human Nature in Politics* (1908). Orthodox psychological economics, launched by W. S. Jevons, the Austrian school, and John Bates Clark, was also founded primarily upon the principles of Bentham. Even Karl Marx was profoundly affected by Bentham's views: his theory of class-consciousness and the class-struggle involved a considerable acceptance of Bentham's attitude respecting the rationality of man. Economic psychology was not freed from the Benthamite incubus until about the time of the World War, when Carlton H. Parker, Wesley C. Mitchell, and others revealed its fundamental fallacies.

Bentham's great error lay in ignoring those subconscious and irrational factors in human nature and social behavior which have been revealed by a subsequent century of study in psychology, psychiatry, and social pschology. He knew little or nothing of the human subconscious and of its influence upon conduct. He did not reckon with custom, convention, tradition, and other partially or wholly irrational factors affecting social behavior. Nor did he take account of crowd or mob psychology. His picture of human behavior was, perhaps, a noble essay in describing how things might be if man and society were different. But his theories do not square with the facts either of human nature or of social behavior.

Although Bentham made many useful contributions to practical reform in the fields of education, poor relief, prison reform, and public-health legislation, few writers in the history of sociological thought have been responsible for such extensive and persistent errors in social thinking. He dominated most of the social psychology of the nineteenth century, and it was a rule of error almost without exception. It generated a false optimism about human nature and made it difficult for social thinkers to come to grips with the grim realities concerning social change.

BIOLOGICAL THEORIES IN SOCIOLOGY

The rise of the theory of evolution and the growing interest in biological science were provocative of a large collection of errors in social thinking. The first of these we shall discuss was the theory of the social organism.[2] This doctrine assumed the essential identity between the nature of the individual biological organism and that of society as a whole. In the individual organism, for example, there is a directive faculty in the brain and nervous system, a nourishing mechanism in the digestive organs and the circulation of the blood, and so on. In society we find comparable systems of organs. Similar to the brain in the individual organism is government in society. Comparable to the digestive organs in the individual are the agricultural and productive classes in society. Most writers contented themselves with pointing out in more or less detail—sometimes most elaborately—the similarities between the individual organism and society. But a few went so far as to hold that society is literally a true organism. The outstanding writers who expounded the doctrine of the social organism were Herbert Spencer, Paul Lilienfeld, Albert Schaeffle, and René Worms.

The extreme members of this group of writers were in error in alleging any identity between the individual organism and society. There are only vague similarities and parallels often achieved by forced interpretations; and what truth there may be in the analogy is of little practical value. Certainly it did not warrant the detailed and voluminous writings that were devoted to its elucidation. The theory has been criticized in constructive fashion by L. T. Hobhouse and others. They have indicated that all there is of value in the idea of a social organism consists in the implication of the desirability of an adequate development of social coördination, the division of labor, and the coöperative principle in society.

[2] The related error of social Darwinism, already noted in connection with the origins of society and the state, applies equally to other applications and belongs in this group.

Another widely and bitterly debated error in sociological thought growing out of biological influences was what we know as Malthusianism. It derived its name from the author of the theory, the English economic philosopher Thomas Robert Malthus (1766-1834), whose *Essay on Population* appeared in 1798. The substance of Malthus' argument was that population tends to increase in a geometric ratio (1, 2, 4, 8, 16, etc.) whereas the food-supply cannot possibly be made to increase in more than an arithmetic ratio (1, 2, 3, 4, 5, etc.) ; hence, population tends always to press upon the underlying means of subsistence. Malthus recognized two kinds of checks to the tendency of population to outrun its food-supply: the first, positive checks—war, pestilence, and starvation; and the second, negative checks—postponement of marriage to a later age and what he described as "moral restraint." In the England of his day Malthus regarded it as inevitable that a considerable part of the population should lead a life of poverty and misery. He feared that, for the immediate future at least, any increase in the means of subsistence would tend to bring about a more than corresponding increase in the population.

Both the vigorous defenders and bitter critics of Malthusianism have usually lacked historical perspective. As applied to conditions as they existed in 1798, the position of Malthus was reasonably sound, but scientific and technological changes since his day have made Malthusian principles in large part an intellectual curiosity in our era. The two or three "industrial revolutions" have completely altered the situation respecting the food-supply. The First Industrial Revolution gave us the reaper, binder, thresher, and other mechanical means of carrying on agriculture, the elements of scientific fertilization, and the beginnings of efficient methods of storing and preserving foods. The Second Industrial Revolution carried the process further by providing the tractor, gang-plow, reaping and threshing combine, ultra-scientific soil-testing and fertilization, scientific canning and refrigeration to further food preservation,.and other comparable advances.

So important are these results that O. W. Willcox estimates that we could produce all the food needed in the United States by the scientific cultivation of one-fifth of the land now under tillage and with one-fifth of the farm labor now employed. Even more dramatic and portentous for the future is the advent of the Third Industrial Revolution, which offers the prospect of the synthetic chemical production of foods in laboratories and factories remote from the farms and with no discernible limits to food production. There are social considerations justifying population limitation, but we can produce food enough today to support any reasonable or probable increase in population.

Another major error in Malthus' theory is related to his assertion that population increase is controlled primarily by the food-supply. It is an interesting fact that the rate of population increase has slowed down notably in the last decade and in the most highly industrialized nations—in other words, at a time when we are uniquely potent in food production and in the countries most competent to feed a growing population. Cultural factors have become more important than physical elements in controlling population.

One of the most serious and in some instances arrogant errors in sociological thought of biological derivation is that represented by the extreme school of writers on the subject of eugenics. This cult was founded by Sir Francis Galton, Karl Pearson, Otto Ammon, Wilhelm Schallmayer and G. Vacher de Lapouge, and was popularized in the United States by C. B. Davenport and Henry Fairfield Osborn. These writers, and their less scientific disciples to an even greater extent, contend that the physical equipment of man is the dominant factor in social life and progress. It is held that a utopian state of society is retarded primarily by the low level of physical stamina and mental ability in the population. The general tendency of such eugenicists is to hold that biological factors are of primary importance in society and to scoff at social reforms and cultural influences. They maintain that no far-reaching improvements can be expected in human society until we create a race of supermen by selective breeding. Social legislation providing for more equitable distribution of wealth and opportunity, for equitable taxation, and for scientific social agencies is condemned as ineffective and demoralizing.

There is no doubt that we would profit by eliminating the physically and mentally unfit and producing an ever higher type of human being. There is also every probability that if the eugenicists were given their way, they could produce such a race of superior beings. It is desirable, therefore, that we should encourage their program in every conceivable way. But it would be very dangerous to surrender to the extreme biological fatalism that is represented by most hundred-percent eugenicists. The obvious fact is that human society will perish or be saved before any eugenics program, granted existing opposition and inertia, can breed a super-race. Our problems will not permit a century or more of evasion and postponement.

Further, the appeal to history is the best answer to the eugenicists' claim that nothing important can be done until the race is improved in a physical and intellectual way. It is conceded by anthropologists and biologists that the native physical and intellectual qualities of the race have not improved in the last thirty thousand years. *Homo sapiens* appeared about that long ago, and man has not increased in native intelligence since, whatever his subsequent accumulations of information. Indeed, some of the eugenically inclined biologists hold that the race is somewhat inferior today compared with what it was at the close of the cave-dwelling age. Yet all that we know as civilization has been achieved by this same race of men—always inferior according to eugenic standards. We have advanced from the cave to the Empire State Building with all of our physical defects. It seems equally probable that with the human race as it is, if we followed sound ideas, we could create a rational social system and intelligently apply our vast scientific knowledge to the comprehensive betterment of mankind.

There is another eugenics fallacy which needs to be silenced forever, namely, the arrogant assumption that our present hierarchy of wealth and social classes rests upon a valid biological basis—with the ablest at the top and the inferiors at the bottom of the pyramid. This illusion has been punctured by a biologist, Herman J. Muller of the University of Texas, who has written: "We must realize that in a society having such glaring inequalities of environment as ours, our tests [that is, social status] are of little account in the determination of individual genetic differences in intelligence, except in some cases where these differences are extreme, or where essential likenesses of both home and outer environment can be proved." There are too many artificial limitations and favoritisms in our present social and economic order to permit the assumption that status is today correlated in any complete manner with innate ability.

Related to the interpretation of progress in terms of superior human types is the effort to associate specific cultural achievements with alleged racial differences and superiorities. A vulgar fallacy, not wholly absent from Aristotelian doctrine, it may be traced back at least as far as romanticism and the reaction against the French Revolution. Then there developed those misleading dogmas of the fickleness and political incompetence of the French as contrasted with the unparalleled political sagacity and capacity for achievement of the Teutonic and Anglo-Saxon peoples. It was further reinforced by the amusing but tragic combination of fallacies in Joseph Arthur de Gobineau's *Essay on the Inequality of the Human Races* (1853-1855), with its eulogy of the "Aryan race." It was solemnly confirmed by the philologists and reached its *reductio ad absurdum* in the nineties in the dithyrambic exultation of Houston Stuart Chamberlain over the cultural supremacy of the Teuton. Even modern biology was drawn upon to support the doctrine of racial superiority, and Sir Francis Galton came forward with an allegation of the vast psycho-physical superiority of the "mythical Greek" over the average member of the intellectual classes of the present day. These racial fallacies have come to possess a good deal of practical significance in our own day because they have been put into practice in the politics of a great state. The Aryanism of Herr Hitler is nothing more than a vulgarized version of the ideas of Gobineau and Chamberlain. It was left for sociological historians such as Fustel de Coulanges and Frederick Seebohm to challenge the romanticist-Teutonic philosophy of history; for W. Z. Ripley and Roland B. Dixon to shatter forever the myth of an Aryan race and to show the hopeless confusion and mixture of races in every leading European state; and for Franz Boas to demonstrate that no sufficient evidence can at present be adduced to prove the biological superiority of any race or sub-race. By the results of these studies the racial interpretation of history has been utterly discredited and can in the future be the refuge only of the uninformed or of the partisan advocate.[3]

CRITIQUE OF SOCIAL-ECONOMIC SYSTEMS

Sociological thought both directs economic systems and derives from them. Especially have the great industrial changes influenced recent thought alike in defense and in criticism of prevailing social orders. Such justification of the *status quo* may readily involve the error of prejudice or partisanship as the systems in question are involved in active interests. The opponents of the system of capitalism regard either its premises or its trends in practice, or both, as "errors" in the legalized regulation of economic forces; its defenders dispute the thesis. The issue becomes controversial, and the word "error" shifts its bearing toward practical benefit to the total social order; it involves concepts of values. A survey of the arguments of defense and opposition to the social order, past and present, illustrates the movements of sociological thought and the resulting liabilities to error.

[3] F. H. Hankins has brought together all of the cogent material in his *Racial Basis of Civilization* (New York, 1926). See also Julian Huxley, A. C. Haddon and A. M. Carr-Saunders, *We Europeans* (New York, 1935).

Among the early defenders of the system commonly called capitalism were Saint-Simon and his younger contemporary J. B. Say. They proposed to hand over the direction of society to the captains of industry and finance. They believed that their aspirations and methods were perfectly designed to promote the well-being of human society. Saint-Simon's distinguished disciple Auguste Comte held that industrial growth and the search for profits helped to promote human progress and social justice. A similar position was taken by James Mill (1773-1836), father of John Stuart Mill, and it persists among some contemporary sociologists, notably Thomas Nixon Carver. Arraigned in opposition are Albion W. Small, L. T. Hobhouse, J. A. Hobson, Beatrice and Sidney Webb, Emile Durkheim, and Thorstein Veblen. They contend that the self-seeking and short-sighted policies of capitalism have brought about "a mixture of lottery and famine." These and many other writers have especially emphasized the antisocial and suicidal character of finance capitalism, which dominates the contemporary economic order.

The new capitalism gave rise to sociological rationalizations in opposition to remedial social legislation. The theory of evolution was drawn upon to justify this philosophy of "rugged individualism," as Newtonian physics had earlier been exploited to justify the free-trade philosophy. It was held that social progress must be achieved in an automatic and natural way, just as biological progress is attained through the operation of the survival of the fittest. For man to interfere with the natural processes of economic evolution would be disastrous. It would impoverish society, impede social progress, and affect adversely the poorer classes whom the humanitarians seek to protect by their ill-advised interference. Unlimited competition is the only road to industrial and social utopia. This position was most effectively set forth in Herbert Spencer's *Social Statics*, in William Graham Sumner's *What Social Classes Owe to Each Other*, and in the many writings of Jacques Novicow, Ludwig Gumplowicz, Gustave LeBon, and Fabian Franklin.

Another group of sociologists and reformers advocate state intervention to protect the masses from exploitation and to secure adequate social planning. Prominent among them are Lester F. Ward, Small, Hobhouse, the Webbs, Hobson, R. H. Tawney, Gustav Schmoller, and Paul H. Douglas. Contemporary opponents add to their destructive critique of the capitalistic premises the evidence of history in the general breakdown of capitalism after nearly a century of relative freedom from social control. They emphasize as especially cogent the economic facts of recent American history: the richest and mostly highly developed of the modern economic states, enjoying the greatest individualism, has been brought face to face with collapse in 1929 and later as result of the very policies that Sumner and his school extolled.

At the other extreme from the protagonists of capitalism and individualism appear the Marxians, advocating a thoroughgoing state control over economic life. Karl Marx (1818-1883) developed his theories in considerable part from preceding writers such as Ricardo, Feuerbach, Rodbertus, and Sismondi. He elaborated, however, a definite economic and social philosophy, based upon the economic interpretation of history, the labor theory of value, the doctrine of surplus value and the exploitation of the workers, class-consciousness and the class struggle; he predicted an ultimate proletarian revolution which would install the workers in the control of society.

There was much validity in the doctrines of Marx. He erred chiefly in overemphasizing a very potent factor in social evolution, namely, the material element. At times it has certainly been a determining factor, but not always. Further, capitalism has made greater contributions to social progress and has been less ruthless than Marx contended. Especially was Marx in error in his assumption of the class-consciousness of the workers. His economic and political psychology was based almost exclusively on Benthamite rationalism, which assumed that man is a cold, calculating being capable of discerning and following his material interests.

Marx, like Bentham, failed to reckon with the effects of custom, tradition, education, cultural lag, and such psychological mechanisms as that of "identification," which leads the workers into psychic associations and loyalties diametrically opposed to their class-interests. Contemporary Marxians, instead of developing a new economic philosophy suited to the times, have been content to parrot the text of their master, who in some respects is as outmoded as Adam Smith or Herbert Spencer. Marx knew nothing of the Second Industrial Revolution or of contemporary finance capitalism. The pronouncement of H. G. Wells may go too far, but it has some validity: "Indeed from first to last the influence of Marx has been an unqualified drag upon the progressive reorganization of human society. We should be far nearer a sanely organized world system today if Karl Marx had never been born." Marx retains a high place as an economist and social prophet; but the slavish loyalty of radicals to Marx is as out of place in the modern scene as the citation of Smith and Spencer by "rugged individualists."

PREMATURE SYSTEMATIZATION

A major error in the growth of sociological thought was the formulation of elaborate sociological systems and the production of highly generalized works on sociology long before the information that would warrant any such achievement was available. What came first should have come last, if, indeed, such formulations are ever possible. Moreover, these generalizations substituted deduction for induction and became a rationalized elaboration of the personal views of the authors. The leading writers to whom this statement applies were Auguste Comte, Herbert Spencer, Lester F. Ward, Albion W. Small, Franklin Henry Giddings, Gustav Ratzenhofer, Frank Oppenheimer, and Leopold von Wiese. It applies with a different emphasis to the most encyclopedic system of the Italian sociologist Vilfredo Pareto, whose "system"—the most ambitious since Bentham—reduces sociology to a natural science of his own peculiar complexion. As Bentham rationalized a faith in pure intellectualism and the capacity of men to govern their conduct by reason, so Pareto's system of sociology becomes a pedantic rationalization of his views of the low rationality of men and the incompetence of the masses. Pareto's major concept is that of "derivations," a by-product equivalent to rationalizations. He exposes the lack of valid scientific foundation of "derivations." Yet his own sociological writings are no more than a vast derivation. What pretends to be an objective treatise on the nature of society is little more than a voluminous argument in terms of abstract concepts in defense of capitalism and the growing trend toward Fascism.

Coincident with the era of sociological systematization was the age of controversies with respect to sociological methods, categories, concepts, and the like. While the formative period of any science must be concerned with such things, the energy devoted to these debates in sociology was out of all proportion to their value and gave the impression that such minute discussions constituted the essence of sociology.

An error of a different order appears in the very useful employment of the statistical technique in the pursuit of sociological problems. It consists in holding too exclusively that sociological data must be confined to those amenable to measurement and thus excluding significant but only partly and subjectively appraisable factors. The movement toward realistic statistics was an inevitable reaction from the older dogmatism of social philosophers and systematizers; but the more ardent advocates of the statistical method have gone to the other extreme and have ruled out the relevance of common sense and logic, as well as the importance of trends and forces clearly discerned but not amenable to quantitative statement.[4]

Moreover, much statistical study of social problems has been little more than a pompous documentation of the obvious, and often it has served as a dignified excuse for

[4] This confusion as to what is measurable (or becomes so by large concessions or assumptions) obtains in other sciences, notably in psychology.

conservatism and neutrality of opinion. Moreover, facts, including those in statistical form, may as readily be assembled in the service of *a priori* prejudices and convictions, drawn from family life, education, class and professional associations, as objectively handled in the interests of truth.

SOCIOLOGICAL APPLICATIONS

As already noted, the varieties of "error" surveyed in these pages are concerned with various applications of social philosophy to the understanding and regulation of the existing social order. There are in this sense "errors" of social psychology, prominent in the development of sociology during the half-century following 1875. Representative writers who introduced psychological motives into sociology included Gustave LeBon, Scipio Sighele, Gabriel Tarde, Emile Durkheim, William McDougall, and Wilfred Trotter. The crowd or herd psychology occupied a major place as the clue to social behavior. LeBon and Sighele ascribed to the crowd abnormal and pathological characteristics dominated by irrationality and bordering on the criminal. Tarde developed a complete system of psychology on the principle or pattern of imitation as universal and dominant. Durkheim emphasized the influence of crowd psychology on the individual, dominating his mental concepts and behavior; social man was ruled by the herd. McDougall laid stress on the variety and potency of the so-called instincts, which he held to direct both individual and social behavior; while in the same trend Trotter wrote engagingly upon the herd instinct as explanatory of human characteristics, in social situations dominantly.

An outstanding weakness of these writers was the fact that few of them, with the exception of McDougall, were trained in psychology or had much technical knowledge of the subject. Most of them were keen observers and clever writers who employed rule-of-thumb and common-sense methods. LeBon and Sighele rationalized their prejudices against democracy and parliamentary government in denouncing the crowd mind. Present-day psychologists doubt the number and specific character of the instincts that McDougall assumed as the foundation of his social psychology, and even question whether the concept of instinct thus employed has a naturalistic basis. While sociologists recognize the remarkably suggestive character of Trotter's work, they question the existence of any specific herd instinct, preferring to record the modification of behavior under the influence of social forces, to which Freud in his clinical approach gave the name of Super-Ego. While modern sociologists are agreed that valid principles are rooted in psychic motivation, the attempts to take over the conclusions of social psychology—in a sense an independent discipline—have not reached a satisfactory stage and reflect the errors of an inadequate social psychology itself.

Among the political problems and institutions on which sociologists have assumed to give us authoritative opinions is that of democracy, which turns upon a view of the extent of the differences among men. Social philosophers such as Le-Bon, W. E. H. Lecky, and Sir Henry Sumner Maine have rationalized their esteem of an aristocratic society into belittling of democracy. Other writers of the eugenics persuasion have assailed the principle of democracy from the standpoint of that of philosophy. They hold that the dominion of society must be turned over to the able individuals, and they stress the mental incompetence or mediocrity of the masses. On the other hand we have enthusiastic defenses of democracy, which are equally a rationalization of the personal leanings and convictions of the writers. Such have been L. T. Hobhouse, W. G. Sumner, Ludwig Stein, Charles H. Cooley, Albion W. Small, and Charles A. Ellwood. Cooley contended that the common people have an unusually shrewd capacity, almost in the manner of *vox populi, vox Dei,* for judging the merit of both persons and public policies.[5]

[5] The fallacy of this dogma, as exposed in actual practice, was devastatingly revealed in Herbert Agar's interesting book, *The People's Choice* (Boston, 1933).

Discriminating sociologists take a middle ground. There are certainly great differences in capacity among men, and the social goal is to put the ablest in positions of leadership. Yet mental ability does not ensure wisdom. Some of the most menacing figures in contemporary life are men of preëminent intellectual capacity. Brain-power is by no means correlated with the possession of sound ideas or a sense for moral values. Moreover, since the masses constitute the majority of those who are directly affected by government, they must surely have some part in its determination. Perhaps the most sensible approach to the whole problem of democracy was that of Giddings. He contended that, whatever the outward form of government, be it monarchy, aristocracy, or democracy, the few will always rule. Ability will inevitably come to the top. Democracy is the preferable form of government because it provides the best opportunity for the able few to rule and rule justly and efficiently. Democracy affords more effective checks upon selfishness and corruption than do other forms of government. With the current trend toward Fascism, a number of sociologists have rationalized expediency into elaborate sociological rationalizations of the Fascist philosophy. This, however, would hardly apply to Pareto in that he evinced definite Fascist leanings before the advent of the present régime in Italy; he stimulated Fascism rather than rationalized existing Fascist practices.

Sociologists have also taken decided positions with respect to imperialism, the expansion of the white race, and the extension of the white man's burden. Two of the most ardent defenders of imperialism and the subjection of inferior races to white dominion have been Franklin Henry Giddings and Benjamin Kidd. They have contended that civilization and progress are promoted by such a process. This position has been sharply attacked by other writers, notably W. G. Sumner and L. T. Hobhouse. These have pointed out the great expense connected with imperialism, its brutalizing tendencies, its unfortunate reactions upon democratic institutions at home, and its stimulation of war. They have also questioned the assumption that the native races are inferior, particularly for the responsibilities of the environment they inhabit. Again the argument turns upon the racial differences of mankind.

A further application of sociological thought in which the prepossessions and traditions of men play large rôles and in so far vitiate objective conclusions is that of morals and religion. Some, like Charles A. Ellwood, have sought justification of Christian ethics as a superior code and ascribed to Christian doctrines the highest social values. Sociologists who approached the subject more objectively have sought the origins of religion in the deification of group values (L. W. Ward and F. H. Hankins) or in the rationalization of crowd stimulation (Durkheim). Historically-minded sociologists trace the evolution of religious forms and beliefs in terms of general culture (Hobhouse) ; others indicate the interrelation between religious beliefs and economic factors (Max Weber, Ernst Troeltsch, and R. H. Tawney). These views with others indicate that there is a growing tendency for sociologists to investigate religion in scientific fashion and to look at morality in a naturalistic and scientific manner, analyzing moral codes and customs with objectivity and impartiality (Leslie Stephen, Durant Drake, James H. Tufts).

It is not possible to include in this survey the application of sociological theories to social work, with its mission to improve the lot of mankind, beyond indicating a significant trend in both directions: to introduce into remedial and corrective measures well-established sociological policies, and to derive from the study of "cases" significant data for sociological generalizations. The largest problem in this domain is that of criminality, which under the influence of scientific disciplines has risen to the proportions of a special science— criminology.

That the course of thought and body of doctrine formulated in recent periods as sociology are replete with error is intelligible. In subject-matter it deals with complicated human relations, themselves the last to yield to the objective approaches and methods of science. Its progress was hindered by emotional attachments to established traditions. Even trained thinkers could not contemplate with detachment or appraise without prejudice the validity of ideas embodied in institutions under which they had grown up with the sentiments of loyalty. In addition, views of the origin and nature of man and his social status were handed down from older patterns of thought which had solved such problems by principles of authority, of dogma, of entrenched conceptions of moral fitness, including also, in a remoter background, supernatural causes and speculative assumptions. In compensation, the recency of sociology gave it the benefit of concepts of the origin of man, of evolutionary progress, of a broadened geographical and ethnological horizon. In its major features sociology reflects the same orders of error—the subjective intrusions and rationalizations especially—as beset the general course of the humanistic disciplines. More than any other department of knowledge it was affected by the vast economic reconstructions of the social structure. In this bearing sociological theory was called upon in the direction of public welfare, and even gross sociological error, as in the case of Aryanism in Germany, has been appealed to in support of public policy. In these implications lies a source of danger for the freedom of the social sciences, in some respects comparable to the dogmatic pressure and taboo that impeded the free investigation of the physical and natural sciences in the past. The firm establishment of a naturalistic basis for the interpretation of man and his works may be counted upon as a stabilizing force in the further progress of sociological thought.

Chapter XII

ERROR IN MEDICINE

By Howard W. Haggard

THE IMPORTANT ERRORS *of medicine are not those in which the physician treating the individual patient fails to apply to best advantage the available knowledge of medicine—when he makes omissions and mistakes. The results of such are insignificant in comparison with those arising from errors in the basic concepts of medicine. The mistakes of the physician affect only the individual patient; the errors of basic concept affect the whole race.*

During the last hundred years the practical achievements of medicine for preventing disease and suffering and prolonging life have been phenomenal. Under their application—even incomplete as it is today—the average length of human life has more than doubled. This beneficial advancement is as much the result of overcoming error as of achieving new knowledge. Indeed, the correction of error is the first step toward acquiring knowledge.

The most persistent error in the field of medicine is complacency—the tendency to accept the prevailing belief as final. This error leads to the subordination of fact to theory; all new findings are interpreted only in the light of the prevailing philosophy; the old is clung to with passionate persistence.

The most important of the erroneous basic concepts that have dominated medicine is that of the supernatural origin of disease. It was founded in prehistoric times, but it remained the guiding thesis of medicine until the sixth, and fifth centuries B.C.; *it is still the belief of uncivilized peoples.*

The error of logic that permitted the long survival of this false philosophy was one which still dominates the medical fallacies, fads, and quackeries of today, and particularly the medical testimonials of modern advertising. It is the fallacy of post hoc, ergo propter hoc—*after it and therefore because of it.*

Science introduces a skepticism which tends to overcome this error of logic. Science raises the question: Did the patient recover because of the remedy or in spite of the remedy? The question then raised is settled by the so-called controlled experiment. This scientific skepticism has not yet permeated the public, and neither has the concept of the controlled experiment. Hence this error is the most insistent one in popular medical beliefs.

But scientific medicine itself is not wholly free from basic errors. Experimentation, which has been the great source of medical advancement, has focused attention on the physical rather than on the emotional aspects of disease. Today medicine is not yet fully a science. The error of overemphasis upon the physical has definitely led to a decline of medicine as an art.

IN LOOKING back over the medicine of the past, the errors that catch and hold attention are those which are really of least importance; they are what may be called the errors of ignorance. They appear as the amazing medical philosophies to which men have devoted their lives, the weird beliefs concerning disease in which they have sought health, and the useless treatments with which they have tortured their flesh. Such ignorance can be pitied; it can even be condoned if with it there is serious effort to find the truth, to remedy ignorance by gaining knowledge. There is a far more serious error of medicine than mere ignorance. It is one which is ingrained in human behavior and which has marked beyond all others the history of medicine. It is the error of *complacency*, the continual tendency to accept what is known as the final, to cease the search for knowledge and to crystallize both fact and fallacy into dogma.

What is spoken of as the progress of medicine, the advancements by which savage medicine over the centuries has been replaced by modern medicine, has not been a continuous progression. Quite the contrary. The usual and hence normal state of medicine has been one of almost complete stagnation and complacency. Century after century has passed in which not one single new fact was gained in the

entire field of medicine—indeed, when the search for the
new was rigidly forbidden.

The progress of medicine has been made by almost ex-
plosive bursts of activity which have shattered the then existing structures of medicine. In the enthusiasm that follows these outbursts definite progress is made. But always
the impetus dies out; the stimulating state of flux is arrested
by the desire for stability. The structure of medicine is then
rebuilt; its pattern is changed, but its form again is rigid.
For a century—perhaps ten centuries—medicine continues
in the comforting state of stability—and sterility—until the
next brief outburst of progress. Such has been the history
of medicine.

PERIODS OF MEDICAL PROGRESS

There have been only two major periods of medical
progress since recorded history, but at least one other must
have occurred before that time. As it is first encountered in
the records of antiquity, medicine is in a phase of complacency. It exists as rigidly organized religious healing such
as that of early Egypt. And Egyptian medicine, as of all
civilization of that period and much later also, is simply an
extension, an amplification, of primitive medicine. The
savage at some far earlier period founded its guiding principles—the belief that disease is due to the action of supernatural forces.

The first recorded outburst of medical progress occurred
in the Greek and Greco-Roman civilizations during a period
extending from the sixth century B.C. to the second century
A.D. The progress during this period was not continuous—
indeed, there were at times definite regressions; but the
episodes of activity were sufficiently close to justify inclusion of these eight centuries as a period of medical readjustment.

The era opened with the overthrow by the Grecian philosophers of the concept of the spiritualistic cause of disease
and the founding of natural science. It reached its peak in the

medical teachings of the school of Hippocrates (died about 357 B.C.), the first science of medicine based upon the fact that knowledge of disease and its treatment is to be gained only by critical observation and careful description of those who are suffering from disease.

At less fundamental levels there were episodes of progress at the school of Alexandria in the fourth and third centuries B.C. Human anatomical and physiological study was founded. In the first and second centuries A.D. there was again definite advancement. The materia medica (to become the herbals of the Middle Ages) was originated by Dioscorides, and obstetrics and gynecology were improved by Soranus of Ephesus. The period closed with the work of Galen. His contribution to progress lay in the fact that he performed experiments particularly in neurophysiology. But in spite of this fact, Galen, perhaps more than any other man in the entire period, was guilty of the error of complacency. He attempted to stabilize the structure of medicine by systematizing every fact and every fallacy. So successful were his efforts in this regard that for nearly a millennium and a half after his death his systematic works were accepted as the irrefutable authority in medicine. He conveniently settled every question of medicine and fully satisfied the complacency of those who came after him.

The second episode of medical progress, the modern, commenced with the revival of learning and the Renaissance. Its beginning was marked by the renewal of observational study after the principle of Hippocrates, best represented by the anatomical studies of Andreas Vesalius in the sixteenth century and the classification of disease by Thomas Sydenham in the seventeenth. Next, the experimental method of Galen was revived and was amplified by the use of mathematical physics, as in the demonstration of the circulation of blood by William Harvey. The progress of medicine from then on followed closely the progress of the natural sciences. The modern epoch had its peak—perhaps it will rise no higher—in the discovery of the bacterial cause of infectious disease in the latter half of the nineteenth century.

Today we still ride on the crest of an epoch of medical progress—the one characterized by the greatest practical heights if not the most fundamental philosophical depths. How intensively successive generations will maintain the present state of movement only the future can reveal, as only it will reveal how soon men will again fall back into the error of complacency and yield to the satisfaction of unprogressive medicine. It is difficult to believe that they can escape the deepest-seated error of medicine.

MEDICINE AS SCIENCE AND AS ART

There are in medicine, more definitely than in any other field save perhaps religion, practical reasons for this repeated tendency towards crystallization.

Medicine is both an art and a science. Its science lies in the impersonal factual knowledge and method learned by observation and experiment—knowledge gained by study at the bedside, in the autopsy-room, and in the laboratory. The accumulated knowledge of medicine is the common property of all physicians. The art of medicine is not factual or universal; it is personal and individual. It is based on the interplay of human personalities, that of the physician on the one hand, and those of his patients on the other. It is manifest as the ability to influence the behavior of other men—sick men—so that they will have complete confidence in the physician, not as an exponent of medical science, but as a man, a personality. It is by art that the physician molds the actions, the beliefs, of his patients so that he can apply with full coöperation the measures of prevention and treatment indicated by medical science. The factual knowledge of medical science deals only with the needs of the flesh; the art of medicine transcends this limitation and deals as well with the psychological aspects of the patient.

The art of medicine is vastly important to the success of the medical practitioner. And art is practised best when the physician has positive belief in the finality of the factual matters at his command. Faith in the practitioner draws forth faith from his patients. When medicine is in that state of flux which we call progress, a severe handicap is put upon the artistry of the physician. When the knowledge gained yesterday becomes error in the light of the knowledge of today, skepticism develops. Skepticism is essential to progress, but it destroys the certainty, the dogmatism, that inspires confidence.

It is notable that the art of medicine has flourished best when science was advancing least, and equally so that those doctors most highly trained in science do not by any means always make the best practitioners.

In the very necessity of practising the art of medicine lies a strong influence tending toward the error of complacency, the error of systematization, which in turn results in the longer sterile phase between the brief epochs of medical progress.

Today unquestionably the art of medicine has declined. The number of physicians engaged in the broad field of general practice has likewise declined; the number who have turned to the more impersonal and more restricted fields of specialization has increased. But the most striking feature of the readjustment to this period of flux is the increasing appeal to intelligent laymen of the simple dogmatizations of the quasi-medical practitioner, the charlatan and the quack.

There is a possibility that the difficulty standing in the way of the continuing progress of modern medicine may yield to discoveries in the field of psychology, and especially psychiatry. Then—though it seems remote at present—the art of medicine itself may be made into a science.

When the physician can evaluate and control the psychological aspects of his patient with the same precision that he can now evaluate and control the symptoms of his disordered flesh, then, and perhaps only then, will the progress persist . . . and the error of complacency disappear.

BASIC CONCEPTS OF IGNORANCE

We turn now to the more striking and more positive errors of medicine, those of ignorance, from which no age, not even the present, has escaped. The mere cataloguing of each known error would fill many volumes, but on examination we should find that most were simply variations of a comparatively few basic errors.

Thus, one may read in eighteenth-century newspapers advertisements for proprietary tooth-powders intended to stop dental decay immediately. In the pseudo-Salernitan documents of the Middle Ages fumigation of the teeth with the smoke of onion-seeds is recommended for this purpose. In still earlier works even more devious methods are suggested, such as eating a mouse or the less disgusting measure of rubbing the teeth with crushed herbs. In spite of the dissimilarity of these remedies, all aimed at one action—killing the worms that were supposed to eat away the substance of a tooth as worms eat away the pulp of an apple. This belief is known to have existed in ancient Babylonia; it probably existed much earlier, and it persisted until the nineteenth century. Every separate one of the multitudinous means that human ingenuity might devise for killing the worms supposed to be in the tooth would, in a classification, constitute a separate error. But the basic error upon which all others depend was the erroneous concept of dental decay. It is the concept that constitutes the important error of ignorance.

With the development of modern bacteriology the fact was demonstrated that the actual destruction of the tooth during dental decay resulted from bacterial action. Bacteria thus replaced the worms of earlier generations. The prevention of decay had then for a time as its theme the use of the toothbrush and dentrifices, particularly antiseptic dentrifices. The slogan expressing the belief that bacterial action was the basic cause of decay was "A clean tooth never decays." The acquisition of further knowledge showed this belief to be in error. "Clean" teeth could decay, and moreover it was impossible to rid, more than momentarily, the teeth and mouth of their bacterial flora by any known dentrifice or antiseptic. The fundamental cause of decay thus receded into the fields of heredity, diet, and local blood-supply. With the change in point of view the exploitation of dentrifices changed correspondingly: the emphasis was no longer put on the prevention of decay, but on preservation of beauty; in the cult for white teeth the inevitable "health" appeal centered on stimulating the gums with the massage of brushing.

Each of the dentrifices previously exploited to rid the teeth of bacteria would, in classification, constitute a separate error. But here again the basic error upon which all of the others depend was the erroneous concept of the primary cause of dental decay.

In dealing here with the errors of ignorance, emphasis will be placed upon such basic concepts rather than upon the separate measures, usually logical and often ingenious, that were dictated by the concepts.

SUPERNATURAL CAUSES OF DISEASE

Most of the fundamental errors of medicine were originated long before recorded history and were mature and well developed when handed on to the earliest civilizations. Very largely it is these same fallacious concepts that have constituted the errors of medicine in all succeeding ages.

Certainly no single error had a more profound influence upon the course of medicine than the belief that disease is due to supernatural causes. It dominated completely the form of all medicine until the sixth and fifth centuries B.C.; it dominates the form of medicine for the greater portion of the inhabitants of the earth today. It is a part of the belief that all of the misfortunes of mankind result from the intervention of supernatural creatures. Variations in the forms attributed to these creatures at different times in history do not alter the basic belief but reflect merely the stages of religious development.

For the savage the forces of misfortune were the simple spirits and demons that inhabited his surroundings, the river, the trees, the rocks. These creatures of evil intent brought illness by tormenting the bodies of men or even by inhabiting them, possessing them. In the subsequent evolution of religion and the development of gods and goddesses, of the one God, of angels and even of the Devil, the general principle still persisted; but the mechanism by which the misfortune was inflicted of necessity altered to correspond to prevailing variations. The Egyptian god Set shed poisoned tears which caused disease in men; Job, at a later period, was tormented with the "arrows of the Almighty."

Belief in the spiritualistic origin of disease was widely held during the Middle Ages and the Renaissance. Possession was the accepted cause of insanity. The belief flowed into such channels as that of witchcraft and magic. It was transmuted gradually into the belief in the harmful effects of night air. Evil spirits and demons were most prevalent during the night, and there was always the likelihood that one might be inhaled. Precautions were taken against this danger. Even after the belief in possession declined, the precautions were continued in this case to avert the ill effects of night air; there was supposed to be some baleful influence from the atmosphere itself. The spirits became the miasms. Mufflers were worn over the mouth after dark, and bedroom windows were hermetically sealed.

The legendary belief, spiritualistic in origin, of the harm from miasms is perpetuated—and not without some justification in this case—in the term malaria, from the Italian words *mala aria*, bad air, referring to the mists from swampy regions—though the mosquitoes from those places, not the air, carried the disease. In the first half of the nineteenth century puerperal infection was still sometimes attributed to corruption of the air; in the latter half of the same century, Joseph Lister, father of modern antisepsis, mentions miasmic causes as one of the theories previously advanced to account for wound infection. Even in the early part of the present century, plumbing codes were rigidly enforced to insure the escape of sewer-gas; for the odor was still believed to be at least a contributing factor in causing infectious diseases.

A natural extension of the spiritualistic conception of disease is involved in the belief that deities have the power not only of inflicting disease, but also of revoking the injury they have inflicted and even of relieving that caused by other supernatural creatures. Thus, the Egyptians appealed to their gods, particularly the deified physician Imhotep, for relief from the machinations of the demons. The Greeks made pilgrimages for similar relief to the temple of their healing god, Æsculapius. The Romans likewise sought recovery from disease by offerings made to their many gods and goddesses. The Christians in turn besought with prayers and sacrifices the healing aid of their Deity either directly or indirectly through the intercession of the saints. Indeed, in time a certain degree of medical specialization was attributed to the saints. Thus Saint Anthony became the patron of those suffering from ergotism, Saint Lazarus of those with leprosy, Saint Appolonia of those with dental diseases, and Saint Vitus of those with chorea. Likewise some of the relics and some of the holy places had powers in alleviating definite diseases. Even in modern times, with the wide recognition of the natural rather than the supernatural cause of disease, efforts at religious healing still persist, as in the miracles anticipated at such shrines as those of Lourdes and Saint Anne de Beaupré.

Certainly the most novel of the delegations of divine healing ability is found in the custom of the royal touch for the king's evil. This disease was scrofula or tuberculosis of the glands, usually of the neck. Over a long period of time the French and English believed that the ability of the king or queen to cure this disease by touch of the hand was evidence of the divine right of kings. Anne was the last of the English sovereigns to practise the custom, but in the nineteenth century it was revived briefly in France.

The separate means by which divine healing virtues have

been invoked are almost infinite in number, but each is merely a mode of satisfying the dictates of a single central erroneous concept.

The origin of the belief in the spiritualistic cause of disease is lost in antiquity. But it seems probable that it grew out of the very nature of the elementary reasoning of ignorant human beings. The savage from the very beginning was faced with accident, disease, and death. Men were struck down by other men; they were torn by animals, and they were crushed by falling trees. Pain, disease, and death followed from these accidents. The various visible forces—the enemy, the beast, the tree—were agencies that caused the misfortune. But there were other men who appeared to exhibit the effects of such accidents, who suffered and died, and yet who had not been acted upon by visible agencies. Nevertheless, the effects of some agency were evident. What, then, was more natural than to reason that there must be invisible agencies? This was a sound deduction. It was in identifying the agencies that the savage made his fundamental error.

The discovery that infection is caused by parasites invisible to the unaided eye came within the lifetime of men still living. Lacking wholly such knowledge as completely as he lacked a microscope, the savage deduced the agents; he created a world of invisible creatures—the spirits. Confusion of life and motion led to the personification of the inanimate; the phantasy of dreams was interpreted as evidence that some essence from the things of the world—even of the dead—could materialize from a distance and then at the moment of waking dematerialize and become again invisible.

FANTASTIC REMEDIES

With the spirits postulated as the cause of disease, the next step was the prevention and treatment of disease by measures intended to keep away or drive away the spirits. Efforts toward these ends have yielded most of the specific errors of ignorance in medicine.

Viewed without the prejudices of familiarity, no practice

in medicine is stranger than that of medication—the swallowing of herbs or mineral or even animal substances, not as food, not to satisfy a natural craving, but specifically to correct altered functions. No practice has filled a greater place in the medicine of the past. Perhaps if the savage had not founded medication, this basic branch of therapeutics might not have been discovered even yet. But in the philosophy of the spiritualistic origin of disease it was both a natural and a logical step.

Human egotism has always been to the foreground of all religious beliefs; the savage was not created in the image of his spirits, but he did create the spirits with his own likes and dislikes. Consequently those things which were unpleasant to the man were unpleasant to the spirits. Experience must have acquainted the savage with many substances of unpleasant taste—astringent berries and barks, bitter and nauseating herbs—which could make the man's body unpleasant for the spirits. Experience too must have revealed other substances that caused salivation or acted as emetics and purges. Visibly the spirits flowed from the body with the secretions or ejections; the accompanying revulsive effects were further evidence of the spirits' departure. It was perhaps from a crude beginning such as this that arose the principle of medication. The number of substances that might serve as medicaments was limited only by the scope of human ingenuity. In the years that followed, everything that grew on the face of the earth, that walked or crawled or flew or swam, even the minerals of the earth itself, found their place in medication.

The nearly universal quality of all medicaments of the past was their unpleasantness; even today (and it was more so a few years ago) the ignorant patient is prone to judge the virtue of his "medicine" in proportion to the vileness of its taste. Castor-oil certainly owes its vogue as a purgative in part at least to its unpleasant taste.

An enumeration of the multitude of medicinal substances recounted in Egyptian papyri, contained in the writings of

the Greeks and Romans, filling the herbals of the Middle
Ages and the medical texts of the Arabs, and comprising the
older pharmacopœias of Europe and of this country, reveals
only the errors of application and the breadth of human
imagination. Excrement even (fly-specks), powdered mummy,
ground-up jewels, insects, unicorn horns, bezoar stones, shoe-
leather, menstrual blood, viper's flesh, saliva, and the moss
scraped from the skull of a man killed by violence—and these
are only a few from many—have all been used as medicaments, with full faith in their curative powers.

Certain medicaments were believed not only to repel
disease, but to carry in addition some of the spiritual
qualities inherent in the source from which they were derived. Thus, the heart of a lion or that of a valiant enemy
gave bravery and fierceness to him who ate them; the lungs
of a fox yielded "long wind" and its brain cunning; the
grease of a bear applied to the scalp made the hair grow,
and that of a snake or earthworm rubbed on the skin made
the muscles supple. Rattlesnake oil was sold by street-corner
"medicine men" as a liniment for rheumatism even within
the twentieth century. The vogue for rare roast beef in the
diet of athletes during training and the belief that raw meat
makes dogs fierce may be traced back to the conception of a
spiritualistic influence of diet. The suggestion came no doubt
from the fact that the fierce carnivora eat meat and eat it
raw; herbivorous animals are universally much gentler.

Along similar lines we find that definite symbolic qualities
—wholly imaginary ones—have been assigned to the organs
and even the secretions of the body. Formerly the liver was
the seat of sexual passion; the heart has now usurped this
function with the somewhat more refined term of the "tender
emotions," and from it comes the symbolization that yields
the bleeding heart, the broken heart, and the sweetheart.
The fact that the blood rather than the genes was formerly
assumed to carry the vital principle of heredity is maintained
in the language in such terms as the "blood line," "bad
blood," and "blood-relations." Of all of the secretions, sweat

has the most varied symbolic significance: the sweat of the brow is the noble fluid by which bread is earned; the sweat of the feet (exactly the same secretion) is wholly disgusting.

An equally symbolic deviation but of teleological origin is found in the belief that plants bear the "signatures" of the diseases for which they have healing virtues. Thus "eye-bright" is remedial for eye disease and the nutmeg for brain disturbances. Needless to say, this ancient thesis had the advantage of convenience in the selection of remedies; its extension was again limited only by the visual imagination of its exponents.

The overthrow of the spiritualistic conception of disease by the Greek physicians and philosophers did not do away with the multitude of useless remedies, even at the hands of physicians who followed the natural sciences. It substituted merely new theories which gave other but equally unsound reasons for continuing the vogue of extensive medication. The general theory of disease accepted by Hippocrates and actively exploited by Galen was Pythagorean in origin. Under this concept all substances, including the human body, possessed in some degree the qualities of heat and cold, wetness and dryness. Disease was a disruption of the balance of these qualities in the body; the disturbance was to be corrected by the administration of substances possessing the particular quality lacking in the body at the time. Herbs and animal substances were in turn classified according to their qualities and administered as needed.

The only virtue of this system lay in the fact that very largely the more poisonous mineral substances were avoided. The herb teas and concoctions did no direct harm even if they did no particular good. This type of medication survives now as the home remedies—as the "sassafras teas" of old women to "cool the blood" and the proprietary "herb remedies" advertised in our newspapers. A harmful quality, indirect but nevertheless serious, is inherent in the use of all such medicaments. Their employment presupposes a diagnosis on the part of a consumer unqualified to make a diagnosis. Attempts at "self-treatment" even with "harmless" remedies often keep the man who suffers from some serious ailment away from the physician until the disease, possibly amenable to prompt treatment, has advanced beyond the possibility of arresting it.

The mineral medicaments, iron, arsenic, mercury, antimony, and the like, were introduced extensively into medical practice at the time of the Renaissance. These (with the exception of antimony) were of real effectiveness in treating certain disease, but they had the great disadvantage of acting, in the doses then often employed, as virulent poisons. For centuries the medical profession was torn by the conflict between the "herbalists" and the "mineralists"—the former with their harmless prescriptions containing a vast array of vegetable substances mainly useless, and the latter with their simple prescriptions for a single highly potent substance. The argument was finally settled only by modern methods of physiological investigation.

It would be unjust to say that none of the older and empirical remedies were beneficial; some were definitely so and are extensively used today. Opium as a sedative, mercury and sulphur for skin diseases, iron as a tonic, quinine for malaria, aloes, senna, castor-oil, and calomel as cathartics, and hemlock buds and scurvy grass as antiscorbutics, were remedies well known and highly valued before the eighteenth century. But as a glance at the pharmacopœia of that period will show, they were the few valuable ones among the many useless ones.

Even hidden in the "brew of the witches' caldron" there was occasionally a possible modicum of value. The broth of boiled toads was used to treat dropsy; we should at once class it as a revolting mess, but modern science has shown that the skin of the toad contains a substance allied to digitalis in its action upon the heart and hence possessing definite curative properties. Again, an old prescription called for burnt sponge in the treatment of simple goiter; iodine is used today, but burnt sponge is rich in iodine. A Chinese prescription calls for dragons' bones in the treatment of convulsions of children, the dragons' bones being those of dinosaurs found buried in desert sands; calcium is used today to relieve the tetany of children, and bones contain calcium. But these remedies are curiosities; they are exceptional because, if correctly applied, they were definitely beneficial.

ILLOGICAL THERAPEUTICS

A great part of the progress in medicinal therapy has been the discarding of the vast collection of useless substances prescribed in the past. But this process of discarding has not been the simple sweeping aside of the older remedies. Rather, each one has been tested and tried and saved or rejected according to the dictates of modern science.

One might assume that 250 centuries or more of dosing human beings constituted in itself a fairly extensive pharmacological investigation which should have solved the problem as to which of the remedies in use were beneficial and which were worthless. Such an assumption is erroneous. Certain basic errors of judgment, which can be removed only by the application of science, have prevented a correct evaluation; these same errors have applied also to other therapeutic measures discussed here.

The errors of evaluation arose in each case from one or more of the following reasons:

1. The confusion of cause and effect.
2. The lack of controlled experiments.
3. The ignorance of the self-reparative ability of the body.
4. The failure to recognize the influence of the mind upon the subjective symptoms of disease, and also the fact that even physical symptoms may have their origin in psychological disturbances.

The first of these, the confusion of cause and effect, is inherent in the *post hoc ergo propter hoc* type of logic. It yields the error of the medical testimonial—the scientifically invalid but nevertheless firm conviction that some form of treatment has "cured" the individual. A man falls ill; a remedy is given, or a course of treatment is prescribed; in time health is regained. In this situation are all of the ingredients ready for the application of false logic: the man was ill, he was treated, he recovered; therefore, he recovered because of the treatment. *Post hoc ergo propter hoc;* after it, therefore because of it. When men are ill, something is usually done for them; the majority recover. Before the days of the controlled experiment the recovery was assumed to be proof that the treatment cured the disease. Almost any treatment applied would, under this type of reasoning, result in a fair percentage of recoveries: hence the multiplicity of remedies.

Modern science teaches in this regard a deep skepticism. It raises the questions: Did the man recover because of the remedy or in spite of it? Might he not have recovered if no remedy was used?

To answer these questions, the physician of today turns to the controlled experiment. In its simplest form, but for obvious reasons a form seldom used, the physician, in testing a remedy, would select a group of patients, all suffering from the same disease, and use them as experimental subjects. The group would be large enough to allow statistical validity. All of the patients would be treated precisely alike except in one single regard. Half of them, the experimental group, would be given the remedy to be tested; the other half, the control group, would not be given the remedy. A comparison of the number of recoveries and the average speed of recovery in the two groups would show at once whether the remedy tested was beneficial, useless, or harmful. As the result of controlled experimentation the majority of the remedies believed in implicitly in the past have been discarded; but what is equally important, the physician has been able to select beneficial remedies and prove their merit.

As simple as the controlled experiment is in principle, its application and interpretation without error require at times the highest scientific skill. A famous instance testified to this fact. Ambroise Paré, the French surgeon of the sixteenth century, was one of the earliest to attempt in a crude way the controlled experiment, but he missed in part the correct interpretation. In treating a powder burn on the face of a soldier he was persuaded by an old woman to use a dressing of chopped-up onion rather than the usual ointment. The wound healed quickly. In consequence Paré performed an experiment to test further, as he believed, the healing virtue of this vegetable. Finding a man who was burnt on both sides of the face, he covered one side with onion and the other with ointment. The side to which the onion was applied healed sooner than the side to which the ointment was applied. Paré, with some apparent justification, concluded that onion was remedial for burns. But his was not a true controlled experiment—merely a comparison between onion and the ointment then in use. A further experiment with one side of the face entirely untreated might have shown that onion was merely *less harmful* than the salve. And no doubt it was; for in those days ointments were often compounded of most dubious ingredients and universally contaminated with bacteria.

A similar conclusion applies also to the so-called weapon-ointments and to the famous powder of sympathy, exploited by Sir Kenelm Digby and patronized by King James I. In using weapon-ointment the wound was simply cleaned and covered with a bandage; the salve was applied copiously to the weapon. In using the powder of sympathy the wound was similarly treated, but instead of anointing the weapon, which was often difficult to obtain, a piece of bloody clothing was dipped in a solution of the powder. In both cases the wounds healed more quickly than they did when salves were applied, not because of any virtue in the weapon-ointment or the powder, but simply because the filthy ointment was not applied to infect the wound.

PSYCHOTHERAPY

The final category of errors in evaluation of remedies—failure to recognize the influence of the mind upon the subjective symptoms of disease, and also the fact that even physical symptoms may have their origin in psychological disturbances—applies less to the use of medicinal substances than to other forms of treatment. It applies especially to psychotherapy, which was the main treatment used by the savage and by the priestly physician of early civilizations, and which is, of course, the basis of all religious and metaphysical healing and a large part also of quackery and charlatanism. Nothing said here is intended to cast disrepute on psychotherapy; properly applied it is as valid and useful a form of treatment as is medication. But as with medication its selection and application need the rigorous control of the scientific method.

It is only now that the modern sciences of psychology and psychiatry are rescuing this form of treatment from the hands of the unqualified and making it into a valuable therapeutic measure for the modern physician. A fair portion of what in an earlier section was called the art of medicine falls in the field of psychotherapy. The wild incantations of the savage, the dignified ceremonies of the priests of Imhotep and of Æsculapius, the miracle of shrines, and the philosophy of the negation of disease, all have elements that find their place in the power of the competent physician to inspire confidence in his patient.

The mind influences the action of body: paralysis, deafness, blindness, constipation, indigestion, palpitation of the heart, may occur as symptoms of disturbances which are purely psychological. When they are of this origin, they may be remedied, temporarily at least, by any form of suggestion that inspires confidence. "Miracles" of recovery have occurred from the mere touching of a holy relic: the lame have thrown away their crutches, the halt have walked, the blind have seen. Similar miracles have been performed with medicinal substances, with metal rods, electrical appliances, mere philosophical discussions, and in fact with any form of treatment that catches and holds the attention and inspires belief in recovery.

The most striking symptoms of physical disease are pain and anxiety. Both are subjective, they are entirely mental. Any form of treatment in which the patient has implicit faith may relieve pain and anxiety. But the mere abolition of symptoms—symptomatic treatment—does not arrest the underlying disease. Morphine and the various anesthetics will also relieve pain and blot out anxiety, but no one would suggest that the mere administration of an anesthetic would remedy an inflamed appendix or a gallstone that might be the cause of the pain, or cure the pneumonia or heart-disease that might be the source of the feeling of anxiety.

The error of psychotherapy, as it was practised by the savage and the priest of the past and as it is still practised by the cultist and quack of today, lies in the fact that while it may relieve symptoms, it arrests no disease unless the disease is entirely of psychological and not of physiological origin. In short, the pain of a toothache can be stopped by suggestion, but an abscess may form in the tooth and systemic disease follow regardless of the presence or absence of the pain. Psychotherapy fills no dental cavities, it removes no cancer, it cures no infectious diseases.

Precise diagnosis by a skilled physician is a prerequisite to the application of psychotherapy, just as much as it is to the application of surgery or medication. The nature of the disturbance must be determined before treatment is instituted. It is self-evident that medication and surgery incorrectly applied can of themselves cause harm; the harm of indiscriminately applied psychotherapy is less evident but none the less real. Unquestionably in modern times in many thousands of instances the victims of cancer, tuberculosis, diphtheria, diabetes, appendicitis—to name only a few diseases—have literally lost their lives because of delay in instituting proper medical treatment while they attempted to treat symptoms by some crude method of psychotherapy, erroneously applied.

DIAGNOSIS

Exact diagnosis is an absolutely essential prerequisite to any form of treatment—it is also the most difficult branch of medical practice. Occasional errors of diagnosis are inevitable, even at the hands of the most highly trained clinicians; but the fundamental error in this field today is not with the physicians, but with the laity who fail to appreciate the importance of diagnosis or who attempt even the impossibility of self-diagnosis.

From the historical point of view the error of concept appears in the field of diagnosis as a confusion of disease and diseases. To the savage there was only one basic disease state —a disturbance resulting from the action of supernatural creatures. The variety of the symptoms that sick men presented was interpreted as due merely to the way in which the spirits applied their torment. A pain in the head and one in the foot had the same cause; in one case the spirits centered their action in the head and in the other in the foot. There was only one cause of disease. Diagnosis was unnecessary.

When the supernatural cause of disease gave place to the natural at the hands of the Greek philosophers, the situation so far as diagnosis was concerned was not greatly changed; the theory that the symptoms of disease were due to an unbalanced state of humors or qualities within the body was substituted for the belief that they were caused by the action of supernatural creatures from the outside. There still remained one general disease state with many different manifestations.

The Greeks were interested primarily in prognosis and not diagnosis. The question raised was not what disease has the patient, but what will the outcome of his illness be? Hippocrates in writing clinical histories was not attempting to differentiate disease, but merely describing and classifying the symptom-complexes arising from a general disease state in order that the course of the particular illness might be predicted. Hippocrates and his followers, in evaluating symptoms for prognosis, studied their patients carefully by every means at their command. But in the years to follow, this rigorous method was largely discarded and replaced by less laborious procedures. Astrological prediction, the casting of horoscopes, the mere visual inspection of urine, and similar means that could be carried out even in the absence of the patient, served as the bases of prognosis and as the guides to therapy.

Modern diagnosis in internal medicine arose in the seventeenth century with the overthrow of the doctrine of the universality of disease at the hands of the English physician Thomas Sydenham. He advanced the now proven fact that instead of one disease there are many. He defined disease entities and started the study of the "natural history of disease." Since his time one of the great advances of medicine has been the development of means for diagnosis. Such now commonplace instruments or measures as the stethoscope, percussion, the clinical thermometer, the counting of the pulse, the X-ray, and all of the seriological tests have been developed and applied since the time of Sydenham.

Today new measures of diagnosis, new means of treatment, and in fact new diseases are being described in rapid succession. The modern period is one of the rare periods of progress in medicine. A strain is put upon the practitioner to keep abreast of the advance of medicine, to adjust his ideas as each new concept and new discovery arises. The patient of the modern physician is too often conscious of the lack of dogmatic positiveness which inevitably must exist in the mind of the modern physician. His tendency too often is to drift into the hands of the charlatan, the cultist, and the faddist, who keep alive the errors of ancient medicine—the errors of ignorance and complacency which permit the practitioner an uncritical state of mind and a confidence inspiring dogmatism. Such quasi-medicine—folk medicine—requires no elaborate diagnosis; it postulates one simple cause of disease which the layman can readily comprehend, and it "cures" just as the medicine of the savage "cured," because the body tends to heal itself in spite of the wrong treatment. It derives its credit for the "cure" not from merit but from the confusion of cause and effect.

Today not only the physician must know the pitfalls that are the errors of medicine, but the patient must also know them if he too will avoid them.

Chapter XIII

ERROR IN PSYCHIATRY

By Abraham Myerson

By DEFINITION *psychiatry is the branch of medicine that deals with the mental diseases. For the orientation of the lay reader it must be emphasized that the* term mental diseases *has a far different significance from the term* insanity. *Insanity is fundamentally a legal concept—that is, at the present day. In earlier days it merely denoted someone whose conduct was so extraordinary that it could be explained only by establishing some possession or some punishment; in other words, a supernatural significance was given to the term. As the world became free from the strangle-hold of theology, as secular interpretations and conceptions gained ground, the legal phases of mental diseases became dominant; and so psychiatry dealt with something which legally was an incapacity to administer and care for one's own affairs and, generally speaking, to be responsible in the social sphere. That is, an insane person was one who could not administer his finances, enter into matrimony, make a will, or be responsible for a criminal act. Mental disease is a far wider concept. It includes those conditions which necessitate the sequestration, called commitment, to an insane hospital; but it also deals with the same individuals long before they become committable or insane—that is, when the disease process is in its earlier phases. Furthermore, it deals with many conditions that never reach the insane hospital at all.*

Take, for example, the neuroses. These are conditions of common incidence—in fact, so prevalent that they form part of the practice of every medical man and probably enter very consequentially into all human relationships. The neurotic is disturbed in his emotional stability. He becomes obsessed by notions of pathological type. His visceral functions become disorganized, even though the most careful examination that we can make at the present time shows no change in the organic structure of the organs of the body. He rarely becomes insane in the legal sense of the term, yet his condition is more of a mental disease than that of the individual who loses his mentality to the point of impotence by reason of a syphilitic infection of the brain.

Because the mental state of man enters as the dominant factor in all his social relations, the abnormal social relations have become part of the field of psychiatry. For example, in the earlier days of human relations the criminal was regarded merely as an individual who chose of his own free will to range himself against society and to commit acts abhorred by the group and punishable by them. Later on, as the idea of cause and effect entered more deeply into the common life, conduct became studied from the standpoint of causation; the criminal act and the criminal himself entered into the domain of science, and since abnormal mentality might be postulated as a cause, the psychiatrist entered vigorously and deeply into the problem of crime. In short, the psychiatrist became a criminologist. We need only mention the fact that the great Lombroso, who gave such a turn to the whole field of crime, was a psychiatrist.

The sexual relations have always been a source of difficulty to man and, in fact, have often obsessed legislator, priest, and thinker. The question whether psychiatry might explain what are defined as abnormal and disharmonious sexual lives became raised when the sexual life became freed, in some measure, from taboo. At the present time the basis of the sexual life, and its difficulties, is one of the fundamental fields of psychoanalysis; and of those who have helped formulate into something like science the relations of the sexes the most important have been psychiatrists, notably such men as Krafft-Ebing and Moll.

The function of the psychiatrist in these social problems has been twofold: most importantly, he has brought into the social consciousness the fact that these fields may be studied without loathing or moral implication—that is, may be studied objectively and without emotion; and he has contributed some results of a scientific nature in the way of laws and formulations.

The psychiatrist and psychiatry have passed from the stage in which the final products of mental disease formed the field of investigation to the stage in which the whole world of mental activity and social relations in part fall within the medical domain. Probably this overextension has dissipated the energy of the psychiatrist and has led him to hasty and rash conclusion. It has, however, marked benefits. While psychiatry may well regret many of its pronunciamentos, it has no reason to be ashamed of the basic effects of its activity.

THE TERMS "ancient" and "modern" have a significance in psychiatry quite different from almost any other field of science. Although mental disease has been known and described from most ancient days, there was no organized psychiatry until early in the eighteenth century and no consequential specialization in the field until the nineteenth century. Hence, discussion of error in psychiatry has its own chronological relations: what would be modern in astronomy would be relatively ancient in psychiatry. We speak of men of the late eighteenth and early nineteenth centuries as *early* psychiatrists, whereas astronomers, for example, of importance appeared over two thousand years ago. The reader, therefore, must keep in mind in this discussion that time, so to speak, is telescoped into a shorter space in this chapter than perhaps in any other in the volume.

TRUTH IN PSYCHIATRY

Working in a field in which the fruits to be gathered are the causes and successful treatment of mental diseases, truth for the psychiatrist is scientific knowledge of the conditions underlying such maladies as epilepsy, feeble-mindedness, the psychoses, and the psychoneuroses, as well as the techniques by which he, the physician, may treat the sufferer. Error, consequently, is that which dissipates his energy, leads him along fruitless ways, and keeps him from his supreme triumph, the understanding by which he can both diagnose the ailment of his patient and cure him.

The candid man who practises in this field of medicine in our day has little enough of these joys by which to replenish his spirit. Few are the conditions that we completely understand, and fewer still are those which we can successfully treat. People become mentally sick, and in the most of instances we know not why. And even when recovery happily occurs, our techniques and laborious efforts in the majority of cases have not contributed in *a provable way* to the recovery. We bring forth rash generalizations, and we embark too enthusiastically in a new and promising field of therapeutics. We write great and learned books and talk, as all specialists do, with dogmatic authority. That the art is long and understanding difficult is most true of the field of psychiatry.

Here we may profitably establish at once a belief in causes and consequently a belief in cures. The psychiatrist is in much the same position as the plumber. The latter is called to a house where there is a leak in the water-pipe. Damage is being done to rugs, as well as to the tempers of the inhabitants. He finds finally a worn place in the pipe, which he establishes as the cause of the leak and which he promptly repairs by one technique or another, or perhaps he introduces a new pipe into the water-system. He does not bother his head with the universe of variables, such as the constitution of water, the nature of pipes, gravitation, ultra-violet and other types of radiation, all of which may be brought into relevancy with the leak in the pipe. He starts with a stable universe which presupposes normal pipes, finds something new has been introduced, which he calls the trouble, and produces a new something which he calls the cure.

Since frankness is the order of the day when one discusses error, it becomes incumbent upon me to declare my prejudices, or, more suavely, my cultural compulsives. I dislike abstractions and generalizations that are reached logically, for I find them too neat for the disorderly phenomena of existence; and I react to the attitude of the promulgators as savoring a little too much of omniscience. I dislike in science great subtlety and especially metaphor and symbol, which is my chief objection to the Freudian doctrine. I am very distrustful of the precise separation of phenomena into the definite and opposing groups that words impose upon us, even though I know it cannot be helped. I like the biological explanation as opposed to the philosophic, and I am unable to separate the structure of an organism from its functions. It is more comforting to me to find a chemical or structural cause than a psychological one; for I can see the former in the test-tube or under the microscope, but there is no way in which I can at present handle a psychological factor. I think the philosophers, mathematicians, and physicists have a right to deny cause and effect in an abstract universe reduced to mathematical formulae. But in the world in which they, as well as we, dwell, one cannot take one step in the manipulation of things-as-they-are towards things-as-we-wish-them without consciously or unconsciously propping up our activities with the belief in cause and effect.

Psychiatry is created by psychiatrists who turn out, on long acquaintance, to be intensely human and to reflect in their attitudes the general attitudes of both the learned and the unlearned people of their times. Hence, the errors of psychiatry are not fundamentally different from the errors that are found in all other scientific effort—to put it more concretely, in all of medical thought and practice. Perhaps they become a little more glaring as the subject-matter of psychiatry itself is more mysterious, more intensely reacted to than the subject-matter of disease in general. This latter fact, to wit: *that mental disease somehow awakens psychological responses which are absent in our reactions to the sickness of other types,* is quite well exemplified by prevailing customs and habits of speech. It is the commonest type of reproach to say "You're crazy," or "He's a nut," as if it were a crime; whereas no one, embodying the same attitude, will say "You have cancer or tuberculosis." On the stage and in private life one of the surest sources of humor, one of the easiest ways of raising a laugh, is to act the boob or the lunatic. It would take us far afield to attempt to uncover the psychology by which the most dreaded of human misfortunes arouses the heartiest laughter.

The words of that mixture of charlatanism and genius known to his contemporaries and posterity as Theophrastus Bombastus Paracelsus (1493-1541) may be quoted as embodying the attitudes of his time and of the previous ages in respect to mental diseases. "That person," he says, "is ill in mind in whom the mortal and the immortal, the irrational and the rational spirit do not appear in the proper proportion and strength. Those who have failed because of weakness of the rational spirit are called imbeciles; fools and senseless, on the contrary, are wrathful, who are mad from an excess of brutish intelligence because they have, as it were, drunk more of the astral wine than they could digest. Both follow solely their native animal mentality nevertheless; with this difference, that the fools reveal themselves as animals with senses unchanged, the mad and senseless on the contrary as deranged beasts. What the fools do is animal cunning, tradition, and good sense; the senseless on the other hand and the deaf to reason give evidence instead of vain confusion." "Lunacy," he stated later, "grows worse at full and new moon because the brain is the microcosmic moon. In salt, sulphur, and mercury are the first beginnings of all diseases; the mercury is sublimated, distilled, or precipitated through heat; the sublimation causes insanity, precipitation gout, and distillation paralysis and melancholy."

In this remarkable statement one finds many of the errors that persist in our own time. First, that the soul, a metaphysical immortal part, is concerned with mind and consequently with madness. As I shall show later, although the soul has disappeared from psychiatric concern, separation of the mind from the body still persists as a formidable obstacle to the development of psychiatry. Second, one sees what appears throughout the writings of the ancients and is part superstition and part poetry, that the celestial bodies are concerned somehow with the welfare of the individual. The anthropomorphic god and the anthropomorphic universe appear conspicuously in all the thinking of ancient times and have not yet disappeared from our thought. The Deity, the planets, and the forces of Nature generally are concerned with the welfare of the individual man, and his conduct brings about their conduct as reward or punishment. This conception appears in all the discussion and study of the physical diseases, but it disappeared as an active force in diseases of the body long before it disappeared in the diagnoses and treatment of the diseases of the mind.[1]

MADNESS AS PUNISHMENT FOR SIN

The doctrine that God punished men by insanity persisted, despite Hippocrates, Galen, and notable successors, throughout the ages as the general etiology of mental disease. That doctrine appears with devastating force and power in the Bible. In *Deuteronomy* XXVIII the Lord announces through Moses that insanity is a punishment for sin and especially the sin of unbelieving and non-worship. At the end of a long list of horrors to be visited upon the sinful and unbelieving, he states, "The Lord shall smite them with madness and blindness and astonishment of heart." The Jews were not alone in this belief. It appears in the Greek legends—for example, when the epilepsy of Hercules is laid to the wrath of a goddess; and it is part of the workaday superstition of all primitive peoples. But since the Bible became the Book of Books to the Christian world, this ancient and barbaric notion governed the thought of Western doctors, priests, and lay folk for centuries and into our own times.

Thus, the madness of Saul came upon him because he had mixed a little mercy, kindness, and human feeling in his dealings with a subjugated foe, against the express commands of one of the harshest men recorded in history, the Prophet

[1] It would, I think, surprise any one, except those acquainted with the facts, to know that there are in every large city in the country at present witches and wizards who practise the cure of the mentally sick. The strata of magic and superstition that still persist in the thought and deed of contemporary humanity need no exposition on my part, for this has been brilliantly done by such men as Frazer.

Samuel, who spoke for God. Saul did not kill the king of the tribe he conquered, and when the conquered man stated "in a delicate spirit" that the time for blood has passed, he was hewed to pieces by an irate prophet; and then King Saul was afflicted with manic-depressive psychosis as his reward for a ray of kindness.

Centuries later that rare old arm-chair psychiatrist and philosopher Robert Burton (1577-1640), who recorded in his inimitable way the psychiatric attitudes of the preceding and contemporary literature, starts off the first partition of his *Anatomy of Melancholy* with the statement that "the impulsive cause of these miseries of man ... was the sin of our first parent, Adam, eating of the forbidden fruit, by devils of instigation and allurement.... To punish, therefore, this blindness and obstinacy of ours as a concomitant cause and principal agent, is God's just judgment in bringing these calamities upon us, to chastise us, I say, for our sins, and to satisfy God's wrath." He goes on to indicate, by citation piled on citation, that the cause of mental diseases that he sums up under the heading of melancholy, although their ostensible cause is one thing or another, is finally punishment for sin.

The practical sense, perhaps the growing secularity, of man rejected this doctrine in so far as bodily sickness was concerned long before it was recoiled from as cause of the mental diseases. And so for centuries the insane were treated by whippings, chastisements of graded horror, cruelties beyond description, on the basis that this treatment accorded with God's will and His technique. It is true that from time to time notable men arose to protest against the barbarities inflicted on the insane. Thus, a fifth-century Roman, Cælius Aurelianus, raises a humane voice in a desert arid of pity and understanding. He says: "They themselves seem to rave rather than to be disposed to cure their patients when they compare them with wild beasts that must be softened by the deprivation of food and the torments of thirst. Misled, doubtless, by the same errors they recommend that they be chained, and that it is more expedient to restrain them by the hand of men than by the weight of iron. They go so far as to advise physical violence by the whip as if by such means to force the return of reason. That deplorable treatment can only aggravate the condition and supply unwelcome memories to salute the return of their intelligence."

Centuries passed. It was finally first in France that the chains of the insane were broken by Philippe Pinel (1745-1826); that they were removed from filth and degradation; in short, that they were treated as sick. Even into our times restraints of an inhuman type were used—the strait-jacket, the leg-irons, the cuffs. It took the twentieth century to bring it about in really civilized communities that the mentally sick were housed in good quarters, ate at tables with fork and knife, had places where they could work, games which they could play, books which they might read; in short, when it became realized that the delicate instrument of man's superiority and humanness, his mind, must be treated with gentleness for recovery to take place.

DEMONIACAL POSSESSION

Linked with this idea of punishment was that of supernatural possession—that devils, demons, witches, and the like were visitants in the bodies and minds of the psychotic. This, too, stems apparently from religion, as when it is stated in the Bible that an evil spirit from God brought about the alienation of Saul. But, undoubtedly, it is more ancient than formal religion and had its origin in the folk superstition. However it arose, it worked havoc with the treatment of the insane.

It is curious how types of cases that appear to have given considerable perturbation of spirit to the ancient writers have disappeared entirely from our modern psychiatry. Lycanthropy, or the transformation of men into wolves, received considerable space in the ancient literature. One rarely sees today patients who believe themselves inhabited by the souls of others; but, as a matter of fact, the insane also reflect the opinions of their times. Indeed, they may create some of the opinions, especially of possession. Thus, the delusion of persecution takes on a modern dress with each advance. Undoubtedly in witchcraft days the dementia præcox patients believed themselves persecuted by witches. In my memory hypnotism was a favorite persecutory tech. nique. This was followed later by electricity, which was being introduced into the body of the patient by his enemies. Then there appeared upon the scene wireless, and now the paranoiac speaks of radio as the instrument by which he is tortured. So it is easy to understand why physicians of a former time gave such thought to witches, wizards, demons, and wolves. Undoubtedly, *their patients* complained of being inhabited by these strange and awful creatures.

To show how powerfully the idea of witches and possession affected the thought of the sixteenth-century physicians and authorities, I quote the following: "The supposed witches and possessed were perhaps never more numerous than in the sixteenth century and belief in the influence of demons in the events of the world has never given rise to so much mischief as precisely in this period. Pope Innocent VIII, therefore, appointed the two Dominicans Institor and Sprenger as inquisitors in Germany to wipe out the crime of sorcery; he even prevailed upon Maximilian I to promise the inquisitors the help of his secular arm. The devastation which these ruthless persons wrought for the honor of the Church was frightful; in the electorate of Trier alone 6,500 persons were put to death within a few years." [2] And the author goes on to state that Luther and Melanchthon—to cite only two of the illustrious persons of the time—ascribed illnesses of all kinds to demons and ghosts. Most of the psychiatric writing of the sixteenth century is pro and con the question of possession. For example, John van Wier (1515-1558) strongly and stoutly stood against the idea of

[2] I am indebted for this quotation from a manuscript by Horst, as well as for other details in this chapter, to a series of papers on "Some Origins in Psychiatry" by Clarence B. Farrar which ran in three issues of the *American Journal of Insanity* in 1908 and 1909.

possession in mental disease and pointed out: "It is usually melancholy and hysterical women with disordered power of imagination who start the rumors and ideas of possession and witchcraft." He was fiercely assailed for this. Scribonius wrote against him. Thomas Erastus exhorted the Christian rulers to rid the world of these monstrous creatures; George Pictorius was especially vehement in his incrimination of witches as producing madness and in his horrified denunciation of those who opposed the idea.

After the sixteenth century, however, the prevalence, amongst physicians at least, of the idea of possession and demoniacal influence began to wane. The center of the stage of psychiatric thought came to be occupied by more "natural" errors, of a type one cannot quarrel with too vigorously because of the sneaking belief that will not down, to wit: that future writers will view the psychiatrists of our time as possessed of error, but doing their best in view of the nature of the thought of their time.

DIAGNOSIS BY COINCIDENCE

A second error arises from the universal and naïve reaction of *post hoc ergo propter hoc*. I label this error as the worship of the great god Coincidence. It is difficult, impossible in fact, for untutored man to discriminate between coincidence and cause. It is easy to say when a man has been disappointed and then raves in mania, that the disappointment causes his disease; just as it is pleasant to do something, whether it is to pray, give electricity, remove tonsils, and to ascribe the recovery of the patient to prayer, electricity, and tonsillectomy. We cannot rest without cause; unless we are mathematically inclined and know about probability, chance, and error, we ascribe the striking event that is antecedent to the illness or the recovery to the series of events that follow. The disposition to praise and blame thus enters largely into the acceptance of coincidence as cause.

Thus, throughout the literature of insanity, we find that disappointment in love is ascribed as a cause of mental disease, and examples are cited in proof. Masturbation, the "solitary sin," appears early as a cause, since the insane masturbate. Even in the writings of the earlier psychiatrists of the nineteenth century this error is conspicuously present. Then it commenced to fade out of the literature as cause and to become merely a phenomenon of the sexual life, sometimes associated with mental disease, but more often appearing as a frustrated or partial sexual reaction. Economic reverses and social reverses of any type stand out as causes, not only in the official literature of all times, but more conspicuously in the lay mind, despite the fact that financial depressions come and go with no increase in the psychoses. Any adverse circumstance occurring to an individual who later falls sick becomes blamed for his mental disease, such as an ancient blow on the head, a frustration, a passing fear. It is curious that only few are the cases where a great joy is said to "unbalance" the mind. The term "unbalance" indicates the general opinion held by both the laity and the specialists, namely, that the mind is something which has an equilibrium, and when that equilibrium is disturbed, it falls into insanity. This relic of anthropomorphism has not yet disappeared, although psychiatrists as a whole have discarded it. But since every court of law is a tribunal which judges men sane or insane, the lay opinion still exerts a profound influence. By that lay opinion the untoward event of whatever type may be blamed for mental disease, so that a trifling injury with subsequent mental disease becomes a matter of litigation; a lay jury of twelve good men and true is supposed to decide the delicate question of cause and effect in relation to the most difficult of human problems.

A third error resides in the disposition so commonly seen throughout the history of human learning, to inflate a few facts into a generalization, to develop neat logical systems, to blind oneself to opposing perceptions and contrary experiences, and to see only that which one wishes to see. The dogmatist incorporates his theory into his egoism, the theory becomes part of his *amour propre*—something to be intruded into the thinking and acting of his contemporaries, something to be defended with debate, that lesser form of battle which beclouds thinking. "All cats look gray in the dark." It is only when daylight comes that one is seen to be black, another gray, still another white, to say nothing of intermediate and diverse colors and shades. Moreover, there is thrust upon the specialist of whatever type the necessity to be dogmatic; people expect it of him. They want to lean upon his assuredness and certainty; they cannot believe that he does not know all about his subject. So, consciously or unconsciously, he plays up to this expectation and talks learnedly, vigorously, and logically, where there is neither certainty, learning, nor logic.

PHYSIOLOGICAL AND PSYCHOLOGICAL CONCEPTIONS

From the Greek days and for centuries the influence of the "humors" inducing mental disease have been stressed; but the admixture with superstition and magic vitiated even this effort at scientific thinking. The term "melancholia" immortalizes the belief that a black humor produces the depression, the lowered energy, and the gloom of one of the great mental diseases. The Hippocratic humoral pathology governed the thought of disease in general and especially of mental disease to the seventeenth century, and it has, in a sense, a lineal descendant in the present-day endocrinal theory of the origin of mental disease, although there is only a limited resemblance, certainly not identity. According to this theory, "man was a miniature embodiment of the universe, the four elements of the latter—earth, water, fire, and air—being represented by the four humors, yellow bile, black bile, blood, and mucus respectively, and these four humors contributing to the body the respective qualities of dryness, moisture, heat, and cold. The various manifestations of insanity were merely the effects, therefore, of an excess of one or the other of these elements, the expression of too much heat or cold, or of too much or too little moisture in the brain."

It would take us too far afield to discuss this notion in

any detail, nor would it add anything to cite the authors who formulated their concepts of mental disease in terms of the humors. The point is that *for centuries no attempt whatever was made to substantiate or to reject this doctrine.* It was derived from authority and from an ingenious identification of man with the universe. The concept of the macrocosmos and the microcosmos ruled the world of medical thought. Any direct study of a single brain would dispel any notion that there was any dryness or overabundance of moisture, any accumulation of bile, mucus, or any of the humors in the melancholic. In fact, the most refined investigations of our present day show nothing to substantiate any such notion. Psychiatry was a matter of logical speculation, inference, and authority. Although men made shrewd observations in respect to the clinical symptoms of mental disease and there appeared from time to time very good descriptions, the attempts at etiology and, consequently, at treatment pathetically illustrate the thesis that man is a herd animal following a leader blindly.

Long ago the father of medicine, Hippocrates, discerned a physiological fact of the greatest importance to mankind. He stated: "And by the same organ [referring to the brain] we become mad and delirious; and fears and terrors assail us, some by night and some by day, and dreams and untimely wanderings, and cares that are not suitable, and ignorance of present circumstances, desuetude and unskilfulness. All these things we endure from the brain when it is not healthy, but is more hot, more cold, more moist, or more dry than natural, or when it suffers from any other preternatural or unusual affection."

Unfortunately, the greater authority of Aristotle entered the field of thought; he sponsored the theory that the heart was the seat of the mind. Moreover, the philosophers and religionists, actuated by various motives, separated the mind from the body; so that diseases of the mind, despite the clinical evidence and common observation that every bodily condition, from fatigue to brain injury, altered mental capacity and activity, were placed in a separate category of events, and a dichotomy was established which is still evident in psychiatry. The literature of the Middle Ages is curiously replete with the grossest materialism, in that the state of the bowels, for example, is intimately linked up with mental activity, but by a curious ambivalence the mind was given by the same observers a spiritual significance and linkage akin to the soul and often identified with it. This, to a certain degree, accounts for the harsh treatment of the insane, since it has been a popular truism and is uneasily present in the thought of most of us, that a man can "help" what he thinks and what he feels, even though he may be helpless against a belly-ache.

This error has definitely permeated the thinking of the psychiatry of our own time. I cite as an example Bernard Hart. This leader of British psychiatry has written a book, *The Psychology of Insanity* (fourth edition, 1931), which has been widely read. He differentiates between the physiological and psychological conceptions of mental disease in the following terms:

The physiological conception admits that the phenomena of insanity are phenomena of consciousness, but it assumes that the mental processes are accompanied by corresponding changes in the brain, and to these brain changes it devotes all its attention. With the phenomena as facts of consciousness it has no immediate concern. The first aim of the physiological conception is to find the actual changes in the brain which occur in insanity and the brain changes correlated with each morbid mental process. Its ultimate aim is the discovery of convenient "laws" which will describe these brain processes in the shortest and most comprehensive manner. These laws will, of course, contain nothing but physiological terms—terms of *consciousness* will find no place therein.

The psychological conception, on the other hand, takes from the outset an altogether different route. It regards the conscious process occurring in insanity as the actual phenomena with which it is called upon to deal. Its ultimate aim is the discovery of convenient "laws" which will shortly and comprehensively describe these *conscious processes*. In this case the laws will contain nothing but psychological terms—terms of *brain* will find no place therein.

It is of the utmost importance that, in the final "laws" obtained by either the physiological or psychological conceptions, there should be no mixing of the terms. The physiological laws must contain no psychological terms, and the psychological laws must contain no physiological terms. Nothing but hopeless confusion can result from the mixture of "brain-cells" and "ideas." The reader must be asked to accept this statement as a dogma. And adequate demonstration of its truth would take us far beyond the limits of this book.

That this is not an isolated attitude is further exemplified by the writings of Paul DuBois, who follows the psychological conception of mental disease, if not through departmental considerations, through the belief that the mind can be treated fundamentally only through mental stimulations in the form of good advice, moral teaching, and philosophic admonition. Unfortunately, the patient who presents himself to me or to any other practising psychiatrist does not come with his mind in one hand and his physiology in the other. Suppose that he presents the clinical picture, as medical men phrase it, of one suffering from a dementia; in other words, his memory is poor, his power of entering into intellectual relations low, his ability to plan and carry out purpose deficient. Suppose, in addition, that he has changed in personality, in that he has lost those ethical reactions which characterize the so-called normal man, and his esthetic conduct has reached the animal stage. Surely, here is someone changed mentally; yet on examination one finds that the disease process involved is syphilis of the brain, and that the entire mental alteration rests upon the disturbed physiology incident to the invasion of the *trepanoma pallida*. The reaction of the body to this invader, whether it appears in liver, heart, or brain, interferes with the function of the organ involved. The function of the organ involved in this case is the expression of those qualities we call mental, although any tyro in physiological psychology knows that every emotion spreads itself out to the smallest blood-vessel and the furthest organ; that the whole chemistry of the organism is changed; that the viscera partake of mind and the expression of mentality in such a way as to make them parts of the mental mechanism of the body. In the particular case cited, the psychology of the situation presents no approach whatever for therapeutics, whereas definite physiological chemical measures offer hope to the patient, as when he is treated with the arsenical preparation introduced by Ehrlich or by the malarial treatment introduced by Wagner-Jauregg.

We may develop this thesis in many clinical directions. Thus, the alcoholic, seeking to escape from his difficulties and his own personality, attempts to wipe out those difficulties and to alter that personality by imbibing a chemical substance, C_2H_5OH. The fact that he does imbibe this drug for the effects that it produces shows that psychology and physiology cannot be separated. Moreover, the result on his organism of the continuous use of alcohol is, for example, to make him abnormally suspicious of his wife, so that he sees and hears her imaginary lovers on every side, and it so alters his ethical responses that he kills her, although when he is sober, he may be the kindest and gentlest of husbands. Dramatic as these examples are, there are numerous others to be taken from the field of clinical psychiatry which make one rather scornful of those nice arm-chair philosophizings by which the mind is separated from the body, and of the profound remarks that by no manner of means can the bridge between the neurone and the idea be crossed. As a matter of fact, no logical bridge can ever be erected between any two phenomena, although experience and experiment may indissolubly link them together.

PSYCHOTHERAPY OF THE NEUROSES

This may bring up the question in the informed reader's mind: What about the whole field of psychotherapy, which, in a formal sense, is a recent development in psychiatry? What about the searching for complexes and personality difficulties as explaining the neuroses? What about the psychoses for which no physiological cause as yet has been found? The answer to all this is that substantially the neuroses are as yet unsolved problems, despite the claim of the psychoanalysts and psychotherapists generally. The neuroses are "cured" by psychoanalysis, Christian Science, osteopathy, the shrine of Ste. Anne de Beaupré, a turn in fortune, nux vomica, and sodium bromide, as well as phenobarbital. I suspect, therefore, that most neuroses are self-limited, that they tend to run their course and get well more or less independently of treatment, just as many physical diseases do; this is borne out by the fact that there are so many methods and so many schools claiming to "cure" the neuroses. When a therapeutic measure or method really cures, it quickly gains the field and is without rival: for example, the treatment of diphtheria by antitoxin, the prevention and treatment of tetanus, the use of the arsenicals in syphilis of liver and its analogues in pernicious anemia, and of quinine in malaria. Where there are many therapeutic approaches, it is probable that none of them is actually curative.

Thus, even where we advocate a psychological technique for the treatment of the neuroses, it is to be remembered that the complaints of the neurotic are largely physiological. That is, he may be the victim of pathological complexes; he may be reacting to a life situation in an inadequate manner; he may need advice, analysis, catharsis, or whatever approach one elects to take to his difficulties. Even then, his physiological state is of importance. Help for his sleeplessness, tonics for his appetite, may be of psychological benefit; for no man is a psychological series of states floating in thin air. Mood, emotion, energy, are anchored in an organism whose physiology it is always important to understand.

Furthermore, the very recent developments in electrical science have shown us that the mental processes give off action-currents of an extraordinary type, that each perceptual function is a differentiated electrical response. We are coming much nearer to the time when we can express, quantitatively and qualitatively, the formulae of the mental states.

The progress of psychiatry depends upon taking into account the whole milieu of the individual; the psychiatrist needs to keep in mind that social pressure and social conditioning are not merely names but actual physiological processes.

HEREDITY AND POLYMORPHISM

I turn now to a pair of errors which were most conspicuous in the psychiatry of the nineteenth century, although they are even yet in evidence. As etiology of the mental diseases, heredity has been given the first place, not only by a few thoughtful observers in the Greek period and the Middle Ages, but especially since the biological conceptions achieved their great publicity and importance. Starting with the observation which is forced upon any man of experience that mental disease "runs in families," there was developed a school of thought which gave to hereditary transmission the first place in the causation of the psychoses and which merely changed its form in the twentieth century by taking on a Mendelian coloring. In stating that there was gross error in the work and in the conclusions of the distinguished proponents of heredity, I do not mean to deny that heredity is an important factor in mental diseases and feeble-mindedness. I merely state that what passed as scientific proof was, in the main, a fine example of *petitio principii*.

There developed in the early nineteenth century a theory, fathered by two French psychiatrists, Auguste Benedicte Morel and Jean Etienne Esquirol, but given its greatest impetus by the celebrated Cesare Lambroso (1836-1909), called the polymorphic theory or polymorphism. On this I quote from an earlier publication.[3]

There is a unitary something in the neuropathic or psychopathic inheritance which makes itself manifest under many forms [thus the *polymorphism*]. All manner of mental diseases, including the organic diseases, are thus labeled, and all psychoneuroses, including as well epilepsy, feeble-mindedness, and

[3] Abraham Myerson, *The Inheritance of Mental Diseases* (Baltimore, 1925), pp. 270-72.

crime. Up to this day this doctrine prevails in France, in England, and to a limited extent in Germany, where attention to the details of mental diseases has side-tracked polymorphism, and in America, where under a new guise, as Mendelism, it has been given the sanction of the foremost writers until very lately. In the excessive development given this theory, headache in an ancestor was given a hereditary value in relation to the mental disease of his descendant. Even headache and fainting-spells in a cousin had a dread significance, while the earlier writers laid great emphasis on tuberculosis, cancer, hemiplegia, gout, "rheumatism," and every chronic or semichronic human ailment. At the present day practically every mental-hospital history in America wastes good paper by including data on these diseases in relation to the ancestors and relatives of patients.

Lombroso and his pupil Nordau gave to the term *degeneracy* the widest significance, so that on the one side the lowest human beings—cretins, idiots, etc.—were linked together with the men and women of genius. His "reasoning" is typical of that of the whole polymorphic school, and I cite a few examples of it, though I could multiply them indefinitely in the writings of Morel, Feré, Dejerine, Clouston, MacPherson, and others who have subscribed to the theory.

Lombroso based much of his theory on his ideas of the stigmata of degeneracy which are both physical and mental. Thus he gives as signs of degeneracy common to both genius and insanity, shortness *and* tallness, rickets (which includes, God save the mark, rachitic, *lame, hunchback,* or clubfooted, thus linking together at least a hundred diverse things under one heading), pallor and emaciation (which as every one knows are secondary to other conditions), stammering, left-handedness, delayed development, precocity, etc., up to a "fondness for certain words"! And this passed for science, and was part of the work of an extraordinary personage. It never occurred, apparently, to Lombroso that all these stigmatas are very common and found almost as frequently among the "normal" as among the insane or the geniuses. He finds it rather sinister that a list of geniuses were weak and sickly in childhood, such personages as Demosthenes, Bacon, Descartes, Newton, etc. It would surely be a blessing if more people were then weak and sickly in childhood, if, alas, mediocrity did not also follow childhood diseases.

In another place he notes the greater capacity of the skull in men of genius, and likens it to the "case reported of an insane man with a large skull." As a matter of fact, the weight of the brain in insane-hospital autopsies is certainly not greater and is generally less than that observed elsewhere. "Men of genius frequently stammer"—I know many stammering mediocrities. It is hard to believe that Lombroso's work found so great a vogue, so patent are its absurdities. Thus he lays stress on the *resemblance* of monomania to the distinguished pertinacity by which men of science devote themselves to one problem!

As a matter of fact, amongst the seven hundred families of the insane at Taunton I found no cases of "genius" and few of high-grade talent. Yet Lombroso concludes "we may confidently affirm that genius is a true degenerative psychosis belonging to the group of moral insanity and may temporarily spring out of other psychoses." All the evidence he produces is to adduce *resemblances* of the flimsiest kind, without any "control" of "normal" families. In fact, his definition of degeneracy as anything deviated from the commonplace must inevitably force genius into degeneracy, since the very essence of genius is that it is peculiarly exuberant in certain qualities. Thus the mistake in logic which nullifies all the work of the polymorphists is seen at its best in Lombroso, that is, to make the premise inevitably include the conclusion, to "prove" what one assumes.

This polymorphic theory underlies even the work of Charles B. Davenport and the American eugenic school. In studying the inheritance of epilepsy he makes up a list of characters all of which are given equal hereditary weight in reaching the conclusion that the patient's condition rests on a hereditary basis. These characters are all vague, may arise from a number of causes; furthermore, their hereditary weight is the point to be proven. Yet with an appalling sang-froid their hereditary potency is assumed. The list starts with *A* (which equals alcoholism), proceeds with apoplexy, blind, Bright's disease, criminalistic, cancerous, deaf, etc., and goes blithely enough through the alphabet to *V* or vagrant. The final criticism of such a list as that which Davenport here uses is this: by extending the list of conditions that may be declared hereditary, one finally extinguishes all differences, comes down to the fact that man is mortal, has sicknesses and weakness, and is sinful. A linking-up thus can be made of the vicissitudes and misfortunes of life and the psychiatric diseases.

This kind of work, which assumes that a sausage-mill will take care of meat, paving-stones, old clothes, and soap and turn out good sausages, is the absolutely untenable basis of the work done not only by Davenport, but by those who have portrayed with gusto and alarm the royal families of the feeble-minded, the Nams, Kallikaks, the hill-folk, Zero tribe, etc.

The errors in this whole theory, which dates back to a period of medical knowledge before there was any definite understanding of brain physiology, spinal-fluid chemistry, reflex pathways, serology, endocrinology, X-ray study, may be thus enumerated:

First, the fallacy of the positive instance dominates. If, for example, it turns out that the non-insane have more cases of apoplexy in their ancestry and collaterals than do the insane, as is the case, then apoplexy in an ancestor is merely coincidence and not related to insanity. If it is assumed that alcoholism in an ancestor or relative is a hereditary psychopathic character, how is the fact explained that the children of the alcoholic Scotch are no worse so far as feeble-mindedness, etc., is concerned than the children of the non-alcoholic Scotch, as shown by the study of Elderton and Pearson? Headache, fainting-spells, epilepsy—all these are terms of varied clinical meaning and etiologic value; they are not specific and may relate to such diverse matters as brain tumor, fractured skull, bad teeth, sinusitis, heart-disease, indigestion, nervous instability, and malingering. At least 90 per cent of the work done on the heredity of mental diseases has been wasted, and a large part of it is merely a bolstering of preconceived theory by specious figures.

Second—and this error is as bad on the other end of the situation as the weighting of the psychopathic ancestors and relatives—any approach to the problems of psychiatry that makes a unity or even a unit of "insanity" or feeble-mindedness or crime is based on a fallacy of the overgeneralization type—so common when little is known about a subject. The "all cats look gray in the dark" axiom aptly fits the case.

Insanity is a legal term and merely means that the person labeled as insane needs incarceration or must have his affairs managed by others. Mental disease has quite a different connotation, though a false unity is implicit in the word mental. What links together, for example, the case of general paresis, due to syphilis, with the alcoholic hallucinosis; or either or both with the case of dementia præcox, which rests very likely on some hereditary disorder; or all of these with senile dementia, which comes to everyone who lives long enough to have his brain disordered before his body dies? Similarly with feeble-mindedness—this is merely the name of a symptom, just as cough is. No more than the cough of pneumonia, tuberculosis, aneuryism of the aorta, nasal spur, hay-fever, hysteria, and embarrassment links these diseases and conditions together does "feeble-mindedness" link together cretinism, mongolism, microcephaly, brain injury, and the perhaps hereditary group of the feeble-minded.

Where little is known in medicine, heredity is invoked as a cause, just as in tuberculosis the early clinicians called it a family or hereditary disease on the same basis that we now speak of the familial psychoses, namely, that brothers and sisters, ancestors and descendants, had tuberculosis. Then came Koch and the discovery of the tubercle bacillus, which knocked out heredity and placed the blame on an environmental organism. The ground then shifted to predisposition —that is, patients acquired tuberculosis because they were predisposed to get it. While it cannot be denied that there may exist a predisposition and that some persons resist forces to which others easily succumb, there is a good bit of *petitio principii* in the general use of the word. A skull can be fractured by a falling brick only if it is thin enough—that is, predisposed to be broken; otherwise it would crack the brick, in which case the predisposition would be with the brick. Predisposition, constitution, these are logically necessary inferences; but they are not facts, nor until the whole world of variables is taken into account can they be shown even mathematically to exist.

The whole subject of the heredity of mental disease needs a fresh start. The nineteenth-century work can be discarded almost completely, and most of the twentieth-century researches make a jump from genetics, which is limited in its application and deals with simple characters of simple creatures, to the mentality of man, which, whatever weight we may give to heredity, is certainly in part a product of social and cultural forces. There is some hereditary background for a few of the mental diseases, apparently for manic-depressive psychosis, perhaps for dementia præcox, and for many of the cases of feeble-mindedness. To infer that these are Mendelian characters is to proceed without the slightest possibility of proof—certainly at the present time.

SOCIAL PSYCHIATRY

A consideration of past errors would have no more than a historical value if it did not help us to guard against repetition of these same errors or of similar ones in our own times. Psychiatry seems to me to have been particularly distinguished by its facility for generalization and for dogmatic statement entirely beyond proof and fact. There has been a trend of modern psychiatry, dating from the time of Lombroso in the latter part of the nineteenth century, which carries with it the grave danger of overgeneralization, of mistaking difficulty in the social relations for psychiatry, and in general of carrying the attitude of the specialist altogether too far. I refer to the whole development of what may, for purposes of convenience and classification, be called *social psychiatry.*

It would be unfair to stigmatize one of the great developments of psychiatry in the past fifty years as error. The extension of the speciality into social matters, into the problems of delinquency, family adjustment, and the care of children, is a logical and natural development. What I here stigmatize as error is that extension of thought and activity by which the social problems that are becoming a new part of the field of psychiatry are discussed as if they were mainly psychiatric problems. The point I shall make is that psychiatry has something to contribute in the discussion as an approach, as a philosophy (if one chooses to use that word), which is important; but basically these problems are not psychiatric. The solution must rest in the hands of those who are not psychiatrists or who, if they happen to be psychiatrists, must in addition become sociologists, economists, and students of the social structure generally. The principal error that has been made is to mistake social maladjustment as necessarily psychopathological.

Again we trace an error to the great Lombroso. The adjective "great" is not here used in any cynical or insincere sense, for a man may be great despite his errors, in that his influence has been directive and powerful. Such individuals are often extremists; they overemphasize and distort, but nevertheless they have modified men's thinking and acting to a great degree. Lombroso started the study of the individual delinquent on a psychiatric basis; but the trouble is that he entirely forgot that crime is a matter of social definition, that he did not contrast or compare his delinquents with the normal population, that somehow he compared a criminal with a hypothetical normal who never has been seen on land or sea. Thus he reached the conclusion that crime was like genius, a sort of degenerative neurosis or psychosis. His idea, namely, that the individual delinquent should be studied rather than the crime, was profound and far-reaching, despite the fact that almost every conclusion he made about the criminal was incorrect.

Since his day there has been, so to speak, a line of psychiatrists working on the idea that crime has its roots in something akin to mental disease. Thus, in the first flush of fervor in the application of the Binet tests, it was quite conclusively proved to the satisfaction of some leading psychiatrists and psychologists that the large majority of criminals were feeble-minded. This misunderstanding—which could have easily been prevented by that *sine qua non* of scientific effort, the control study—received a lethal blow with the mental examination of the Army during the World War, the results of which should have taught psychiatrists that their prime approach to any of their problems included a study of the so-called normal man. At about the same time a little bit of circular reasoning was indulged in by several important psychiatrists interested in criminology, when they discovered that most criminals were psychopathic personalities. This conclusion they reached, first, by defining certain types of psychopathic personalities as chronic criminals, and then stating that these chronic criminals were psychopathic personalities. No control studies have been made as to the amount of psychopathic personality defined in other terms than criminality found amongst college professors, school-teachers, mechanics, and even the learned psychiatrists themselves.

If by a psychopathic personality is meant serious emotional difficulty or impaired capacity of the individual to adjust to others without outstanding peculiarity or great strain, independent of the direction or manifestation, then it will be found, I venture to say, that there is an extraordinary amount of psychopathic personality in the world. Samuel Johnson was a psychopathic person, and so is some shallow sex exhibitionist confined in a county jail. To drink too much is psychopathic, and to drink not at all is also psychopathic. The over-good are psychopathic, as are the over-sinful. Anybody who is not easily classified can be stigmatized as psychopathic. Except as applied to a few limited groups, the term is used without any real meaning and constitutes one of the darkest areas in psychiatry. Our difficulties will not be cleared up by any blanket terms that cover ignorance.

Such complex phenomena as crime are not easily settled by any unitary explanation, nor must we make the mistake of judging all deviations of conduct from a legal or moral standard as psychopathic. The field of character has not yet been subjected to laws and formulae. Furthermore, in the adjustment of any individual to the complex social structure in which he lives, there are at least two sets of factors of importance—first, the nature of the individual, and second, the particular set of social forces playing upon him. The fault in maladjustment may be the "fault" of the social structure and not the "fault" of the individual. In other words, what psychopathology there may be may rest with the traditions, customs, methods of teaching, economics, and legislation of the times. The individual by himself may conceivably be excellent in another setting, yet he may be hopelessly disabled by what has happened and what is happening to him. So the abnormality with which the psychiatrist may find himself compelled to deal may be due to the psychopathology of society rather than the psychopathology of the individual. This social psychopathology may render hopeless a situation which, if the social structure were flexible enough and had any thought for the individual, might well be curable or, to put it more broadly, adjustable. There is a definite tendency on the part of psychiatrists to talk and act as if the social structure were all right—were, so to speak, normal—whereas any consideration of social life as it is reveals its ignorance, brutality, and inadequacy.

Social psychopathology involves great difficulties, as any student of the social sciences knows. The capacity to find cause and effect in the ocean of social life, in the currents of all kinds that sweep on and through the individual, is probably at the present time beyond human power. All the psychiatrist can hope to do is to select from the milieu those factors which, in his best judgment and as a result of his experience and that of others, seem to be basic in creating individual maladjustment. Thus, the home, that very complicated social product, imposes strains and burdens on its occupants which, though they are in the majority of cases compensated for by the advantages, may in many other cases bring no such compensation. Again, the sexual life, including marriage, may reasonably seem a fundamental source of emotional tension, fatigue, obsession, and fixation. The economic structure with its competitiveness, with its burdens of the impress of fear, the sense of inferiority and failure, may, on examination, appear to be the important variable that compresses and depresses the individual.

Nevertheless, while the psychiatrist may hypothetically set up any social factor as important and work as though he had discovered the source of maladjustment, he should always keep in mind that this may be mere coincidence, that his etiological conclusion may be as naïve as those which were reached by Paracelsus and his contemporaries. He must avoid, at least from the scientific angle, a doctrinaire position and the bold building up of social and individual theory of his subject-matter. In the present state of its knowledge, psychiatry needs research, research, and research. It also needs the humble, diligent spirit.

INDEX

Abnormality, *see* Insanity
Adams, J. C., 56
Agassiz, Louis, 81, 82, 213
Agricola, Georgius, 66, 67
Albertus Magnus, 97, 325
Alchemy, 8, 9, 126, 173 *ff.*, 231
Alcmæon, 324
Altamira, 287
Anatomy, comparative, 256, 257
Anaxagoras, 324
Animals, fabulous, 200, 201, 202
Anthropology, 292 *ff.*; physical, 300; *see also* Man, science of
Arbuthnot, John A., 368
Archæology, classical, 306, 307; cultural, 306 *ff.*; prehistoric, 307
Archeus, see Spirits, 231
Archimedes, 177
Aristotle, 50, 96, 130, 131, 169, 173, 201 *ff.*, 223, 227, 251, 252, 324, 325, 329, 333, 425
Association, 342
Astrology, 6, 7, 8, 231; and mental disorder, 417
Astronomy, 26, 39 *ff.*; naïve, 40, 41; sidereal, 58, 59
Atlantic, 98 *ff.*
Atom, 150; theory of, 185 *ff.*
Attraction, 11
Australis, 105
Avicenna, 70, 178
Avogado, Amadeo, 188, 189, 190

Bacon, Francis, 16, 17, 166
Bacon, Roger, 16, 102
Bagehot, Walter, 351
Bentham, Jeremy, 372, 373
Berringer, Johannes, 71, 72
Berzelius, J. J., 191
Binet tests, 436
Black, Joseph, 238
Blas, 232 *ff.*
Blumenbach, J. F., 300
Boas, Frank, 303
Bodin, Jean, 368

Bohr, Niels, 159
Bonnet, Charles, 208
Borelli, Giovanni, 235
Boyle, Robert, 165, 169, 181, **236**
Bradley, James, 53, 54
Brahe, Tycho, 47, 48, 49
Brain, 255, 324 *ff.*, 336 *ff.*; *see also* mind; *also* nervous system
Buckle, Henry Thomas, 369
Buffon, G. L., 300
Burton, Robert, 419

Campanella, Tommaso, **360**
Cannon, W. B., 260
Capitalism, 379 *ff.*
Cardan, Jerome, 9
Carrel, Dr. Alexis, 261
Cause and effect, 29; search for, 136 *ff.*
Charms, 10
Chemistry, 165 *ff.*
Cheselden, William, **331**
Child, C. M., 258
Christian Science, 429
Circulation, 226 *ff.*
Civilization, the procession of, 271 *ff.*
Coghill, G. E., 254
Columbus, Christopher, 98 *ff.*
Combustion, 238-240; *see also* Phlogiston
Cook, Captain, 108
Copernicus, Nicolaus, 45-47
Criminality, 437; *see also* Insanity
Cro-Magnan, 287
Cultural compulsive, 285-286
Culture, 289; diffusion of, 317 *ff.*; evolution of, 312-317
Cuvier, 212, 286-287, 308

D'Ailly, Pierre, 97, 98
Dalton, John, 185 *ff.*
Darwin, Charles, 211 *ff.*, 257, 272, 273, 286, 299 *ff.*, 351, 352, 353, 374
Davenport, Charles B., 432-433
Da Vinci, Leonardo, 70, 298, 325